GREENHOUSE GAS: SCIENCE AND TECHNOLOGY

Edited by

Judith Rosales

ArclerPress

Greenhouse Gas: Science and Technology

Edited by Judith Rosales

Edition 2018

ArclerPress

Published by
Arcler Press LLC
708 3rd Avenue, 6th Floor
New York
NY 10017
United States of America
www.arclerpress.com

ISBN: 978-1-68094-625-3 (Hardback)

Printed in United States of America

Library of Congress Control Number: 2017933217

About the Editor

Judith Rosales

Prof. Judith is an Ecologist and Environmental Specialist with more than 30 years of academic and consulting experience. She holds a PhD in geography from the University of Birmingham UK and wrote the PhD and Master Academic Programs of Environmental Sciences for the Universidad de Guayana, Venezuela.

Table of Contents

2 HALOCARBON COMPOUNDS

3 GREENHOUSE EFFECT

4 CARBON CAPTURE AND STORAGE

7 GLOBAL WARMING AND CLIMATE CHANGE

8 ADAPTATION TO ENVIRONMENT

List of Abbreviations

ACE-CRC	Antarctic Climate and Ecosystems Cooperative Research Centre
BECS	Bio-energy with carbon storage
CCS	Carbon capture and storage
CRC	Carboniferous Rainforest Collapse
CM	Carbon monoxide
CURC	Coal Utilization Research Council
EDP	Environmentally displaced person
EPICA	European Project for Ice Coring in Antarctica
FGD	Flue-gas desulfurization
GWP	Global warming potential
GHG	Greenhouse gas
IPCC	Intergovernmental Panel on Climate Change
IOM	International Organization for Migration
IPICS	International Partnerships in Ice Core Sciences,
LCA	Life cycle assessment
LDV	Light duty vehicles
NMVOCs	Non-methane volatile organic compounds
ONC	Ocean Nourishment Corporation
PC	Pulverized coal
TCE	Trichloroethylene
UNFCCC	UN Framework Convention on Climate Change
UNFCCC	United Nations Framework Convention on Climate Change

WAIS	West Antarctic ice sheet
WMO	World Meteorological Organization

Preface

On Earth, the atmosphere is warmed by absorption of infrared thermal radiation from the underlying surface, absorption of shorter wavelength radiant energy from the sun, and convective heat fluxes from the surface. Greenhouse gases in the atmosphere radiate energy, some of which is directed to the surface and lower atmosphere. The mechanism that produces this difference between the actual surface temperature and the effective temperature is due to the atmosphere and is known as the greenhouse effect. Earth's natural greenhouse effect is critical to supporting life. Human activities, primarily the burning of fossil fuels and clearing of forests, have intensified the natural greenhouse effect, causing global warming. Greenhouse gases allow direct sunlight (relative shortwave energy) to reach the Earth's surface unimpeded. As the shortwave energy heats the surface, longer-wave (infrared) energy (heat) is reradiated to the atmosphere. Greenhouse gases absorb this energy, thereby allowing less heat to escape back to space, and 'trapping' it in the lower atmosphere. Many greenhouse gases occur naturally in the atmosphere, such as carbon dioxide, methane, water vapor, and nitrous oxide, while others are synthetic. Those that are man-made include the chlorofluorocarbons (CFCs), hydrofluorocarbons (HFCs) and Perfluorocarbons (PFCs), as well as sulfur hexafluoride (SF6). Atmospheric concentrations of both the natural and man-made gases have been rising over the last few centuries due to the industrial revolution. As the global population has increased and our reliance on fossil fuels (such as coal, oil and natural gas) has been firmly solidified, so emissions of these gases have risen. While gases such as carbon dioxide occur naturally in the atmosphere, through our interference with the carbon cycle (through burning forest lands, or mining and burning coal), we artificially move carbon from solid storage to its gaseous state, thereby increasing atmospheric concentrations. Despite their relative scarcity, the so-called greenhouse gases play an important role in the regulation of the Earth's energy balance. Greenhouse gases in the atmosphere trap infrared heat energy trying to escape back to space. In doing so they raise the temperature of the lower atmosphere and the Earth's surface in contact with it. This warming process is called the natural greenhouse effect, but during the last 200 years, mankind's pollution of the atmosphere with extra greenhouse gases has enhanced this natural greenhouse effect that may be contributing to global warming. In addition to the man-made increase of naturally occurring greenhouse gases, mankind has released some completely new chemicals into the atmosphere, including the CFCs or chlorofluorocarbons. Although these have now been banned in an attempt to save the ozone layer, they will remain in the atmosphere for at least another 50 years. Although their abundance in the atmosphere is very low, molecule for molecule they can be thousands of times better at absorbing heat than carbon dioxide, and consequently contribute significantly to the enhanced greenhouse effect.

Content Coverage

Chapters One and *Two* present an introduction of greenhouse gases and halocarbon compounds, correspondingly. Human activities since the beginning of the Industrial Revolution have produced a 40% increase in the atmospheric concentration of carbon dioxide, from 280 ppm in 1750 to 400 ppm in 2016. This increase has occurred despite the uptake of a large portion of the emissions by various natural "sinks" involved in the carbon cycle.

Chapters Three and *Four* present a coverage of greenhouse effect and the process of capturing carbon, consistently. The greenhouse effect is a biological process in Earth's atmosphere in which certain gases bind together and form what is essentially a layer of insulation, however the growing concerns about rising carbon dioxide levels in the early 21st century led to some interest in carbon capture as a technique which could be used to manage carbon emissions.

Chapters Five and *Six* are aimed to focus on human impact on the environment and climate change, separately. Human impact on the environment includes impacts on biophysical environments, biodiversity, and other resources. Climate change occurs when the Earth's average temperature changes dramatically over time.

Chapters Seven and *Eight* are intended to global warming and adaptation process to environment, separately. Global warming is concerned with the rise in the average temperature of the Earth's climate system and its related effects. The developmental norm of reaction for any given trait is essential to the correction of adaptation as it affords a kind of biological insurance or resilience to varying environments.

CHAPTER 1

Greenhouse Gases: An Introduction

Chapter Contents

- **Introduction**
- **Gases in Earth's Atmosphere**
- **Impacts on the Overall Greenhouse Effect**
- **Natural and Anthropogenic Sources**
- **Anthropogenic Greenhouse Gases**
- **Role of Water Vapor**
- **Direct Greenhouse Gas Emissions**
- **Life-Cycle Greenhouse-Gas Emissions of Energy Sources**
- **History of Scientific Research**

Greenhouse gases are gases that work to keep heat in the Earth's atmosphere. By doing so, they help keep the temperature on Earth within a livable range for human beings, plant life, and animals. Greenhouse gases are so named because they act in a similar manner to the glass ceiling and walls of a greenhouse. They allow light to penetrate Earth's atmosphere, but prevent a significant amount of heat from leaving it.

A person may first learn about greenhouse gases during a discussion of the greenhouse effect. Essentially, the greenhouse effect means that temperatures on the Earth increase because of gases that filter energy from the sun, much like the panels of a greenhouse. If these gases did not behave in this manner, more of the heat from the sun would move back out of the Earth's atmosphere and into space. Scientists theorize that this would leave the Earth considerably colder, with a temperature that averaged around 60 degrees Fahrenheit.

INTRODUCTION

A greenhouse gas (abbrev. GHG) is a gas in an atmosphere that absorbs and emits radiation within the thermal infrared range. This process is the fundamental cause of the greenhouse effect. The primary greenhouse gases in Earth's atmosphere are water vapor, carbon dioxide, methane, nitrous oxide, and ozone. Without greenhouse gases, the average temperature of Earth's surface would be about −18 °C (0 °F), rather than the present average of 15 °C (59 °F). In the Solar System, the atmospheres of Venus, Mars and Titan also contain gases that cause a greenhouse effect.

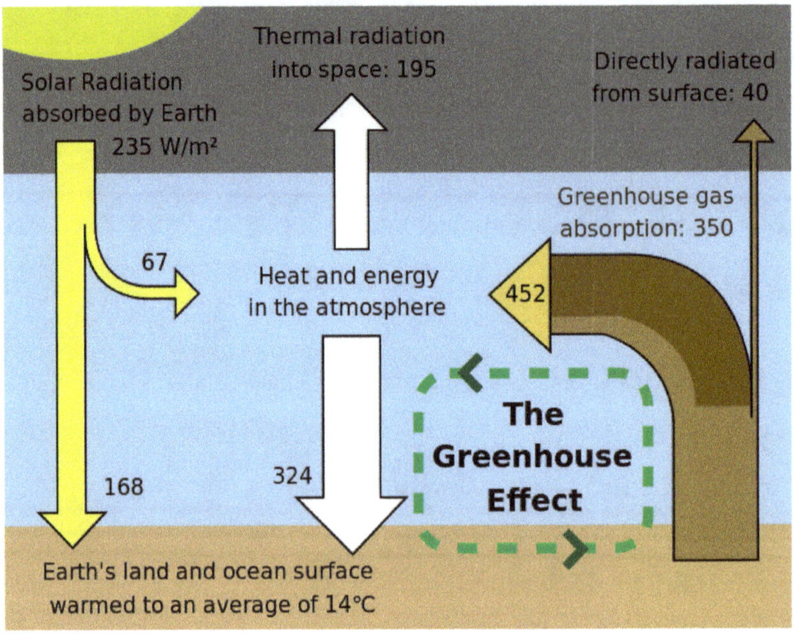

Greenhouse effect schematic showing energy flows between space, the atmosphere, and Earth's surface. Energy influx and emittance are expressed in watts per square meter (W/m^2).

Human activities since the beginning of the Industrial Revolution (taken as the year 1750) have produced a 40% increase in the atmospheric concentration of carbon dioxide, from 280 ppm in 1750 to 400 ppm in 2016. This increase has occurred despite the uptake of a large portion of the emissions by various natural "sinks" involved in the carbon cycle. Anthropogenic carbon dioxide (CO_2) emissions (i.e. emissions produced by human activities) come from combustion of carbon-based fuels, principally coal, oil, and natural gas, along with deforestation, soil erosion and animal agriculture.

It has been estimated that if greenhouse gas emissions continue at the present rate, Earth's surface temperature could exceed historical values as early as 2047, with potentially harmful effects on ecosystems, biodiversity and the livelihoods of people worldwide. Recent estimates suggest that on the current emissions trajectory the Earth could pass a threshold of 2 °C global warming, which the United Nations' IPCC designated as the upper limit for "dangerous" global warming, by 2036.

GASES IN EARTH'S ATMOSPHERE

Greenhouse Gases

Atmospheric absorption and scattering at different wavelengths of electromagnetic waves. The largest absorption band of carbon dioxide is in the infrared.

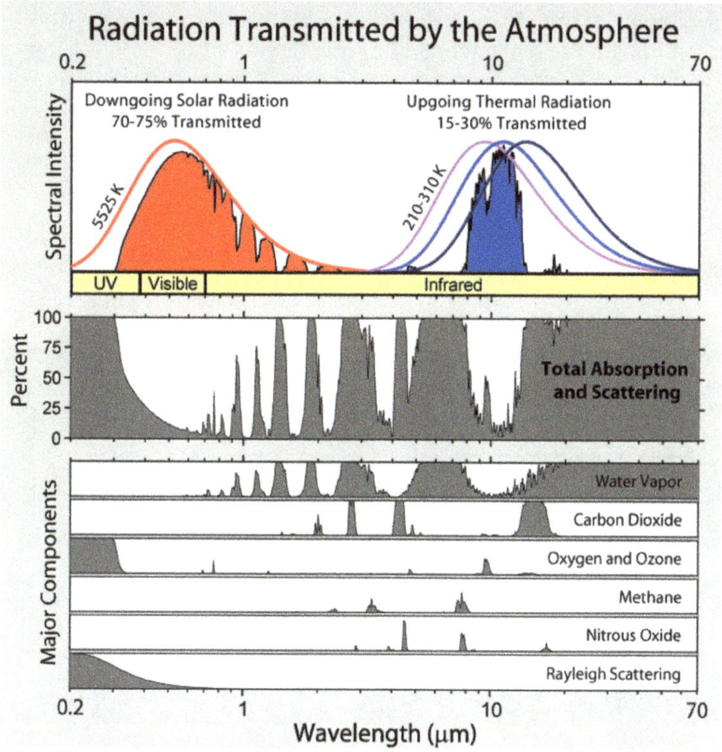

Greenhouse gases are those that absorb and emit infrared radiation in the wavelength range emitted by Earth. In order, the most abundant greenhouse gases in Earth's atmosphere are:

- Water vapor (H_2O)
- Carbon dioxide (CO_2)
- Methane (CH_4)
- Nitrous oxide (N_2O)
- Ozone (O_3)
- Chlorofluorocarbons (CFCs)

Atmospheric concentrations of greenhouse gases are determined by the balance between sources (emissions of the gas from human activities and natural systems) and sinks (the removal of the gas from the atmosphere by conversion to a different chemical compound). The proportion of an emission remaining in the atmosphere after a specified time is the "airborne fraction" (AF). More precisely, the annual airborne fraction is the ratio of the atmospheric increase in a given year to that year's total emissions. Over the last 50 years (1956–2006) the airborne fraction for CO_2 has been increasing at 0.25 ± 0.21%/year.

Non-Greenhouse Gases

The major atmospheric constituents, nitrogen (N_2), oxygen (O_2), and argon (Ar), are not greenhouse gases because molecules containing two atoms of the same element such as N_2 and O_2 and monatomic molecules such as argon (Ar) have no net change in the distribution of their electrical charges when they vibrate and hence are almost totally unaffected by infrared radiation. Although molecules containing two atoms of different elements such as carbon monoxide (CO) or hydrogen chloride (HCl) absorb infrared radiation, these molecules are short-lived in the atmosphere owing to their reactivity and solubility. Therefore, they do not contribute significantly to the greenhouse effect and usually are omitted when discussing greenhouse gases.

Indirect Radiative Effects

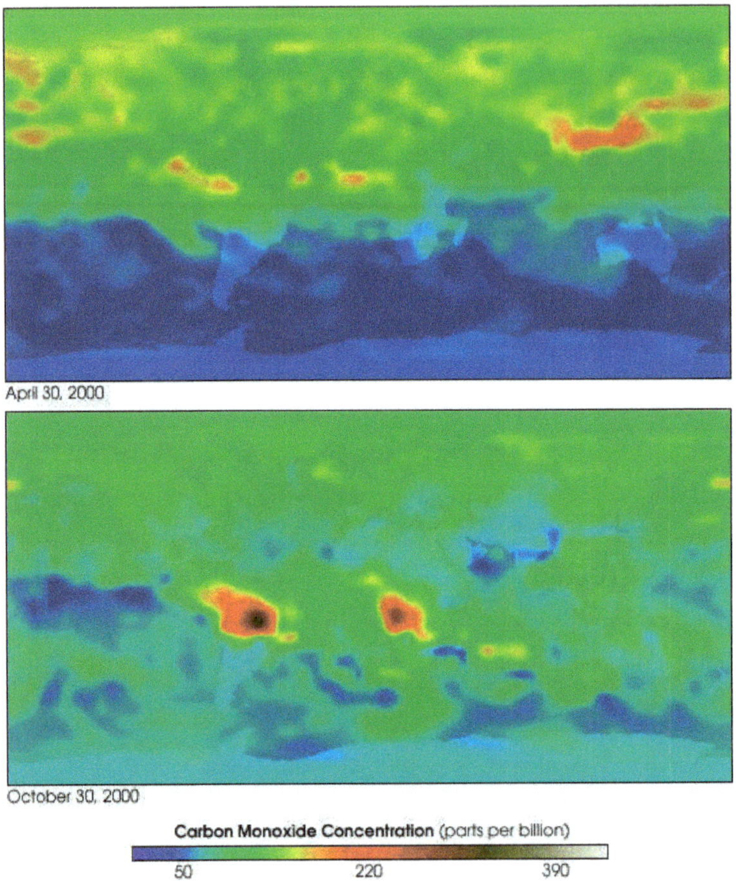

The false colors in this image represent concentrations of carbon monoxide in the lower atmosphere, ranging from about 390 parts per billion (dark brown pixels), to 220 parts per billion (red pixels), to 50 parts per billion (blue pixels).

Some gases have indirect radiative effects (whether or not they are greenhouse gases themselves). This happens in two main ways. One way is that when they break down in the atmosphere they produce another greenhouse gas. For example, methane and carbon monoxide (CO) are oxidized to give carbon dioxide (and methane oxidation also produces water vapor; that will be considered below). Oxidation of CO to CO_2 directly produces an unambiguous increase in radiative forcing although the reason is subtle. The peak of the thermal IR emission from Earth's surface is very close to a strong vibrational absorption band of CO_2(667 cm^{-1}). On the other hand, the single CO vibrational band only absorbs IR at much higher frequencies (2145 cm^{-1}), where the ~300 K thermal emission of the surface is at least a factor of ten lower. On the other hand, oxidation of methane to CO_2, which requires reactions with the OH radical, produces an instantaneous reduction, since CO_2 is a weaker greenhouse gas than methane; but it has a longer lifetime. As described below this is not the whole story, since the oxidations of CO and CH 4 are intertwined by both consuming OH radicals. In any case, the calculation of the total radiative effect needs to include both the direct and indirect forcing. A second type of indirect effect happens when chemical reactions in the atmosphere involving these gases change the concentrations of greenhouse gases. For example, the destruction of non-methane volatile organic compounds (NMVOCs) in the atmosphere can produce ozone. The size of the indirect effect can depend strongly on where and when the gas is emitted.

Methane has a number of indirect effects in addition to forming CO_2. Firstly, the main chemical that destroys methane in the atmosphere is the hydroxyl radical (OH). Methane reacts with OH and so more methane means that the concentration of OH goes down. Effectively, methane increases its own atmospheric lifetime and therefore its overall radiative effect. The second effect is that the oxidation of methane can produce ozone. Thirdly, as well as making CO_2 the oxidation of methane produces water; this is a major source of water vapor in the stratosphere, which is otherwise very dry. CO and NMVOC also produce CO_2 when they are oxidized. They remove OH from the atmosphere and this leads to higher concentrations of methane. The surprising effect of this is that the global warming potential of CO is three times that of CO_2. The same process that converts NMVOC to carbon dioxide can also lead to the formation of tropospheric ozone. Halocarbons have an indirect effect because they destroy stratospheric ozone. Finally hydrogen can lead to ozone production and CH4 increases as well as producing water vapor in the stratosphere.

Contribution of Clouds to Earth's Greenhouse Effect

The major non-gas contributor to Earth's greenhouse effect, clouds, also absorb and emit infrared radiation and thus have an effect on radiative properties of the greenhouse gases. Clouds are water droplets or ice crystals suspended in the atmosphere.

IMPACTS ON THE OVERALL GREENHOUSE EFFECT

Schmidt *et al.* (2010) analysed how individual components of the atmosphere contribute to the total greenhouse effect. They estimated that water vapor accounts for about 50% of Earth's greenhouse effect, with clouds contributing 25%, carbon dioxide 20%, and the minor greenhouse gases and aerosols accounting for the remaining 5%. In the study, the reference model atmosphere is for 1980 conditions. Image credit: NASA.

The contribution of each gas to the greenhouse effect is affected by the characteristics of that gas, its abundance, and any indirect effects it may cause. For example, the direct radiative effect of a mass of methane is about 72 times stronger than the same mass of carbon dioxide over a 20-year time frame but it is present in much smaller concentrations so that its total direct radiative effect is smaller, in part due to its shorter atmospheric lifetime. On the other hand, in addition to its direct radiative impact, methane has a large, indirect radiative effect because it contributes to ozone formation. Shindell *et al.* (2005) argue that the contribution to climate change from methane is at least double previous estimates as a result of this effect.

When ranked by their direct contribution to the greenhouse effect, the most important are:

Compound	Formula	Concentration in atmosphere(ppm)	Contribution (%)
Water vapor and clouds	H_2O	10–50,000[(A)]	**36–72%**
Carbon dioxide	CO_2	~400	**9–26%**
Methane	CH_4	~1.8	**4–9%**
Ozone	O_3	2–8[(B)]	**3–7%**
notes: (A) Water vapor strongly varies locally (B) The concentration in stratosphere. About 90% of the ozone in Earth's atmosphere is contained in the stratosphere.			

In addition to the main greenhouse gases listed above, other greenhouse gases include sulfur hexafluoride, hydrofluorocarbons and perfluorocarbons (see IPCC list of greenhouse gases). Some greenhouse gases are not often listed. For example, nitrogen trifluoride has a high global warming potential (GWP) but is only present in very small quantities.

Proportion of Direct Effects at a Given Moment

It is not possible to state that a certain gas causes an exact percentage of the greenhouse effect. This is because some of the gases absorb and emit radiation at the same frequencies as others, so that the total greenhouse effect is not simply the sum of the influence of each gas. The higher ends of the ranges quoted are for each gas alone; the lower ends account for overlaps with the other gases. In addition, some gases such as methane are known to have large indirect effects that are still being quantified.

Atmospheric Lifetime

Aside from water vapor, which has a residence time of about nine days, major greenhouse gases are well mixed and take many years to leave the atmosphere. Although it is not easy to know with precision how long it takes greenhouse gases to leave the atmosphere, there are estimates for the principal greenhouse gases. Jacob (1999) defines the lifetime τ of an atmospheric species X in a one-box model as the average time that a molecule of X remains in the box. Mathematically τ can be defined as the ratio of the mass m (in kg) of X in the box to its removal rate, which is the sum of the flow of X out of the box (F_{out}), chemical loss of X_L, and deposition of $\tau = \frac{m}{F_{out} + L + D}$ If one stopped pouring any of this gas into the box, then after a time {\displaystyle \tau }, its concentration would be about halved.

The atmospheric lifetime of a species therefore measures the time required to restore equilibrium following a sudden increase or decrease in its concentration in the atmosphere. Individual atoms or molecules may be lost or deposited to sinks such as the soil, the oceans and other waters, or vegetation and other biological systems, reducing the excess to background concentrations. The average time taken to achieve this is the mean lifetime.

Carbon dioxide has a variable atmospheric lifetime, and cannot be specified precisely. The atmospheric lifetime of CO_2 is estimated of the order of 30–95 years. This figure accounts for CO_2molecules being removed from the atmosphere by mixing into the ocean, photosynthesis, and other processes. However, this excludes the balancing fluxes of CO_2 into the atmosphere from the geological reservoirs, which have slower characteristic rates. Although more than half of the CO_2 emitted is removed from the atmosphere within a century, some fraction (about 20%) of emitted CO_2 remains in the atmosphere for many thousands of years. Similar issues apply to other greenhouse gases, many of which have longer mean lifetimes than CO_2. E.g., N_2O has a mean atmospheric lifetime of 114 years.

Radiative Forcing

Earth absorbs some of the radiant energy received from the sun, reflects some of it as light and reflects or radiates the rest back to space as heat. Earth's surface temperature depends on this balance between incoming and outgoing energy. If this energy balance is shifted, Earth's surface could become warmer or cooler, leading to a variety of changes in global climate.

A number of natural and man-made mechanisms can affect the global energy balance and force changes in Earth's climate. Greenhouse gases are one such mechanism. Greenhouse gases in the atmosphere absorb and re-emit some of the outgoing energy radiated from Earth's surface, causing that heat to be retained in the lower atmosphere. As explained above, some greenhouse gases remain in the atmosphere for decades or even centuries, and therefore can affect Earth's energy balance over a long time period. Factors that influence Earth's energy balance can be quantified in terms of "radiative climate forcing." Positive radiative forcing indicates warming (for example, by increasing incoming energy or decreasing the amount of energy that escapes to space), whereas negative forcing is associated with cooling.

Global Warming Potential

The global warming potential (GWP) depends on both the efficiency of the molecule as a greenhouse gas and its atmospheric lifetime. GWP is measured relative to the same **mass** of CO_2 and evaluated for a specific timescale. Thus, if a gas has a high (positive) radiative forcing but also a short lifetime, it will have a large GWP on a 20-year scale but a small one on a 100-year scale. Conversely, if a molecule has a longer atmospheric lifetime than CO_2 its GWP will increase with the timescale considered. Carbon dioxide is defined to have a GWP of 1 over all time periods.

Methane has an atmospheric lifetime of 12 ± 3 years. The 2007 IPCC report lists the GWP as 72 over a time scale of 20 years, 25 over 100 years and 7.6 over 500 years. A 2014 analysis, however, states that although methane's initial impact is about 100 times greater than that of CO_2, because of the shorter atmospheric lifetime, after six or seven decades, the impact of the two gases is about equal, and from then on methane's relative role continues to decline. The decrease in GWP at longer times is because methane is degraded to water and CO_2 through chemical reactions in the atmosphere.

Examples of the atmospheric lifetime and GWP relative to CO_2 for several greenhouse gases are given in the following table:

Atmospheric lifetime and GWP relative to CO_2 at different time horizon for various greenhouse gases.

Gas name	Chemical formula	Lifetime (years)	Global warming potential (GWP) for given time horizon		
			20-yr	100-yr	500-yr
Carbon dioxide	CO_2	30–95	1	1	1
Methane	CH_4	12	72	25	7.6
Nitrous oxide N 2O 114 289				298	153
CFC-12	CCl_2F_2	100	11 000	10 900	5 200
HCFC-22	$CHClF_2$	12	5 160	1 810	549
Tetrafluoromethane	CF_4	50 000	5 210	7 390	11 200
Hexafluoroethane	C_2F_6	10 000	8 630	12 200	18 200
Sulfur hexafluoride	SF_6	3 200	16 300	22 800	32 600
Nitrogen trifluoride	NF_3	740	12 300	17 200	20 700

The use of CFC-12 (except some essential uses) has been phased out due to its ozone depleting properties. The phasing-out of less active HCFC-compounds will be completed in 2030.

Carbon dioxide in Earth's atmosphere if *half* of global-warming emissions are *not* absorbed. (NASA simulation; 9 November 2015)

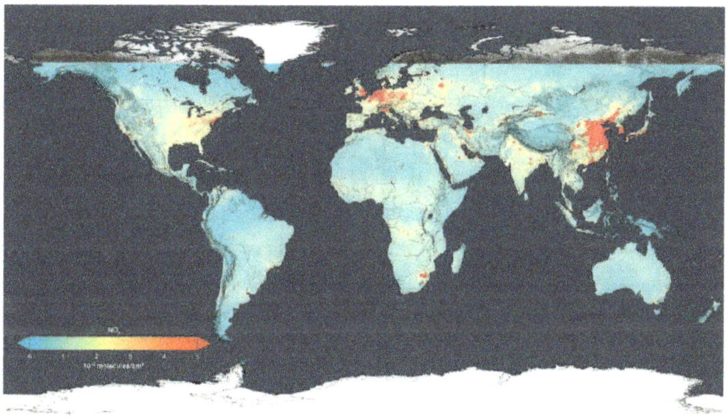

Nitrogen dioxide 2014 – global air quality levels(released 14 December 2015).

NATURAL AND ANTHROPOGENIC SOURCES

This diagram shows a simplified representation of the contemporary global carbon cycle. Changes are measured in gigatons of carbon per year (GtC/y). Canadell *et al.* (2007) estimated the growth rate of global average atmospheric CO_2 for 2000–2006 as 1.93 parts-per-million per year (4.1 petagrams of carbon per year).

Aside from purely human-produced synthetic halocarbons, most greenhouse gases have both natural and human-caused sources. During the pre-industrial Holocene, concentrations of existing gases were roughly constant. In the industrial era, human activities have added greenhouse gases to the atmosphere, mainly through the burning of fossil fuels and clearing of forests.

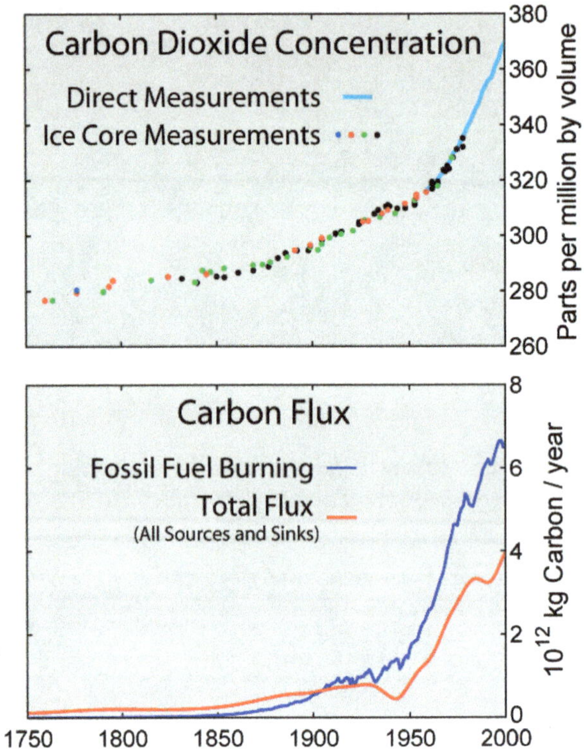

Top: Increasing atmospheric carbon dioxide levels as measured in the atmosphere and reflected in ice cores. Bottom: The amount of net carbon increase in the atmosphere, compared to carbon emissions from burning fossil fuel.

The 2007 Fourth Assessment Report compiled by the IPCC (AR4) noted that "changes in atmospheric concentrations of greenhouse gases and aerosols, land cover and solar radiation alter the energy balance of the climate system", and concluded that "increases in anthropogenic greenhouse gas concentrations is very likely to have caused most of the increases in global average temperatures since the mid-20th century". In AR4, "most of" is defined as more than 50%.

Abbreviations used in the two tables below: ppm = parts-per-million; ppb = parts-per-billion; ppt = parts-per-trillion; W/m^2 = watts per square metre

Current greenhouse gas concentrations					
Gas	Pre-1750 tropospheric concentration	Recent tropospheric concentration	Absolute increase since 1750	Percentage increase since 1750	Increased radiative forcing (W/m^2)
Carbon dioxide (CO_2)	280 ppm	395.4 ppm	115.4 ppm	41.2%	1.88

Methane (CH_4)	700 ppb	1893 ppb / 1762 ppb	1193 ppb / 1062 ppb	170.4% / 151.7%	0.49
Nitrous oxide (N_2O)	270 ppb	326 ppb / 324 ppb	56 ppb / 54 ppb	20.7% / 20.0%	0.17
Tropospheric ozone (O_3)	237 ppb	337 ppb	100 ppb	42%	0.4

Relevant to radiative forcing and/or ozone depletion; all of the following have no natural sources and hence zero amounts pre-industrial		
Gas	**Recent tropospheric concentration**	**Increased radiative forcing (W/m²)**
CFC-11 (trichlorofluoromethane) (CCl_3F)	236 ppt / 234 ppt	0.061
CFC-12 (CCl_2F_2)	527 ppt / 527 ppt	0.169
CFC-113 ($Cl_2FC-CClF_2$)	74 ppt / 74 ppt	0.022
HCFC-22 ($CHClF_2$)	231 ppt / 210 ppt	0.046
HCFC-141b (CH_3CCl_2F)	24 ppt / 21 ppt	0.0036
HCFC-142b (CH_3CClF_2)	23 ppt / 21 ppt	0.0042
Halon 1211 ($CBrClF_2$)	4.1 ppt / 4.0 ppt	0.0012
Halon 1301 ($CBrClF_3$)	3.3 ppt / 3.3 ppt	0.001
HFC-134a (CH_2FCF_3)	75 ppt / 64 ppt	0.0108
Carbon tetrachloride (CCl_4)	85 ppt / 83 ppt	0.0143
Sulfur hexafluoride (SF_6)	7.79 ppt / 7.39 ppt	0.0043
Other halocarbons	Varies bysubstance	collectively0.02
Halocarbons in total		0.3574

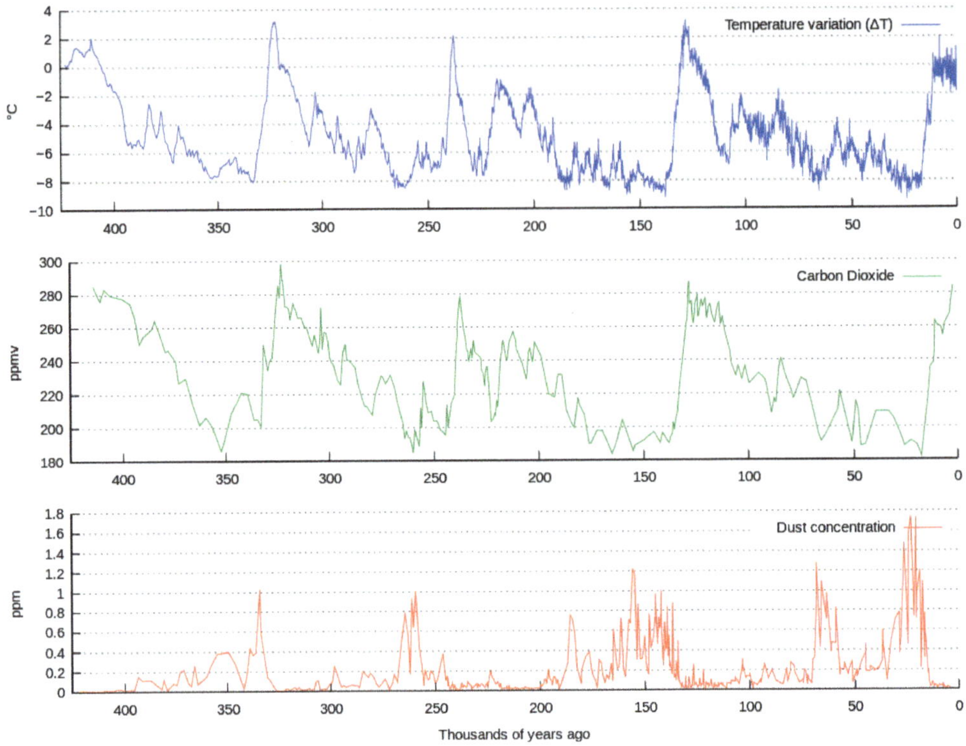

400,000 years of ice core data.

Ice cores provide evidence for greenhouse gas concentration variations over the past 800,000 years (see the following section). Both CO_2 and CH_4 vary between glacial and interglacial phases, and concentrations of these gases correlate strongly with temperature. Direct data does not exist for periods earlier than those represented in the ice core record, a record that indicates CO_2 mole fractions stayed within a range of 180 ppm to 280 ppm throughout the last 800,000 years, until the increase of the last 250 years. However, various proxies and modeling suggests larger variations in past epochs; 500 million years ago CO_2 levels were likely 10 times higher than now. Indeed, higher CO_2 concentrations are thought to have prevailed throughout most of the Phanerozoic eon, with concentrations four to six times current concentrations during the Mesozoic era, and ten to fifteen times current concentrations during the early Palaeozoic era until the middle of the Devonian period, about 400 Ma. The spread of land plants is thought to have reduced CO_2 concentrations during the late Devonian, and plant activities as both sources and sinks of CO_2 have since been important in providing stabilising feedbacks. Earlier still, a 200-million year period of intermittent, widespread glaciation extending close to the equator (Snowball Earth) appears to have been ended suddenly, about 550 Ma, by a colossal volcanic outgassing that raised the CO_2 concentration of the atmosphere abruptly to 12%, about 350 times modern levels, causing extreme greenhouse conditions and carbonate deposition as limestone at the rate of about 1 mm per day. This episode marked the close of the Precambrian eon, and was succeeded by the generally warmer conditions of the Phanerozoic, during which multicellular animal and plant life evolved. No volcanic carbon dioxide emission of comparable scale has occurred since. In the modern era, emissions to the atmosphere from volcanoes are only about 1% of emissions from human sources.

Ice Cores

Measurements from Antarctic ice cores show that before industrial emissions started atmospheric CO_2mole fractions were about 280 parts per million (ppm), and stayed between 260 and 280 during the preceding ten thousand years. Carbon dioxide mole fractions in the atmosphere have gone up by approximately 35 percent since the 1900s, rising from 280 parts per million by volume to 387 parts per million in 2009. One study using evidence from stomata of fossilized leaves suggests greater variability, with carbon dioxide mole fractions above 300 ppm during the period seven to ten thousand years ago, though others have argued that these findings more likely reflect calibration or contamination problems rather than actual CO_2 variability. Because of the way air is trapped in ice (pores in the ice close off slowly to form bubbles deep within the firn) and the time period represented in each ice sample analyzed, these figures represent averages of atmospheric concentrations of up to a few centuries rather than annual or decadal levels.

Changes since the Industrial Revolution

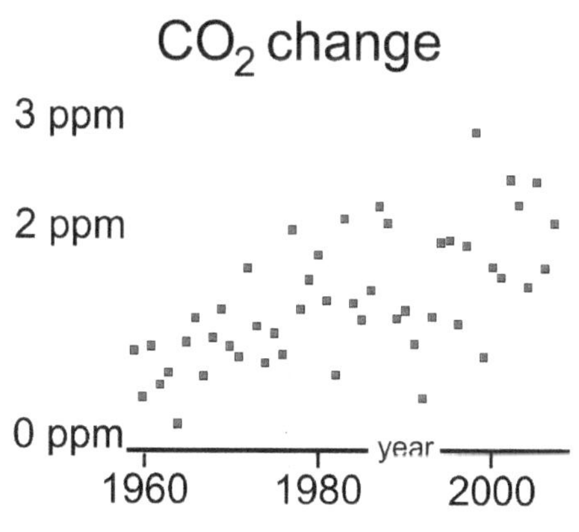

Recent year-to-year increase of atmospheric CO_2.

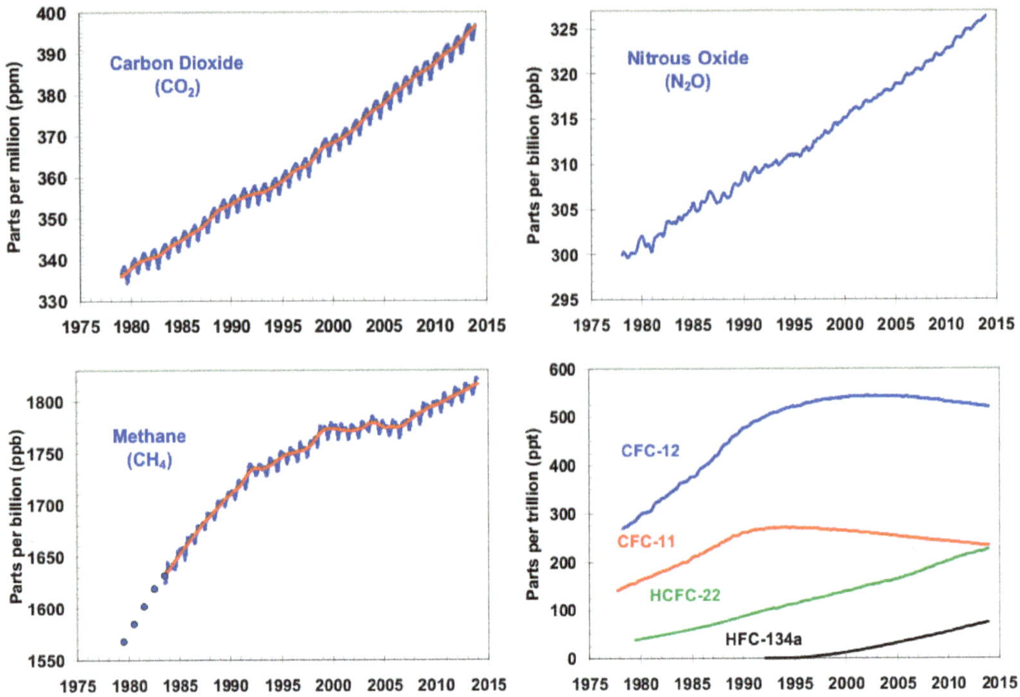

Major greenhouse gas trends.

Since the beginning of the Industrial Revolution, the concentrations of most of the greenhouse gases have increased. For example, the mole fraction of carbon dioxide has increased from 280 ppm by about 36% to 380 ppm, or 100 ppm over modern pre-industrial levels. The first 50 ppm increase took place in about 200 years, from the start of the Industrial Revolution to around 1973.; however the next 50 ppm increase took place in about 33 years, from 1973 to 2006.

Recent data also shows that the concentration is increasing at a higher rate. In the 1960s, the average annual increase was only 37% of what it was in 2000 through 2007.

Today, the stock of carbon in the atmosphere increases by more than 3 million tonnes per annum (0.04%) compared with the existing stock. This increase is the result of human activities by burning fossil fuels, deforestation and forest degradation in tropical and boreal regions.

The other greenhouse gases produced from human activity show similar increases in both amount and rate of increase. Many observations are available online in a variety of Atmospheric Chemistry Observational Databases.

ANTHROPOGENIC GREENHOUSE GASES

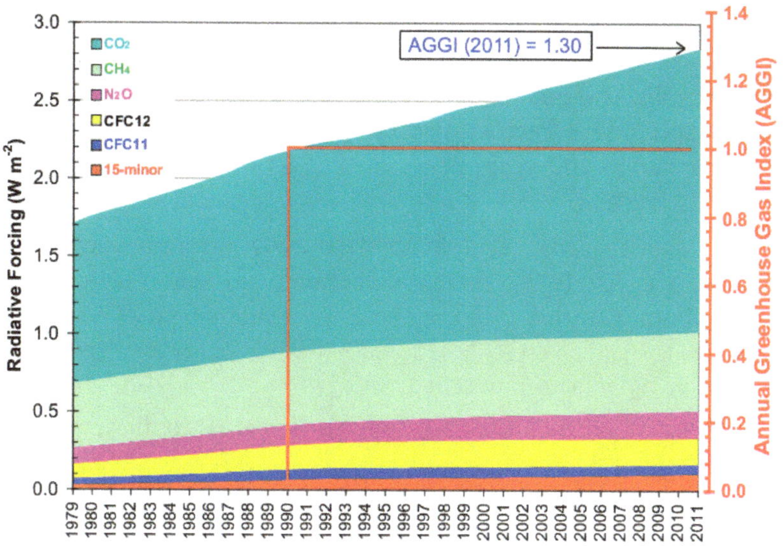

This graph shows changes in the annual greenhouse gas index (AGGI) between 1979 and 2011. The AGGI measures the levels of greenhouse gases in the atmosphere based on their ability to cause changes in Earth's climate.

This bar graph shows global greenhouse gas emissions by sector from 1990 to 2005, measured in carbon dioxide equivalents.

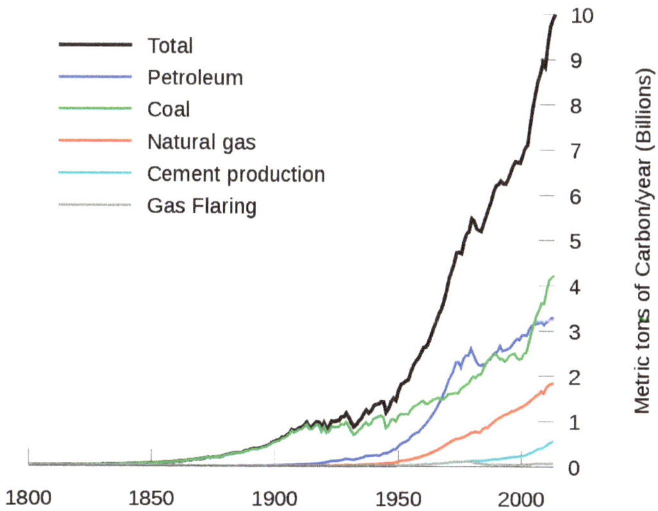

Modern global CO2 emissions from the burning of fossil fuels.

Since about 1750 human activity has increased the concentration of carbon dioxide and other greenhouse gases. Measured atmospheric concentrations of carbon dioxide are currently 100 ppm higher than pre-industrial levels. Natural sources of carbon dioxide are more than 20 times greater than sources due to human activity, but over periods longer than a few years natural sources are closely balanced by natural sinks, mainly photosynthesis of carbon compounds by plants and marine plankton. As a result of this balance, the atmospheric mole fraction of carbon dioxide remained between 260 and 280 parts per million for the 10,000 years between the end of the last glacial maximum and the start of the industrial era.

It is likely that anthropogenic (i.e., human-induced) warming, such as that due to elevated greenhouse gas levels, has had a discernible influence on many physical and biological systems. Future warming is projected to have a range of impacts, including sea level rise,increased frequencies and severities of some extreme weather events, loss of biodiversity, and regional changes in agricultural productivity.

The main sources of greenhouse gases due to human activity are:

- burning of fossil fuels and deforestation leading to higher carbon dioxide concentrations in the air. Land use change (mainly deforestation in the tropics) account for up to one third of total anthropogenic CO_2emissions.
- livestock enteric fermentation and manure management, paddy rice farming, land use and wetland changes, man-made lakes, pipeline losses, and covered vented landfill emissions leading to higher methane atmospheric concentrations. Many of the newer style fully vented septic systems that enhance and target the fermentation process also are sources of atmospheric methane.
- use of chlorofluorocarbons (CFCs) in refrigeration systems, and use of CFCs and halons in fire suppression systems and manufacturing processes.
- agricultural activities, including the use of fertilizers, that lead to higher nitrous oxide (N 2O) concentrations.

The seven sources of CO_2 from fossil fuel combustion are (with percentage contributions for 2000–2004):

Seven main fossil fuel combustion sources	Contribution (%)
Liquid fuels (e.g., gasoline, fuel oil)	36%
Solid fuels (e.g., coal)	35%
Gaseous fuels (e.g., natural gas)	20%
Cement production	3 %
Flaring gas industrially and at wells	< 1%
Non-fuel hydrocarbons	< 1%
"International bunker fuels" of transport not included in national inventories	4 %

Carbon dioxide, methane, nitrous oxide (N_2O) and three groups of fluorinated gases (sulfur hexafluoride (SF6), hydrofluorocarbons (HFCs), and perfluorocarbons (PFCs)) are the major anthropogenic greenhouse gases,:[147] and are regulated under the Kyoto Protocol international treaty, which came into force in 2005. Emissions limitations specified in the Kyoto Protocol expired in 2012. The Cancún agreement, agreed in 2010, includes voluntary pledges made by 76 countries to control emissions. At the time of the agreement, these 76 countries were collectively responsible for 85% of annual global emissions.

Although CFCs are greenhouse gases, they are regulated by the Montreal Protocol, which was motivated by CFCs' contribution to ozone depletion rather than by their contribution to global warming. Note that ozone depletion has only a minor role in greenhouse warming though the two processes often are confused in the media. On 15 October 2016, negotiators from over 170 nations meeting at the summit of the United Nations Environment Programme reached a legally binding accord to phase out hydrofluorocarbons (HFCs) in an amendment to the Montreal Protocol.

Sectors

Tourism

According to UNEP global tourism is closely linked to climate change. Tourism is a significant contributor to the increasing concentrations of greenhouse gases in the atmosphere. Tourism accounts for about 50% of traffic movements. Rapidly expanding air traffic contributes about 2.5% of the production of CO_2. The number of international travelers is expected to increase from 594 million in 1996 to 1.6 billion by 2020, adding greatly to the problem unless steps are taken to reduce emissions.

Road Haulage

The road haulage industry plays a part in production of CO_2, contributing around 20% of the UK's total carbon emissions a year, with only the energy industry having a larger impact at around 39%. Average carbon emissions within the haulage industry are falling—in the thirty-year period from 1977 to 2007, the carbon emissions associated with a 200-mile journey fell by 21 percent; NOx emissions are also down 87 percent, whereas journey times have fallen by around a third. Due to their size, HGVs often receive criticism regarding their CO2 emissions; however, rapid development in engine technology and fuel management is having a largely positive effect.

ROLE OF WATER VAPOR

Water vapor accounts for the largest percentage of the greenhouse effect, between 36% and 66% for clear sky conditions and between 66% and 85% when including clouds. Water vapor concentrations fluctuate regionally, but human activity does not significantly affect water vapor concentrations except at local scales, such as near irrigated fields. The atmospheric concentration of vapor is highly variable and depends largely on temperature, from less than 0.01% in extremely cold regions up to 3% by mass in saturated air at about 32 °C. (See Relative humidity#other important facts.)

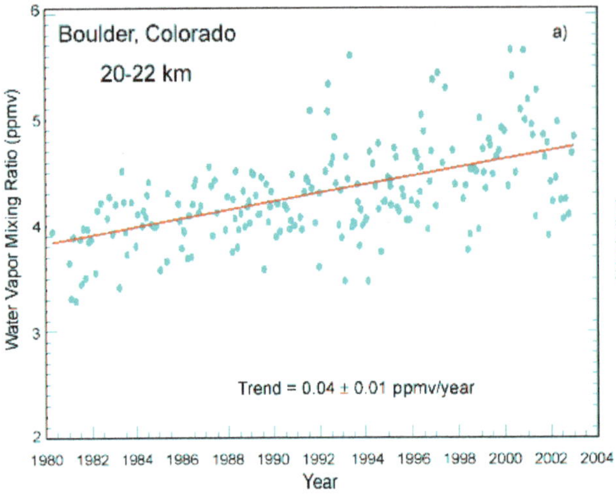

Increasing water vapor in the stratosphere at Boulder, Colorado.

The average residence time of a water molecule in the atmosphere is only about nine days, compared to years or centuries for other greenhouse gases such as CH 4 and CO_2. Thus, water vapor responds to and amplifies effects of the other greenhouse gases. The Clausius–Clapeyron relation establishes that more water vapor will be present per unit volume at elevated temperatures. This and other basic principles indicate that warming associated with increased concentrations of the other greenhouse gases also will increase the concentration of water vapor (assuming that the relative humidity remains approximately constant; modeling and observational studies find that this is indeed so). Because water vapor is a greenhouse gas, this results in further warming and so is a "positive feedback" that amplifies the original warming. Eventually other earth processes offset these positive feedbacks, stabilizing the global temperature at a new equilibrium and preventing the loss of Earth's water through a Venus-like runaway greenhouse effect.

DIRECT GREENHOUSE GAS EMISSIONS

Between the period 1970 to 2004, GHG emissions (measured in CO_2-equivalent) increased at an average rate of 1.6% per year, with CO_2 emissions from the use of fossil fuels growing at a rate of 1.9% per year. Total anthropogenic emissions at the end of 2009 were estimated at 49.5 gigatonnes CO_2-equivalent.[15] These emissions include CO_2 from fossil fuel use and from land use, as well as emissions of methane, nitrous oxide and other GHGs covered by the Kyoto Protocol.

At present, the primary source of CO_2 emissions is the burning of coal, natural gas, and petroleum for electricity and heat.

Regional and National Attribution of Emissions

This figure shows the relative fraction of anthropogenic greenhouse gases coming from each of eight categories of sources, as estimated by the Emission Database for Global

Atmospheric Research version 3.2, fast track 2000 project . These values are intended to provide a snapshot of global annual greenhouse gas emissions in the year 2000. The top panel shows the sum over all anthropogenic greenhouse gases, weighted by their global warming potential over the next 100 years. This consists of 72% carbon dioxide, 18% methane, 8% nitrous oxide and 1% other gases. Lower panels show the comparable information for each of these three primary greenhouse gases, with the same coloring of sectors as used in the top chart. Segments with less than 1% fraction are not labeled.

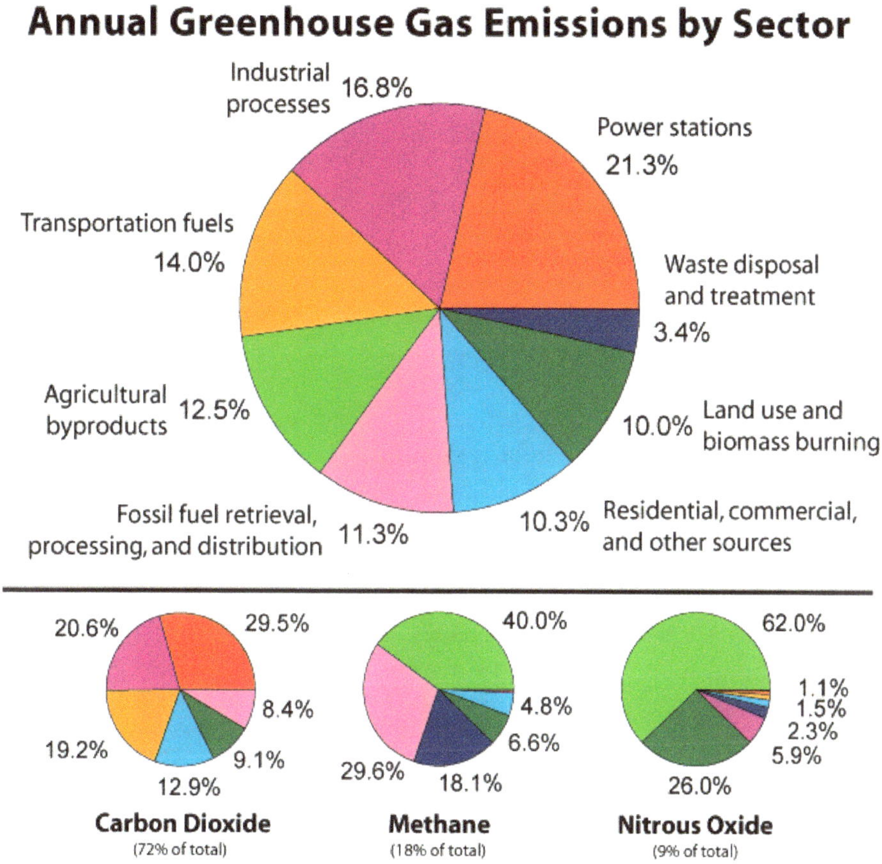

There are several different ways of measuring GHG emissions, for example, see World Bank (2010) for tables of national emissions data. Some variables that have been reported include:

- Definition of measurement boundaries: Emissions can be attributed geographically, to the area where they were emitted (the territory principle) or by the activity principle to the territory produced the emissions. These two principles result in different totals when measuring, for example, electricity importation from one country to another, or emissions at an international airport.
- Time horizon of different GHGs: Contribution of a given GHG is reported as a CO_2 equivalent. The calculation to determine this takes into account how long that gas remains in the atmosphere. This is not always known accurately and calculations must be regularly updated to reflect new information.

- What sectors are included in the calculation (e.g., energy industries, industrial processes, agriculture etc.): There is often a conflict between transparency and availability of data.
- The measurement protocol itself: This may be via direct measurement or estimation. The four main methods are the emission factor-based method, mass balance method, predictive emissions monitoring systems, and continuous emissions monitoring systems. These methods differ in accuracy, cost, and usability.

These different measures are sometimes used by different countries to assert various policy/ethical positions on climate change (Banuri *et al.*, 1996, p. 94). This use of different measures leads to a lack of comparability, which is problematic when monitoring progress towards targets. There are arguments for the adoption of a common measurement tool, or at least the development of communication between different tools.

Emissions may be measured over long time periods. This measurement type is called historical or cumulative emissions. Cumulative emissions give some indication of who is responsible for the build-up in the atmospheric concentration of GHGs (IEA, 2007, p. 199).

The national accounts balance would be positively related to carbon emissions. The national accounts balance shows the difference between exports and imports. For many richer nations, such as the United States, the accounts balance is negative because more goods are imported than they are exported. This is mostly due to the fact that it is cheaper to produce goods outside of developed countries, leading the economies of developed countries to become increasingly dependent on services and not goods. We believed that a positive accounts balance would means that more production was occurring in a country, so more factories working would increase carbon emission levels.(Holtz-Eakin, 1995, pp.;85;101).

Emissions may also be measured across shorter time periods. Emissions changes may, for example, be measured against a base year of 1990. 1990 was used in the United Nations Framework Convention on Climate Change (UNFCCC) as the base year for emissions, and is also used in the Kyoto Protocol (some gases are also measured from the year 1995). A country's emissions may also be reported as a proportion of global emissions for a particular year.

Another measurement is of per capita emissions. This divides a country's total annual emissions by its mid-year population. Per capita emissions may be based on historical or annual emissions (Banuri *et al.*, 1996, pp. 106–107).

Land-use Change

Land-use change, e.g., the clearing of forests for agricultural use, can affect the concentration of GHGs in the atmosphere by altering how much carbon flows out of the atmosphere into carbon sinks. Accounting for land-use change can be understood as an attempt to measure "net" emissions, i.e., gross emissions from all GHG sources minus the removal of emissions from the atmosphere by carbon sinks (Banuri *et al.*, 1996, pp. 92–93).

There are substantial uncertainties in the measurement of net carbon emissions. Additionally, there is controversy over how carbon sinks should be allocated between different regions and over time (Banuri *et al.*, 1996, p. 93). For instance, concentrating on more recent changes in carbon sinks is likely to favour those regions that have deforested earlier, e.g., Europe.

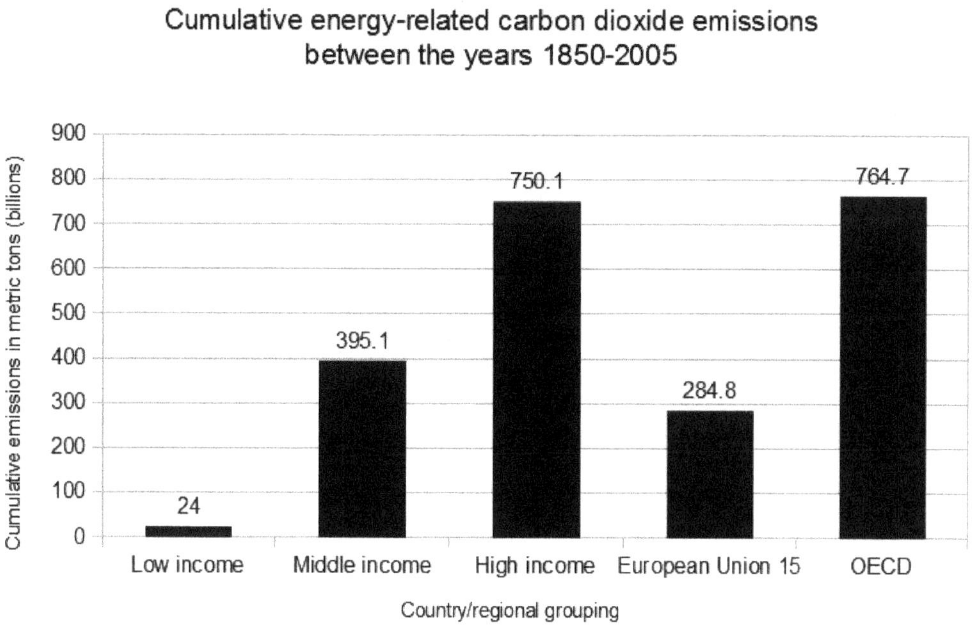

Greenhouse gas emissions from agriculture, forestry and other land use, 1970–2010.

Greenhouse gas intensity

Greenhouse gas intensity is a ratio between greenhouse gas emissions and another metric, e.g., gross domestic product (GDP) or energy use. The terms "carbon intensity" and "emissions intensity" are also sometimes used. GHG intensities may be calculated using market exchange rates (MER) or purchasing power parity (PPP) (Banuri *et al.*, 1996, p. 96). Calculations based on MER show large differences in intensities between developed and developing countries, whereas calculations based on PPP show smaller differences.

Cumulative and Historical Emissions

Cumulative energy-related CO_2emissions between the years 1850–2005 grouped into low-income, middle-income, high-income, the EU-15, and the OECD countries.

Cumulative energy-related CO_2emissions between the years 1850–2005 for individual countries.

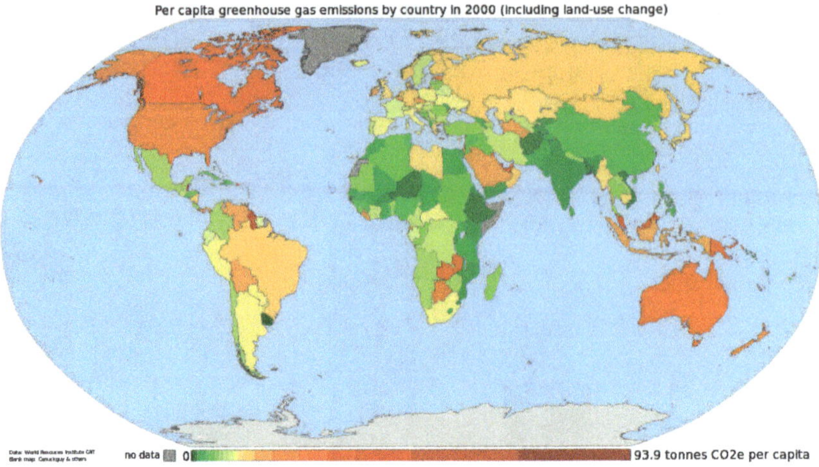

Map of cumulative per capita anthropogenic atmospheric CO_2 emissions by country. Cumulative emissions include land use change, and are measured between the years 1950 and 2000.

trends in annual CO_2emissions from fuel combustion between 1971 and 2009.

Regional trends in annual per capita CO_2 emissions from fuel combustion between 1971 and 2009.

Cumulative anthropogenic (i.e., human-emitted) emissions of $CO_2$from fossil fuel use are a major cause of global warming, and give some indication of which countries have contributed most to human-induced climate change.[:15]

Top-5 historic CO_2 contributors by region over the years 1800 to 1988 (in %)		
Region	**Industrial CO_2**	**Total CO_2**
OECD North America	33.2	29.7
OECD Europe	26.1	16.6
Former USSR	14.1	12.5
China	5.5	6.0
Eastern Europe	5.5	4.8

The table above to the left is based on Banuri *et al.* (1996, p. 94).Overall, developed countries accounted for 83.8% of industrial CO_2emissions over this time period, and 67.8% of total CO_2 emissions. Developing countries accounted for industrial CO_2 emissions of 16.2% over this time period, and 32.2% of total CO_2 emissions. The estimate of total CO_2 emissions includes biotic carbon emissions, mainly from deforestation. Banuri *et al.* (1996, p. 94) calculated per capita cumulative emissions based on then-current population. The ratio in per capita emissions between industrialized countries and developing countries was estimated at more than 10 to 1.

Including biotic emissions brings about the same controversy mentioned earlier regarding carbon sinks and land-use change (Banuri *et al.*, 1996, pp. 93–94). The actual calculation of net emissions is very complex, and is affected by how carbon sinks are allocated between regions and the dynamics of the climate system. Non-OECD countries accounted for 42% of cumulative energy-related CO_2 emissions between 1890 and 2007.[:179–180] Over this time period, the US accounted for 28% of emissions; the EU, 23%; Russia, 11%; China, 9%; other OECD countries, 5%; Japan, 4%; India, 3%; and the rest of the world, 18%.[:179–180]

Changes since a Particular Base Year

Between 1970 and 2004, global growth in annual CO_2 emissions was driven by North America, Asia, and the Middle East. The sharp acceleration in CO_2 emissions since 2000 to more than a 3% increase per year (more than 2 ppm per year) from 1.1% per year during the 1990s is attributable to the lapse of formerly declining trends in carbon intensity of both developing and developed nations. China was responsible for most of global growth in emissions during this period. Localised plummeting emissions associated with the collapse of the Soviet Union have been followed by slow emissions growth in this region due to more efficient energy use, made necessary by the increasing proportion of it that is exported. In comparison, methane has not increased appreciably, and N 2O by 0.25% y^{-1}.

Using different base years for measuring emissions has an effect on estimates of national contributions to global warming.[:17–18] This can be calculated by dividing a country's highest contribution to global warming starting from a particular base year, by that country's minimum contribution to global warming starting from a particular base year. Choosing between different base years of 1750, 1900, 1950, and 1990 has a significant effect for most countries.[:17–18] Within the G8 group of countries, it is most significant for the UK, France and Germany. These countries have a long history of CO_2emissions (see the section on Cumulative and historical emissions).

Annual Emissions

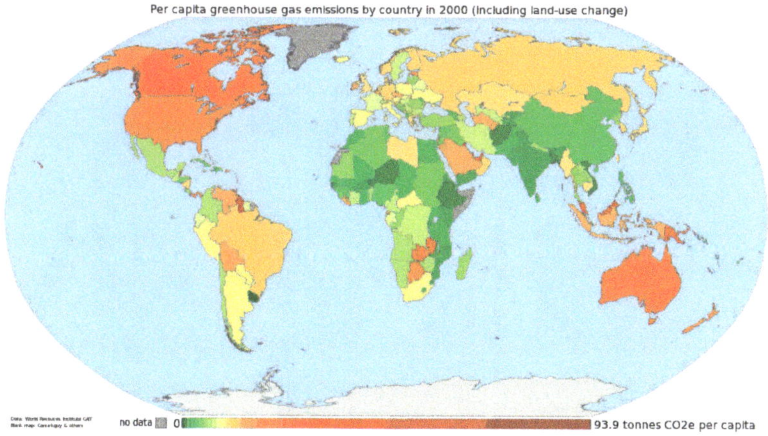

Per capita anthropogenic greenhouse gas emissions by country for the year 2000 including land-use change.

Annual per capita emissions in the industrialized countries are typically as much as ten times the average in developing countries.:144 Due to China's fast economic development, its annual per capita emissions are quickly approaching the levels of those in the Annex I group of the Kyoto Protocol (i.e., the developed countries excluding the USA).Other countries with fast growing emissions are South Korea, Iran, and Australia (which apart from the oil rich Persian Gulf states, now has the highest percapita emission rate in the world). On the other hand, annual per capita emissions of the EU-15 and the USA are gradually decreasing over time. Emissions in Russia and Ukraine have decreased fastest since 1990 due to economic restructuring in these countries.

Energy statistics for fast growing economies are less accurate than those for the industrialized countries. For China's annual emissions in 2008, the Netherlands Environmental Assessment Agency estimated an uncertainty range of about 10%.

The GHG footprint, or greenhouse gas footprint, refers to the amount of GHG that are emitted during the creation of products or services. It is more comprehensive than the commonly used carbon footprint, which measures only carbon dioxide, one of many greenhouse gases.

2015 was the first year to see both total global economic growth and a reduction of carbon emissions.

Top Emitter Countries

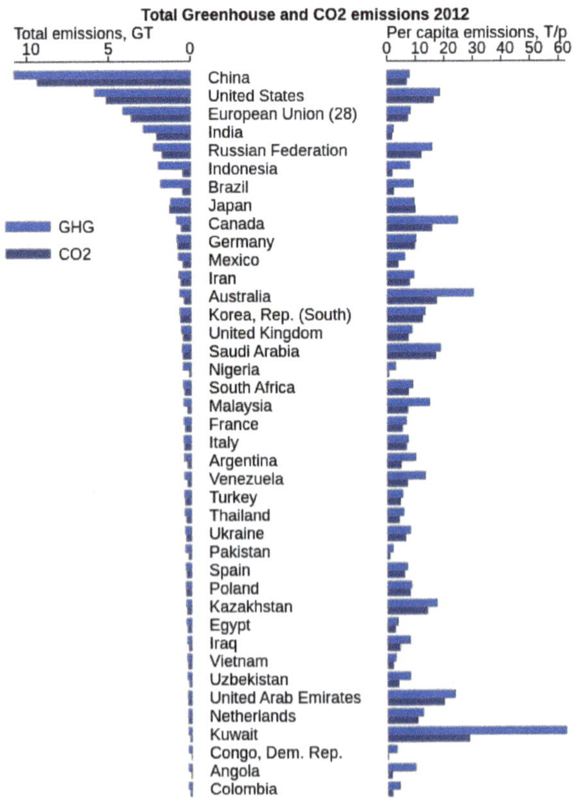

The top 40 countries emitting all greenhouse gases, showing both that derived from all sources including land clearance and forestry and also the CO2 component excluding those sources. Per capita figures are included. Data taken from World Resources Institute, Washington. Note that Indonesia and Brazil show very much higher than on graphs simply showing fossil fuel use.

Annual

In 2009, the annual top ten emitting countries accounted for about two-thirds of the world's annual energy-related CO_2emissions.

Top-10 annual energy-related CO_2emitters for the year 2009

Country	% of global total annual emissions	Tonnes of GHG per capita
People's Rep. of China	23.6	5.1
United States	17.9	16.9
India	5.5	1.4
Russian Federation	5.3	10.8
Japan	3.8	8.6
Germany	2.6	9.2
Islamic Rep. of Iran	1.8	7.3
Canada	1.8	15.4
South Korea	1.8	10.6
United Kingdom	1.6	7.5

Cumulative

Top-10 cumulative energy-related CO_2 emitters between 1850 and 2008

Country	% of world total	Metric tonnes CO_2 per person
United States	28.5	1,132.7
China	9.36	85.4
Russian Federation	7.95	677.2
Germany	6.78	998.9
United Kingdom	5.73	1,127.8

Top-10 cumulative energy-related CO_2 emitters between 1850 and 2008		
Country	**% of world total**	**Metric tonnes CO_2 per person**
Japan	3.88	367
France	2.73	514.9
India	2.52	26.7
Canada	2.17	789.2
Ukraine	2.13	556.4

Embedded Emissions

One way of attributing greenhouse gas (GHG) emissions is to measure the embedded emissions (also referred to as "embodied emissions") of goods that are being consumed. Emissions are usually measured according to production, rather than consumption. For example, in the main international treaty on climate change (the UNFCCC), countries report on emissions produced within their borders, e.g., the emissions produced from burning fossil fuels.[:179:1] Under a production-based accounting of emissions, embedded emissions on imported goods are attributed to the exporting, rather than the importing, country. Under a consumption-based accounting of emissions, embedded emissions on imported goods are attributed to the importing country, rather than the exporting, country.

Davis and Caldeira (2010)[:4] found that a substantial proportion of CO_2 emissions are traded internationally. The net effect of trade was to export emissions from China and other emerging markets to consumers in the US, Japan, and Western Europe. Based on annual emissions data from the year 2004, and on a per-capita consumption basis, the top-5 emitting countries were found to be (in tCO_2 per person, per year): Luxembourg (34.7), the US (22.0), Singapore (20.2), Australia (16.7), and Canada (16.6).[:5] Carbon Trust research revealed that approximately 25% of all CO_2 emissions from human activities 'flow' (i.e. are imported or exported) from one country to another. Major developed economies were found to be typically net importers of embodied carbon emissions — with UK consumption emissions 34% higher than production emissions, and Germany (29%), Japan (19%) and the USA (13%) also significant net importers of embodied emissions.

Effect of Policy

Governments have taken action to reduce GHG emissions (climate change mitigation). Assessments of policy effectiveness have included work by the Intergovernmental Panel on Climate Change,International Energy Agency, and United Nations Environment Programme. Policies implemented by governments have included national and regional targets to reduce emissions, promoting energy efficiency, and support for renewable energy such as Solar energy as an effective use of renewable energy because solar uses energy from the sun and does not release pollutants into the air.

Countries and regions listed in Annex I of the United Nations Framework Convention on Climate Change (UNFCCC) (i.e., the OECD and former planned economies of the Soviet Union) are required to submit periodic assessments to the UNFCCC of actions they are taking to address climate change.[3] Analysis by the UNFCCC (2011)[8] suggested that policies and measures undertaken by Annex I Parties may have produced emission savings of 1.5 thousand Tg CO_2-eq in the year 2010, with most savings made in the energy sector. The projected emissions saving of 1.5 thousand Tg CO_2-eq is measured against a hypothetical "baseline" of Annex I emissions, i.e., projected Annex I emissions in the absence of policies and measures. The total projected Annex I saving of 1.5 thousand CO_2-eq does not include emissions savings in seven of the Annex I Parties.[8]

Projections

A wide range of projections of future GHG emissions have been produced. Rogner *et al.* (2007) assessed the scientific literature on GHG projections. Rogner *et al.* (2007) concluded that unless energy policies changed substantially, the world would continue to depend on fossil fuels until 2025–2030. Projections suggest that more than 80% of the world's energy will come from fossil fuels. This conclusion was based on "much evidence" and "high agreement" in the literature. Projected annual energy-related CO_2 emissions in 2030 were 40–110% higher than in 2000, with two-thirds of the increase originating in developing countries. Projected annual per capita emissions in developed country regions remained substantially lower (2.8–5.1 tonnes CO_2) than those in developed country regions (9.6–15.1 tonnes CO_2). Projections consistently showed increase in annual world GHG emissions (the "Kyoto" gases, measured in CO_2-equivalent) of 25–90% by 2030, compared to 2000.

Relative CO_2 Emission from Various Fuels

One liter of gasoline, when used as a fuel, produces 2.32 kg (about 1300 liters or 1.3 cubic meters) of carbon dioxide, a greenhouse gas. One US gallon produces 19.4 lb (1,291.5 gallons or 172.65 cubic feet)

Mass of carbon dioxide emitted per quantity of energy for various fuels

Fuel name	CO_2 emitted (lbs/10^6 Btu)	CO_2 emitted (g/MJ)	CO_2 emitted (g/KWh)
Natural gas	117	50.30	181.08
Liquefied petroleum gas	139	59.76	215.14
Propane	139	59.76	215.14
Aviation gasoline	153	65.78	236.81

Mass of carbon dioxide emitted per quantity of energy for various fuels			
Fuel name	**CO_2 emitted ($lbs/10^6$ Btu)**	**CO_2 emitted (g/MJ)**	**CO_2 emitted (g/KWh)**
Automobile gasoline	156	67.07	241.45
Kerosene	159	68.36	246.10
Fuel oil	161	69.22	249.19
Tires/tire derived fuel	189	81.26	292.54
Wood and wood waste	195	83.83	301.79
Coal (bituminous)	205	88.13	317.27
Coal (sub-bituminous)	213	91.57	329.65
Coal (lignite)	215	92.43	332.75
Petroleum coke	225	96.73	348.23
Tar-sand Bitumen			
Coal (anthracite)	227	97.59	351.32

LIFE-CYCLE GREENHOUSE-GAS EMISSIONS OF ENERGY SOURCES

A literature review of numerous energy sources CO_2 emissions by the IPCC in 2011, found that, the CO_2emission value that fell within the 50th percentile of all total life cycle emissions studies conducted was as follows.

Lifecycle greenhouse gas emissions by electricity source.		
Technology	**Description**	**50th percentile (g CO_2/kWh_e)**
Hydroelectric	reservoir	4
Ocean Energy	wave and tidal	8
Wind	onshore	12
Nuclear	various generation II reactor types	16
Biomass	various	18
Solar thermal	parabolic trough	22
Geothermal	hot dry rock	45
Solar PV	Polycrystalline silicon	46

Lifecycle greenhouse gas emissions by electricity source.		
Technology	**Description**	**50th percentile ($g\ CO_2/kWh_e$)**
Natural gas	various combined cycle turbines without scrubbing	469
Coal	various generator types without scrubbing	1001

Removal from the atmosphere ("sinks")

Natural Processes

Greenhouse gases can be removed from the atmosphere by various processes, as a consequence of:

- a physical change (condensation and precipitation remove water vapor from the atmosphere).
- a chemical reaction within the atmosphere. For example, methane is oxidized by reaction with naturally occurring hydroxyl radical, OH• and degraded to CO_2 and water vapor (CO_2 from the oxidation of methane is not included in the methane Global warming potential). Other chemical reactions include solution and solid phase chemistry occurring in atmospheric aerosols.
- a physical exchange between the atmosphere and the other compartments of the planet. An example is the mixing of atmospheric gases into the oceans.
- a chemical change at the interface between the atmosphere and the other compartments of the planet. This is the case for CO_2, which is reduced by photosynthesis of plants, and which, after dissolving in the oceans, reacts to form carbonic acid and bicarbonate and carbonate ions (see ocean acidification).
- a photochemical change. Halocarbons are dissociated by UV light releasing Cl• and F• as free radicals in the stratosphere with harmful effects on ozone (halocarbons are generally too stable to disappear by chemical reaction in the atmosphere).

Negative Emissions

A number of technologies remove greenhouse gases emissions from the atmosphere. Most widely analysed are those that remove carbon dioxide from the atmosphere, either to geologic formations such as bio-energy with carbon capture and storage and carbon dioxide air capture, or to the soil as in the case with biochar. The IPCC has pointed out that many long-term climate scenario models require large-scale manmade negative emissions to avoid serious climate change.

HISTORY OF SCIENTIFIC RESEARCH

In the late 19th century scientists experimentally discovered that N_2 and O_2 do not absorb infrared radiation (called, at that time, "dark radiation"), while water (both as true vapor and condensed in the form of microscopic droplets suspended in clouds) and CO_2 and other poly-atomic gaseous molecules do absorb infrared radiation. In the early 20th century researchers realized that greenhouse gases in the atmosphere made Earth's overall temperature higher than it would be without them. During the late 20th century, a scientific consensus evolved that increasing concentrations of greenhouse gases in the atmosphere cause a substantial rise in global temperatures and changes to other parts of the climate system, with consequences for the environment and for human health.

REFERENCES

1. "IPCC AR4 SYR Appendix Glossary" (PDF). Retrieved 14 December 2008.
2. NASA GISS: Science Briefs: Greenhouse Gases: Refining the Role of Carbon Dioxide". www.giss.nasa.gov. Retrieved 2016-04-26.
3. Karl TR, Trenberth KE (2003). "Modern global climate change". Science. 302 (5651): 1719–23. Bibcode:2003Sci...302.1719K. doi:10.1126/science.1090228. PMID 14657489.
4. Le Treut H.; Somerville R.; Cubasch U.; Ding Y.; Mauritzen C.; Mokssit A.; Peterson T.; Prather M. (2007). Historical overview of climate change science. In: Climate change 2007: The physical science basis. Contribution of Working Group I to the Fourth Assessment Report of the Intergovernmental Panel on Climate Change (Solomon S., Qin D., Manning M., Chen Z., Marquis M., Averyt K. B., Tignor M. and Miller H. L., editors) (PDF). Cambridge University Press. Retrieved 14 December 2008.
5. NASA Science Mission Directorate article on the water cycle". Nasascience.nasa.gov. Retrieved 2010-10-16.
6. From non-copyrighted source: Blasing, T. J. (February 2013), Current Greenhouse Gas Concentrations, doi:10.3334/CDIAC/atg.032, on CDIAC 2013. Details on copyright status: Frequently Asked Global Change Questions, Q34: I would like to use a diagram, image, graph, table, or other materials from the CDIAC Web site. How can I obtain permission? Are there copyright restrictions?, retrieved 2012-09-26, on CDIAC 2013.
7. The most recent preliminary estimate of global monthly mean CO_2 concentration (as of May 2013) is 396.71 ppm: (Ed Dlugokencky and Pieter Tans, NOAA/ ESRL ([1])
8. Frequently asked global change questions". Carbon Dioxide Information Analysis Center.
9. ESRL Web Team (14 January 2008). "Trends in carbon dioxide". Esrl.noaa.gov. Retrieved 2011-09-11.
10. AR4 SYR Synthesis Report Summary for Policymakers – 2 Causes of change". ipcc.ch.
11. Mora, C (2013). "The projected timing of climate departure from recent variability". Nature. 502 (7470): 183–187. Bibcode:2013Natur.502..183M. doi:10.1038/nature12540.
12. Mann, Michael E. (2014-04-01). "Earth Will Cross the Climate Danger Threshold by 2036". Scientific American. Retrieved 30 August 2016.
13. Chapter 7: Couplings Between Changes in the Climate System and Biogeochemistry" (PDF). IPCC WG1 AR4 Report. IPCC. 2007. p. FAQ 7.1; report page 512; pdf page 14. Retrieved 11 July 2011.
14. Canadell, J. G.; Le Quere, C.; Raupach, M. R.; Field, C. B.; Buitenhuis, E. T.; Ciais, P.; Conway, T. J.; Gillett, N. P.; Houghton, R. A.; Marland, G. (2007). "Contributions to accelerating atmospheric CO_2growth from economic activity, carbon intensity, and

efficiency of natural sinks". Proc. Natl. Acad. Sci. U.S.A. 104 (47): 18866–70. Bibcode:2007PNAS..10418866C. doi:10.1073/pnas.0702737104. PMC 2141868. PMID 17962418.

15. http://earthobservatory.nasa.gov/Library/RemoteSensingAtmosphere/remote_sensing6.htmlArchived 20 September 2008 at the Wayback Machine.

16. Forster, P.; et al. (2007). "2.10.3 Indirect GWPs". Changes in Atmospheric Constituents and in Radiative Forcing. Contribution of Working Group I to the Fourth Assessment Report of the Intergovernmental Panel on Climate Change. Cambridge University Press. Retrieved 2012-12-02.

17. MacCarty, N. "Laboratory Comparison of the Global-Warming Potential of Six Categories of Biomass Cooking Stoves" (PDF). Approvecho Research Center.

18. Kiehl, J.T.; Kevin E. Trenberth (1997). "Earth's annual global mean energy budget" (PDF). Bulletin of the American Meteorological Society. 78 (2): 197–208. Bibcode:1997BAMS...78..197K. doi:10.1175/1520-0477(1997)078<0197:EAGMEB>2.0.CO;2. Archived from the original (PDF) on 30 March 2006. Retrieved 1 May 2006.

19. "Water vapour: feedback or forcing?". RealClimate. 6 April 2005. Retrieved 1 May 2006.

20. Schmidt, G. A.; R. Ruedy; R. L. Miller; A. A. Lacis (2010), "The attribution of the present-day total greenhouse effect" (PDF), J. Geophys. Res., 115, Bibcode:2010JGRD..11520106S, doi:10.1029/2010JD014287, D20106. Web page for paper.

21. Lacis, A. (October 2010), NASA GISS: CO_2: The Thermostat that Controls Earth's Temperature, New York: NASA GISS

22. IPCC Fourth Assessment Report, Table 2.14, Chap. 2, p. 212

23. Shindell, Drew T. (2005). "An emissions-based view of climate forcing by methane and tropospheric ozone". Geophysical Research Letters. 32 (4): L04803. Bibcode:2005GeoRL..3204803S. doi:10.1029/2004GL021900.

24. Methane's Impacts on Climate Change May Be Twice Previous Estimates". Nasa.gov. 30 November 2007. Retrieved 2010-10-16.

25. http://www3.epa.gov/climatechange/science/indicators/ghg/ghg-concentrations.html. Missing or empty |title= (help)

26. Wallace, John M. and Peter V. Hobbs. Atmospheric Science; An Introductory Survey. Elsevier. Second Edition, 2006. ISBN 978-0-12-732951-2. Chapter 1

27. Prather, Michael J.; J Hsu (2008). "NF3, the greenhouse gas missing from Kyoto". Geophysical Research Letters. 35 (12): L12810. Bibcode:2008GeoRL..3512810P. doi:10.1029/2008GL034542.

28. Isaksen, Ivar S. A.; Michael Gauss; Gunnar Myhre; Katey M. Walter Anthony; Carolyn Ruppel (20 April 2011). "Strong atmospheric chemistry feedback to climate warming from Arctic methane emissions" (PDF). Global Biogeochemical Cycles. 25 (2). Bibcode:2011GBioC..25B2002I. doi:10.1029/2010GB003845. Retrieved 29 July 2011.

29. AGU Water Vapor in the Climate System". Eso.org. 27 April 1995. Retrieved 2011-09-11.

30. Betts (2001). "6.3 Well-mixed Greenhouse Gases". Chapter 6 Radiative Forcing of Climate Change. Working Group I: The Scientific Basis IPCC Third Assessment Report — Climate Change 2001. UNEP/GRID-Arendal — Publications. Retrieved 2010-10-16.

31. Jacob, Daniel (1999). Introduction to atmospheric chemistry. Princeton University Press. pp. 25–26. ISBN 0-691-00185-5.
32. How long will global warming last?". RealClimate. Retrieved 2012-06-12.
33. Jacobson, MZ (2005). "Correction to "Control of fossil-fuel particulate black carbon and organic matter, possibly the most effective method of slowing global warming."". J. Geophys. Res. 110. pp. D14105. doi:10.1029/2005JD005888.
34. Archer, David (2009). "Atmospheric lifetime of fossil fuel carbon dioxide". Annual Review of Earth and Planetary Sciences. 37. pp. 117–134. doi:10.1146/annurev.earth.031208.100206.
35. Meehl, G. A. (2007). "Frequently Asked Question 10.3: If emissions of greenhouse gases are reduced, how quickly do their concentrations in the atmosphere decrease?". In S. Solomon; et al. Chapter 10: Global Climate Projections. Climate Change 2007: The Physical Science Basis. Contribution of Working Group I to the Fourth Assessment Report of the Intergovernmental Panel on Climate Change. Cambridge University Press (CUP), Cambridge, United Kingdom and New York, USA.: Print version:CUP. This version: IPCC website. Retrieved 2011-06-01.
36. See also: Archer, David (2005). "Fate of fossil fuel CO_2 in geologic time" (PDF). Journal of Geophysical Research. 110 (C9): C09S05.1–6. Bibcode:2005JGRC..11009S05A. doi:10.1029/2004JC002625. Retrieved 27 July 2007.
37. See also: Caldeira, Ken; Wickett, Michael E. (2005). "Ocean model predictions of chemistry changes from carbon dioxide emissions to the atmosphere and ocean" (PDF). Journal of Geophysical Research. 110 (C9): C09S04.1–12. Bibcode:2005JGRC..11009S04C. doi:10.1029/2004JC002671. Archived from the original (PDF) on 10 August 2007. Retrieved 27 July 2007.
38. Edited quote from public-domain source: "Climate Change Indicators in the United States". U.S. Environmental Protection Agency (EPA). 2010. Greenhouse Gases: Figure 1. The Annual Greenhouse Gas Index, 1979–2008: Background.. This publication is also available as a PDF (page 18).
39. David L. Chandler, How to count methane emissions, MIT News, April 25, 2014 (Accessed Jan. 15, 2015). Referenced paper is Jessika Trancik and Morgan Edwards, Climate impacts of energy technologies depend on emissions timing, Nature Climate Change, Volume 4, April 25, 2014, p. 347 (Accessed Jan. 15, 2015).
40. St. Fleur, Nicholas (10 November 2015). "Atmospheric Greenhouse Gas Levels Hit Record, Report Says". New York Times. Retrieved 11 November 2015.
41. Ritter, Karl (9 November 2015). "UK: In 1st, global temps average could be 1 degree C higher". AP News. Retrieved 11 November 2015.
42. Cole, Steve; Gray, Ellen (14 December 2015). "New NASA Satellite Maps Show Human Fingerprint on Global Air Quality". NASA. Retrieved 14 December 2015.
43. Canadell, J. G.; et al. (20 November 2007), "Contributions to Accelerating Atmospheric CO2 Growth from Economic Activity, Carbon Intensity, and Efficiency of Natural Sinks (Results and Discussion: Growth in Atmospheric CO_2)", Proceedings of the National Academy of Sciences of the United States of America, 104 (47): 18866–18870, Bibcode:2007PNAS..10418866C, doi:10.1073/pnas.0702737104, PMC 2141868, PMID 17962418
44. Ehhalt, D.; et al., "Ch 4. Atmospheric Chemistry and Greenhouse Gases", Table 4.1,

in IPCC TAR WG1 2001, pp. 244–245. Referred to by: Blasing, T. J. (February 2013),

45. Forster, P.; et al., "Ch 2: Changes in Atmospheric Constituents and in Radiative Forcing", Table 2.1, in IPCC AR4 WG1 2007, p. 141. Referred to by: Blasing, T. J. (February 2013), Current Greenhouse Gas Concentrations, doi:10.3334/CDIAC/atg.032, on CDIAC 2013. For the latest updates, see the NOAA Annual Greenhouse Gas Index at: [2].

46. Prentice, I. C.; et al., "Ch 3. The Carbon Cycle and Atmospheric Carbon Dioxide", Executive summary, in IPCC TAR WG1 2001, p. 185. Referred to by: Blasing, T. J. (February 2013), Current Greenhouse Gas Concentrations, doi:10.3334/CDIAC/atg.032, on CDIAC 2013

47. The pre-1750 value for N2O is consistent with ice-core records from 10,000 B.C.E. through 1750 C.E.: Summary for policymakers", Figure SPM.1, IPCC, in IPCC AR4 WG1 2007, p. 3. Referred to by: Blasing, T. J. (February 2013), Current Greenhouse Gas Concentrations, doi:10.3334/CDIAC/atg.032, on CDIAC 2013

48. Changes in stratospheric ozone have resulted in a decrease in radiative forcing of 0.05 W/m^2: Forster, P.; et al., "Ch 2: Changes in Atmospheric Constituents and in Radiative Forcing", Table 2.12, in IPCC AR4 WG1 2007, p. 204. Referred to by: Blasing, T. J. (February 2013), Current Greenhouse Gas Concentrations, doi:10.3334/CDIAC/atg.032, on CDIAC 2013

49. Berner, Robert A. (January 1994). "GEOCARB II: a revised model of atmospheric CO_2 over Phanerozoic time" (PDF). American Journal of Science. 294 (1): 56–91. doi:10.2475/ajs.294.1.56.

50. Royer, D. L.; R. A. Berner; D. J. Beerling (2001). "Phanerozoic atmospheric CO_2 change: evaluating geochemical and paleobiological approaches". Earth-Science Reviews. 54 (4): 349–92. Bibcode:2001ESRv...54..349R. doi:10.1016/S0012-8252(00)00042-8.

51. Berner, Robert A.; Kothavala, Zavareth (2001). "GEOCARB III: a revised model of atmospheric CO_2over Phanerozoic time" (PDF). American Journal of Science. 301 (2): 182–204. doi:10.2475/ajs.301.2.182.

52. Beerling, D. J.; Berner, R. A. (2005). "Feedbacks and the co-evolution of plants and atmospheric CO_2". Proc. Natl. Acad. Sci. U.S.A. 102 (5): 1302–5. Bibcode:2005PNAS..102.1302B. doi:10.1073/pnas.0408724102. PMC 547859. PMID 15668402.

53. Hoffmann, PF; AJ Kaufman; GP Halverson; DP Schrag (1998). "A neoproterozoic snowball earth". Science. 281 (5381): 1342–6. Bibcode:1998Sci...281.1342H. doi:10.1126/science.281.5381.1342. PMID 9721097.

54. Gerlach, TM (1991). "Present-day CO_2 emissions from volcanoes". Transactions of the American Geophysical Union. 72 (23): 249–55. Bibcode:1991EOSTr..72..249.. doi:10.1029/90EO10192.

55. Flückiger, Jacqueline (2002). "High-resolution Holocene N2O ice core record and its relationship with CH4 and CO_2". Global Biogeochemical Cycles. 16: 1010. Bibcode:2002GBioC..16a..10F. doi:10.1029/2001GB001417.

56. Friederike Wagner; Bent Aaby; Henk Visscher (2002). "Rapid atmospheric CO_2 changes associated with the 8,200-years-B.P. cooling event". Proc. Natl. Acad. Sci. U.S.A. 99 (19): 12011–4. Bibcode:2002PNAS...9912011W. doi:10.1073/pnas.182420699. PMC 129389. PMID 12202744.

57. Andreas Indermühle; Bernhard Stauffer; Thomas F. Stocker (1999). "Early Holocene Atmospheric CO_2 Concentrations". Science. 286 (5446): 1815. doi:10.1126/science.286.5446.1815a. Early Holocene atmospheric CO_2 concentrations". Science. Retrieved 26 May 2005.

58. H. J. Smith; M. Wahlen; D. Mastroianni (1997). "The CO_2 concentration of air trapped in GISP2 ice from the Last Glacial Maximum-Holocene transition". Geophysical Research Letters. 24 (1): 1–4. Bibcode:1997GeoRL..24....1S. doi:10.1029/96GL03700.

59. Monthly Average Carbon Dioxide Concentration, Mauna Loa Observatory" (PDF). Carbon Dioxide Information Analysis Center. 2005. Retrieved 14 December 2008.

60. Dr. Pieter Tans (3 May 2008) Annual CO_2 mole fraction increase (ppm)" for 1959–2007 National Oceanic and Atmospheric Administration Earth System Research Laboratory, Global Monitoring Division (additional details; see also K. A. Masarie; P. P. Tans (1995). "Extension and integration of atmospheric carbon dioxide data into a globally consistent measurement record". J. Geophys. Research. 100: 11593–610. Bibcode:1995JGR...10011593M. doi:10.1029/95JD00859.

61. Dumitru-Romulus Târziu; Victor-Dan Păcurar (Jan 2011). "Pădurea, climatul energia". Rev. pădur. (in Romanian). 126 (1): 34–39. ISSN 1583-7890. 16720. Retrieved 2012-06-11.(webpage has a translation button)

62. Climate Change Indicators in the United States". NOAA. 2012. Figure 4. The Annual Greenhouse Gas Index, 1979–2011.

63. Climate Change Indicators in the United States". U.S. Environmental Protection Agency (EPA). 2010. Figure 2. Global Greenhouse Gas Emissions by Sector, 1990–2005.

64. Climate Change 2001: Working Group I: The Scientific Basis: figure 6-6". Retrieved 1 May 2006.

65. The present carbon cycle — Climate Change". Grida.no. Retrieved 2010-10-16.

66. Solomon, S.; D. Qin; M. Manning; Z. Chen; M. Marquis; K. B. Averyt; M. Tignor; H. L. Miller, eds. (2007).

67. H. Steinfeld, P. Gerber, T. Wassenaar, V. Castel, M. Rosales, C. de Haan (2006) Livestock's long shadow. Environmental issues and options. FAO Livestock, Environment and Development (LEAD) Initiative.

68. Ciais, Phillipe; Sabine, Christopher. "Climate Change 2013: The Physical Science Basis: Chapter 6: Carbon and Other Biogeochemical Cycles". IPCC. Chapter 6: IPCC. p. 473.

69. Raupach, M. R.; et al. (2007). "Global and regional drivers of accelerating CO_2 emissions"(PDF). Proc. Natl. Acad. Sci. U.S.A. 104 (24): 10288–93. Bibcode:2007PNAS..10410288R. doi:10.1073/pnas.0700609104. PMC 1876160. PMID 17519334.

70. Schrooten, L; De Vlieger, Ina; Int Panis, Luc; Styns, R. Torfs, K; Torfs, R (2008). "Inventory and forecasting of maritime emissions in the Belgian sea territory, an activity based emission model". Atmospheric Environment. 42 (4): 667–676. doi:10.1016/j.atmosenv.2007.09.071.

71. Lerner & K. Lee Lerner, Brenda Wilmoth (2006). "Environmental issues: essential primary sources". Thomson Gale. Retrieved 11 September 2006.

72. "Kyoto Protocol". United Nations Framework Convention on Climate Change. Home > Kyoto Protocol.

73. King, D.; et al. (July 2011), "Copenhagen and Cancún", International climate change negotiations: Key lessons and next steps, Oxford, UK: Smith School of Enterprise and the Environment, University of Oxford, p. 12, doi:10.4210/ssee.pbs.2011.0003 PDF version is also available

74. Chris Johnston, Oliver Milman, John Vidal and agencies, Climate change: global deal reached to limit use of hydrofluorocarbons", The Guardian, Saturday 15 October 2016 (page visited on 15 October 2016).

75. Climate change: ‹Monumental' deal to cut HFCs, fastest growing greenhouse gases". BBC News. 15 October 2016. Retrieved 15 October 2016.

76. Nations, Fighting Powerful Refrigerant That Warms Planet, Reach Landmark Deal". New York Times. 15 October 2016. Retrieved 15 October 2016.

77. Environmental Impacts of Tourism – Global Level UNEP

78. search engine optimisation manchester, web design agency london, e commerce manchester". freightbestpractice.org.uk. Retrieved 13 September 2015.

79. Evans, Kimberly Masters (2005). "The greenhouse effect and climate change". The environment: a revolution in attitudes. Detroit: Thomson Gale. ISBN 0-7876-9082-1.

80. Inventory of U.S. Greenhouse Gas Emissions and Sinks: 1990–2010" (PDF). U.S. Environmental Protection Agency. 15 April 2012. p. 1.4. Retrieved 2 June 2012.

81. Held, I. M.; Soden, B. J. (2000). "Water Vapor Feedback and Global Warming1". Annual Review of Energy and the Environment. 25: 441–475. doi:10.1146/annurev.energy.25.1.441.

82. Rogner, H.-H., D. Zhou, R. Bradley. P. Crabbé, O. Edenhofer, B. Hare, L. Kuijpers, M. Yamaguchi (2007). "Executive Summary". In B. Metz; O. R. Davidson; P. R. Bosch; R. Dave; L. A. Meyer. Introduction. Climate Change 2007: Mitigation. Contribution of Working Group III to the Fourth Assessment Report of the Intergovernmental Panel on Climate Change. Cambridge University Press. ISBN 978-0-521-88011-4.

83. Rogner, H.-H., D. Zhou, R. Bradley. P. Crabbé, O. Edenhofer, B.Hare, L. Kuijpers, M. Yamaguchi (2007). "1.3.1 Review of the last three decades". In B. Metz; O. R. Davidson; P. R. Bosch; R. Dave; L. A. Meyer. Introduction. Climate Change 2007: Mitigation. Contribution of Working Group III to the Fourth Assessment Report of the Intergovernmental Panel on Climate Change. Cambridge University Press. ISBN 978-0-521-88011-4. This citation clarifies the time period (1970–2004) for the observed emissions trends

84. Bridging the Emissions Gap: A UNEP Synthesis Report (PDF), Nairobi, Kenya: United Nations Environment Programme (UNEP), November 2011, ISBN 978-92-807-3229-0 UNEP Stock Number: DEW/1470/NA

85. Global Greenhouse Gas Emissions Data". EPA. Retrieved 4 March 2014. The burning of coal, natural gas, and oil for electricity and heat is the largest single source of global greenhouse gas emissions.

86. "Selected Development Indicators" (PDF). World Development Report 2010: Development and Climate Change (PDF). Washington, D.C., USA: The International Bank for Reconstruction and Development / The World Bank. 2010. Tables A1 and A2. doi:10.1596/978-0-8213-7987-5. ISBN 9780821379875.

87. Bader, N.; Bleichwitz, R. (2009). "Measuring urban greenhouse gas emissions: The

challenge of comparability. S.A.P.I.EN.S. 2 (3)". Sapiens.revues.org. Retrieved 2011-09-11.

88. Banuri, T. (1996). Equity and social considerations. In: Climate change 1995: Economic and social dimensions of climate change. Contribution of Working Group III to the Second Assessment Report of the Intergovernmental Panel on Climate Change (J.P. Bruce et al. Eds.) (PDF). This version: Printed by Cambridge University Press, Cambridge, UK, and New York, USA. PDF version: IPCC website. doi:10.2277/0521568544. ISBN 978-0-521-56854-8.

89. World energy outlook 2007 edition – China and India insights. International Energy Agency (IEA), Head of Communication and Information Office, 9 rue de la Fédération, 75739 Paris Cedex 15, France. 2007. p. 600. ISBN 978-92-64-02730-5. Retrieved 2010-05-04.

90. Holtz-Eakin, D. (1995). "Stoking the fires? CO_2 emissions and economic growth". Journal of Public Economics. 57 (1): 85–101. doi:10.1016/0047-2727(94)01449-X. Retrieved 2011-04-20.

91. B. Metz; O. R. Davidson; P. R. Bosch; R. Dave; L. A. Meyer, eds. (2007). "Annex I: Glossary J-P". Climate Change 2007: Mitigation. Contribution of Working Group III to the Fourth Assessment Report of the Intergovernmental Panel on Climate Change. Print version: Cambridge University Press, Cambridge, United Kingdom and New York, USA. This version: IPCC website. ISBN 978-0-521-88011-4. Retrieved 2011-04-11.

92. Markandya, A. (2001). "7.3.5 Cost Implications of Alternative GHG Emission Reduction Options and Carbon Sinks". In B. Metz; et al. Costing Methodologies. Climate Change 2001: Mitigation. Contribution of Working Group III to the Third Assessment Report of the Intergovernmental Panel on Climate Change. Print version: Cambridge University Press, Cambridge, U.K., and New York, N.Y. This version: GRID-Arendal website. doi:10.2277/0521015022. ISBN 978-0-521-01502-8. Retrieved 2011-04-11.

93. Herzog, T. (November 2006). Yamashita, M. B., ed. Target: intensity — an analysis of greenhouse gas intensity targets (PDF). World Resources Institute. ISBN 1-56973-638-3. Retrieved 2011-04-11.

94. Botzen, W. J. W.; et al. (2008). "Cumulative CO_2 emissions: shifting international responsibilities for climate debt". Climate Policy. 8: 570. doi:10.3763/cpol.2008.0539.

95. , N.; et al. (24 September 2010). "Contributions of individual countries' emissions to climate change and their uncertainty" (PDF). Climatic Change. Springer Science+Business Media B.V. 106: 359–391. doi:10.1007/s10584-010-9930-6.

96. World Energy Outlook 2009 (PDF), Paris, France: International Energy Agency (IEA), 2009, pp. 179–180, ISBN 978-92-64-06130-9

97. Rogner, H.-H., D. Zhou, R. Bradley. P. Crabbé, O. Edenhofer, B. Hare, L. Kuijpers, M. Yamaguchi (2007), "1.3.1 Review of the last three decades", in B. Metz; O. R. Davidson; P. R. Bosch; R. Dave; L. A. Meyer, Introduction, Climate Change 2007: Mitigation. Contribution of Working Group III to the Fourth Assessment Report of the Intergovernmental Panel on Climate Change, Cambridge University Press, ISBN 978-0-521-88011-4

98. "Global CO_2 emissions: annual increase halves in 2008". Netherlands Environmental Assessment Agency (PBL) website. 25 June 2009. Retrieved 2010-05-05.

99. Global Carbon Mechanisms: Emerging lessons and implications (CTC748)". Carbon

Trust. March 2009. p. 24. Retrieved 2010-03-31.

100. Vaughan, Adam (2015-12-07). "Global emissions to fall for first time during a period of economic growth". The Guardian. ISSN 0261-3077. Retrieved 2016-12-23.

101. CO_2 Emissions From Fuel Combustion: Highlights (2011 edition), Paris, France: International Energy Agency (IEA), 2011, p. 9

102. Helm, D.; et al. (10 December 2007). Too Good To Be True? The UK's Climate Change Record(PDF). p. 3.

103. Davis, S. J.; K. Caldeira (8 March 2010). "Consumption-based Accounting of CO_2 Emissions"(PDF). Proceedings of the National Academy of Sciences of the United States of America. 107 (12): 5687–5692. Bibcode:2010PNAS..107.5687D. doi:10.1073/pnas.0906974107. Retrieved 2011-04-18.

104. International Carbon Flows". Carbon Trust. May 2011. Retrieved 12 November 2012.

105. International Energy Agency — Energy Policy". Paris, France: International Energy Agency (IEA). 2012.

106. IEA Publications Bookshop: IEA Publications on ‹Energy Policy'". Paris, France: Organization for Economic Co-operation and Development (OECD) / International Energy Agency (IEA). 2012.

107. Bridging the Emissions Gap: A UNEP Synthesis Report (PDF), Nairobi, Kenya: United Nations Environment Programme (UNEP), November 2011, ISBN 978-92-807-3229-0 UNEP Stock Number: DEW/1470/NA

108. 4. Energizing development without compromising the climate" (PDF). World Development Report 2010: Development and Climate Change (PDF). Washington, D.C., USA: The International Bank for Reconstruction and Development / The World Bank. 2010. p 192, Box 4.2: Efficient and clean energy can be good for development. doi:10.1596/978-0-8213-7987-5. ISBN 9780821379875.

109. Sixth compilation and synthesis of initial national communications from Parties not included in Annex I to the Convention. Note by the secretariat. Executive summary. (PDF). Geneva, Switzerland: United Nations Framework Convention on Climate Change (UNFCCC). 2005. pp. 10–12.

110. Compilation and synthesis of fifth national communications. Executive summary. Note by the secretariat. (PDF). Geneva (Switzerland): United Nations Framework Convention on Climate Change (UNFCCC). 2011. pp. 9–10.

111. Fisher, B.; et al. (2007). "3.1 Emissions scenarios". In B. Metz; O. R. Davidson; P. R. Bosch; R. Dave; L. A. Meyer. Chapter 3: Issues related to mitigation in the long-term context. Climate Change 2007: Mitigation. Contribution of Working Group III to the Fourth Assessment Report of the Intergovernmental Panel on Climate Change. Cambridge University Press. ISBN 978-0-521-88011-4.

112. Rogner, H.-H., D. Zhou, R. Bradley. P. Crabbé, O. Edenhofer, B. Hare, L. Kuijpers, M. Yamaguchi (2007). "1.3.2 Future outlook". In B. Metz; O. R. Davidson; P. R. Bosch; R. Dave; L. A. Meyer. Introduction. Climate Change 2007: Mitigation. Contribution of Working Group III to the Fourth Assessment Report of the Intergovernmental Panel on Climate Change. Cambridge University Press. ISBN 978-0-521-88011-4.

113. Rogner, H.-H., D. Zhou, R. Bradley. P. Crabbé, O. Edenhofer, B. Hare, L. Kuijpers, M. Yamaguchi (2007). "1.3.2.4 Total GHG emissions". In B. Metz; O. R. Davidson; P. R.

Bosch; R. Dave; L. A. Meyer. Introduction. Climate Change 2007: Mitigation. Contribution of Working Group III to the Fourth Assessment Report of the Intergovernmental Panel on Climate Change. Cambridge University Press. ISBN 978-0-521-88011-4.

114. Greenhouse Gas Emissions from a Typical Passenger Vehicle, US Environment Protection Agency" (PDF). Epa.gov. Retrieved 2011-09-11.

115. Engber, Daniel (1 November 2006). "How gasoline becomes CO_2, Slate Magazine". Slate Magazine. Retrieved 2011-09-11.

116. Volume calculation for carbon dioxide". Icbe.com. Retrieved 2011-09-11.

117. Voluntary Reporting of Greenhouse Gases Program". Energy Information Administration. Retrieved 21 August 2009.

118. Moomaw, W., P. Burgherr, G. Heath, M. Lenzen, J. Nyboer, A. Verbruggen (2011). "Annex II: Methodology" (PDF). IPCC Special Report on Renewable Energy Sources and Climate Change Mitigation: 10.

119. Obersteiner M; Azar C; Kauppi P; et al. (October 2001). "Managing climate risk". Science. 294 (5543): 786–7. doi:10.1126/science.294.5543.786b. PMID 11681318.

120. Azar, C., Lindgren, K., Larson, E.D. and Möllersten, K. (2006). "Carbon capture and storage from fossil fuels and biomass – Costs and potential role in stabilising the atmosphere" (PDF). Climatic Change. 74: 47–79. doi:10.1007/s10584-005-3484-7.

121. Geoengineering the climate: science, governance and uncertainty". The Royal Society. 2009. Archived from the original on 7 September 2009. Retrieved 12 September 2009.

122. Fischer, B.S., N. Nakicenovic, K. Alfsen, J. Corfee Morlot, F. de la Chesnaye, J.-Ch. Hourcade, K. Jiang, M. Kainuma, E. La Rovere, A. Matysek, A. Rana, K. Riahi, R. Richels, S. Rose, D. van Vuuren, R. Warren, (2007)»Issues related to mitigation in the long term context", In Climate Change 2007: Mitigation. Contribution of Working Group III to the Fourth Assessment Report of the Inter-governmental Panel on Climate Change [B. Metz, O.R. Davidson, P.R. Bosch, R. Dave, L.A. Meyer (eds)], Cambridge University Press, Cambridge.

123. Cook, J.; Nuccitelli, D.; Green, S. A.; Richardson, M.; Winkler, B. R.; Painting, R.; Way, R.; Jacobs, P.; Skuce, A. (2013). "Quantifying the consensus on anthropogenic global warming in the scientific literature". Environmental Research Letters. 8 (2): 024024. doi:10.1088/1748-9326/8/2/024024.

CHAPTER 2

Halocarbon Compounds

Chapter Contents

- **Introduction**
- **Chemical Families**
- **History and Context**
- **Organoiodine Compounds, Including Biological Derivatives**
- **Uses**
- **Hazards**

A compound, such as a fluorocarbon, that consists of carbon combined with one or more halogens. Halocarbons are typically nonflammable and nonreactive, though some halocarbons are broken down by ultraviolet radiation in the upper atmosphere, and this process releases free halogen atoms that damage the ozone layer. Some halocarbons have also been implicated as greenhouse gases.

INTRODUCTION

Halocarbon compounds are chemicals in which one or more carbon atoms are linked by covalent bonds with one or more halogen atoms (fluorine, chlorine, bromine or iodine – group 17) resulting in the formation of organofluorine compounds, organochlorine compounds, organobromine compounds, and organoiodine compounds. Chlorine halocarbons are the most common and are called organochlorides.

Many synthetic organic compounds such as plastic polymers, and a few natural ones, contain halogen atoms; they are known as *halogenated* compounds or *organohalogens*. Organochlorides are the most common industrially used organohalides, although the other organohalides are used commonly in organic synthesis. Except for extremely rare cases, organohalides are not produced biologically, but many pharmaceuticals are organohalides. Notably, many pharmaceuticals such as Prozac have trifluoromethyl groups.

For information on inorganic halide chemistry, see halide.

CHEMICAL FAMILIES

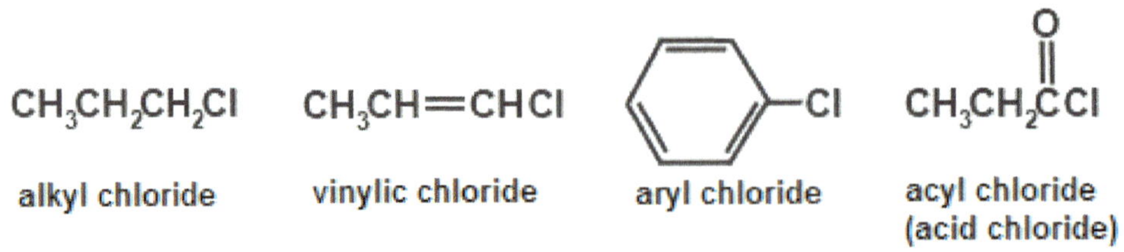

Examples of organohalogens-chlorides.

Halocarbons are typically classified in the same ways as the similarly structured organic compounds that have hydrogen atoms occupying the molecular sites of the halogen atoms in halocarbons. Among the chemical families are:

- haloalkanes—compounds with carbon atoms linked by single bonds
- haloalkenes—compounds with one or more double bonds between carbon atoms
- haloaromatics—compounds with carbons linked in one or more aromatic rings with a delocalised donut shaped pi cloud.

The halogen atoms in halocarbon molecules are often called "substituents," as though those atoms had been substituted for hydrogen atoms. However halocarbons are prepared in many ways that do not involve direct substitution of halogens for hydrogens.

HISTORY AND CONTEXT

A few halocarbons are produced in massive amounts by microorganisms. For example, several million tons of methyl bromide are estimated to be produced by marine organisms annually. Most of the halocarbons encountered in everyday life – solvents, medicines, plastics – are man-made. The first synthesis of halocarbons was achieved in the early 1800s. Production began accelerating when their useful properties as solvents and anesthetics were discovered. Development of plastics and synthetic elastomers has led to greatly expanded scale of production. A substantial percentage of drugs are halocarbons.

Natural Halocarbons

A large amount of the naturally occurring halocarbons are created by wood fire, dioxine for example, or volcanic activities. A second large source are marine algae which produce several chlorinated methane and ethane containing compounds. There are several thousand complex halocarbons known, produced mainly by marine species. Although chlorine compounds are the majority of the discovered compounds, bromides, iodides and fluorides have also been found. The tyrian purple, which is a dibromoindigo, is representative of the bromides, while the thyroxine secreted from the thyroid gland, is an iodide, and the highly toxic fluoroacetate is one of the rare organofluorides. These three representatives, thyroxine from humans, tyrian purple from snails and fluoroacetate from plants, also show that unrelated species use halocarbons for several purposes.

ORGANOIODINE COMPOUNDS, INCLUDING BIOLOGICAL DERIVATIVES

Organoiodine compounds, called **organic iodides**, are similar in structure to organochlorine and organobromine compounds, but the C-I bond is weaker. Many organic iodides are known, but few are of major industrial importance. Iodide compounds are mainly produced as nutritional supplements.

The thyroxin hormones are essential for human health, hence the usefulness of iodized salt.

Six mg of iodide a day can be used to treat patients with hyperthyroidism due to its ability to inhibit the organification process in thyroid hormone synthesis, the so-called Wolff-Chaikoff Effect. Prior to 1940, iodides were the predominant antithyroid agents. In large doses, iodides inhibit proteolysis of thyroglobulin, which permits TH to be synthesized and stored in colloid, but not released into the bloodstream.

This treatment is seldom used today as a stand-alone therapy despite the rapid improvement of patients immediately following administration. The major disadvantage of iodide treatment lies in the fact that excessive stores of TH accumulate, slowing the onset of action of thioamides (TH synthesis blockers). In addition, the functionality of iodides fades after the initial treatment period. An "escape from block" is also a concern, as extra stored TH may spike following discontinuation of treatment.

USES

The first halocarbon commercially used was Tyrian purple a natural organobromide of the *Murex brandaris* marine snail.

Common uses for halocarbons have been as solvents, pesticides, refrigerants, fire-resistant oils, ingredients of elastomers, adhesives and sealants, electrically insulating coatings, plasticizers, and plastics. Many halocarbons have specialized uses in industry. One halocarbon, sucralose, is a sweetener.

Before they became strictly regulated, the general public often encountered haloalkanes as paint and cleaning solvents such as trichloroethane (1,1,1-trichloroethane) and carbon tetrachloride (tetrachloromethane), pesticides like 1,2-dibromoethane (EDB, ethylene dibromide), and refrigerants like Freon-22 (duPont trademark for chlorodifluoromethane). Some haloalkanes are still widely used for industrial cleaning, such as methylene chloride (dichloromethane), and as refrigerants, such as R-134a (1,1,1,2-tetrafluoroethane).

Haloalkenes have also been used as solvents, including perchloroethylene (Perc, tetrachloroethene), widespread in dry cleaning, and trichloroethylene (TCE, 1,1,2-trichloroethene). Other haloalkenes have been chemical building blocks of plastics such as polyvinyl chloride ("vinyl" or PVC, polymerized chloroethene) and Teflon (duPont trademark for polymerized tetrafluoroethene, PTFE). Haloaromatics include the former Aroclors (Monsanto Company trademark for polychlorinated biphenyls, PCBs), once widely used in power transformers and capacitors and in building caulk, the former Halowaxes (Union Carbide trademark for polychlorinated naphthalenes, PCNs), once used for electrical insulation, and the chlorobenzenes and their derivatives, used for disinfectants, pesticides such as dichloro-diphenyl-trichloroethane (DDT, 1,1,1-trichloro-2,2-bis(p-chlorophenyl)ethane), herbicides such as 2,4-D (2,4-dichlorophenoxyacetic acid), askarel dielectrics (mixed with PCBs, no longer used in most countries), and chemical feedstocks.

A few halocarbons, including acid halides like acetyl chloride, are highly reactive; these are rarely found outside chemical processing. The widespread uses of halocarbons were often driven by observations that most of them were more stable than other substances. They may be less affected by acids or alkalis; they may not burn as readily; they may not be attacked by bacteria or molds; or they may not be affected as much by sun exposure.

HAZARDS

The stability of halocarbons tended to encourage beliefs that they were mostly harmless, although in the mid-1920s physicians reported workers in polychlorinated naphthalene manufacturing suffering from chloracne (Teleky 1927), and by the late 1930s it was known that workers exposed to PCNs could die from liver disease (Flinn & Jarvik 1936) and that DDT would kill mosquitos and other insects (Müller 1948). By the 1950s, there had been several reports and investigations of workplace hazards. In 1956, for example, after testing hydraulic oils containing PCBs, the U.S. Navy found that skin contact caused fatal liver disease in animals and rejected them as "too toxic for use in a submarine" (Owens v. Monsanto 2001).

In 1962 a book by U.S. biologist Rachel Carson (Carson 1962) started a storm of concerns about environmental pollution, first focused on DDT and other pesticides, some of them also halocarbons. These concerns were amplified when in 1966 Swedish chemist Soren Jensen reported widespread residues of PCBs among Arctic and sub-Arctic fish and birds (Jensen 1966). In 1974, mexican chemist Mario Molina and U.S. chemist Sherwood Rowland predicted that common halocarbon refrigerants, the chlorofluorocarbons (CFCs), would accumulate in the upper atmosphere and destroy protective ozone (Molina & Rowland 1974). Within a few years, ozone depletion was being observed above Antarctica, leading to bans on production and use of chlorofluorocarbons in many countries. In 2007, the Intergovernmental Panel on Climate Change (IPCC) said halocarbons were a direct cause of global warming.

Since the 1970s there have been longstanding, unresolved controversies over potential health hazards of trichloroethylene (TCE) and other halocarbon solvents that had been widely used for industrial cleaning (Anderson v. Grace 1986) (Scott & Cogliano 2000) (U.S. National Academies of Science 2004) (United States 2004). More recently perfluorooctanoic acid (PFOA), a precursor in the most common manufacturing process for Teflon and also used to make coatings for fabrics and food packaging, became a health and environmental concern starting in 2006 (United States 2010 (begun in 2006)), suggesting that halocarbons, though thought to be among the most inert, may also present hazards.

Halocarbons, including those that might not be hazards in themselves, can present waste disposal issues. Because they do not readily degrade in natural environments, halocarbons tend to accumulate. Incineration and accidental fires can create corrosive byproducts such as hydrochloric acid and hydrofluoric acid, and poisons like halogenated dioxins and furans. Species of Desulfitobacterium are being investigated for their potential in the bioremediation of halogenic organic compounds.

REFERENCES

1. Anderson v. Grace (1986), 628 F. Supp. 1219, Massachusetts, USA, settled between the parties, reviewed in Harr, J., Ed.; Asher, M., Ed. (1996), A Civil Action, Minneapolis, MN, USA: Sagebrush Education Resources
2. Carson, R. (1962), Silent Spring, Boston, MA, USA: Houghton Mifflin
3. Flinn, F.B.; Jarvik, N.E. (1936), "Action of certain chlorinated naphthalenes on the liver", Proceedings of the Society for Experimental Biology and Medicine, 35: 118–120, doi:10.3181/00379727-35-8879p
4. Jensen, S. (1966), "Report of a new chemical hazard", New Scientist, 32: 612
5. Molina, M.J.; Rowland, F.S. (1974), "Stratospheric sink for chlorofluoromethanes: chlorine atom-catalysed destruction of ozone", Nature, 249: 810–812, Bibcode:1974Natur.249..810M, doi:10.1038/249810a0
6. Müller, P.H. (1948), "Dichloro-diphenyl-trichloroethane and newer insecticides" (PDF), Nobel Lecture
7. Owens v. Monsanto (2001), 96-CV-440, Exhibit 3A03F (PDF), Alabama, USA, cited in Chemical Industry Archives, Anniston Case, by Environmental Working Group, Washington, DC, 2002
8. Scott, C.S., Ed.; Cogliano, V.J., Ed. (2000), "Trichloroethylene Health Risks--State of the Science", Environmental Health Perspectives, 108 (S2): 159–60, doi:10.1289/ehp.00108s2159, PMC 1637768, PMID 10928830
9. Teleky, L. (1927), "Die pernakrankheit", Klinische Wochenschrift, Berlin: Springer, Jahrgänge 6: 845
10. U.S. National Academies of Science, Current Projects System (2004), Assessing the Human Health Risks of Trichloroethylene
11. United States, Environmental Protection Agency (2004), Integrated Risk Information System, Trichloroethylene (CASRN 79-01-6)
12. United States, Environmental Protection Agency (2010), PFOA Stewardship Program (begun in 2006)

CHAPTER 3

Greenhouse Effect

Chapter Contents

- **Introduction**
- **History**
- **Mechanism**
- **Greenhouse Gases**
- **Role In Climate Change**
- **Real Greenhouses**
- **Bodies Other Than Earth**
- **Paleoclimatology**

The greenhouse effect is a biological process in Earth's atmosphere in which certain gases bind together and form what is essentially a layer of insulation. This insulation traps warmth and solar radiation. It is a natural process that is widely believed to be essential to life on the planet, since without it heat from the sun would escape and the planet could grow frigidly cold as a result. Just the same, certain "greenhouse gases" have gotten a bad reputation in recent years for the negative ways in which they interact with the process. Chemical emissions and pollution can cause the insulation to be thicker or thinner than normal, for instance, and might even be able to cause holes or perforations. Greenhouse gases are often blamed for the phenomenon known as "global warming," and the greenhouse effect is certainly a part of this — but the process isn't usually viewed as a problem in and of itself.

INTRODUCTION

The **greenhouse effect** is the process by which radiation from a planet's atmosphere warms the planet's surface to a temperature above what it would be without its atmosphere.

If a planet's atmosphere contains radiatively active gases (i.e., greenhouse gases) the atmosphere will radiate energy in all directions. Part of this radiation is directed towards the surface, warming it. The downward component of this radiation – that is, the strength of the greenhouse effect – will depend on the atmosphere's temperature and on the amount of greenhouse gases that the atmosphere contains.

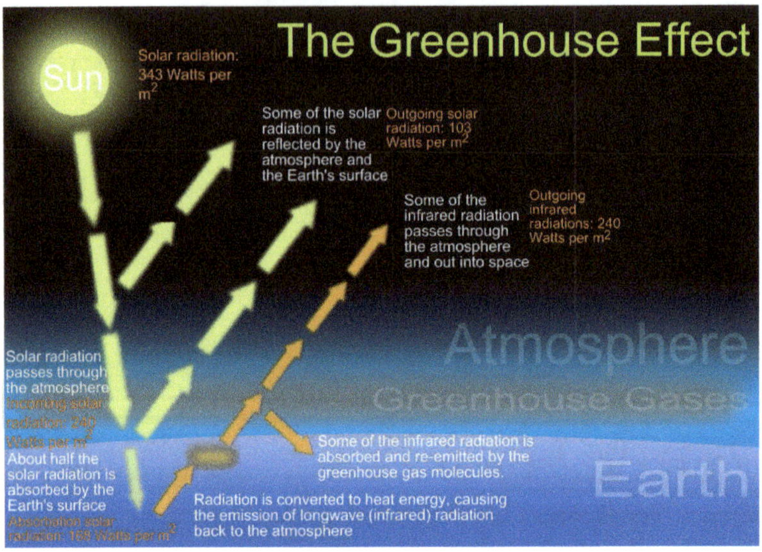

Another diagram of the greenhouse effect.

On Earth, the atmosphere is warmed by absorption of infrared thermal radiation from the underlying surface, absorption of shorter wavelength radiant energy from the sun, and convective heat fluxes from the surface. Greenhouse gases in the atmosphere radiate energy, some of which is directed to the surface and lower atmosphere. The mechanism that produces this difference between the actual surface temperature and the effective temperature is due to the atmosphere and is known as the greenhouse effect.

Earth's natural greenhouse effect is critical to supporting life. Human activities, primarily the burning of fossil fuels and clearing of forests, have intensified the natural greenhouse effect, causing global warming.

The mechanism is named after a faulty analogy with the effect of solar radiation passing through glass and warming a greenhouse. The way a greenhouse retains heat is fundamentally different, as a greenhouse works by reducing airflow and retaining warm air inside the structure.

HISTORY

The existence of the greenhouse effect was argued for by Joseph Fourier in 1824. The argument and the evidence were further strengthened by Claude Pouillet in 1827 and 1838 and reasoned from experimental observations by John Tyndall in 1859. The effect was more fully quantified by Svante Arrhenius in 1896. However, the term "greenhouse" was not used to refer to this effect by any of these scientists; the term was first used in this way by Nils Gustaf Ekholm in 1901.

In 1917 Alexander Graham Bell wrote "[The unchecked burning of fossil fuels] would have a sort of greenhouse effect", and "The net result is the greenhouse becomes a sort of hot-house." Bell went on to also advocate the use of alternate energy sources, such as solar energy.

MECHANISM

Earth receives energy from the Sun in the form of ultraviolet, visible, and near-infrared radiation. Of the total amount of solar energy available at the top of the atmosphere, about 26% is reflected to space by the atmosphere and clouds and 19% is absorbed by the atmosphere and clouds. Most of the remaining energy is absorbed at the surface of Earth. Because the Earth's surface is colder than the photosphere of the Sun, it radiates at wavelengths that are much longer than the wavelengths that were absorbed. Most of this thermal radiation is absorbed by the atmosphere, thereby warming it. In addition to the absorption of solar and thermal radiation, the atmosphere further gains heat by sensible and latent heat fluxes from the surface. The atmosphere radiates energy both upwards and downwards; the part radiated downwards is absorbed by the surface of Earth. This leads to a higher equilibrium temperature than if the atmosphere were absent.

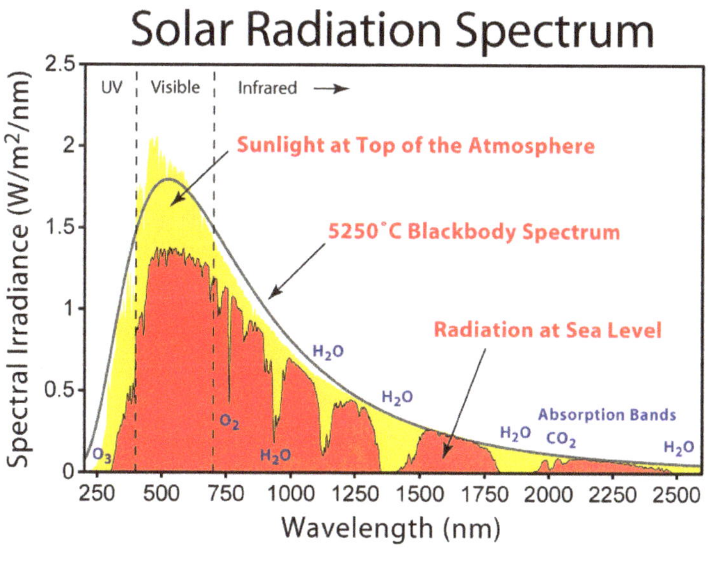

The solar radiation spectrum for direct light at both the top of Earth's atmosphere and at sea level.

An ideal thermally conductive blackbody at the same distance from the Sun as Earth would have a temperature of about 5.3 °C. However, because Earth reflects about 30% of the incoming sunlight, this idealized planet's effective temperature (the temperature of a blackbody that would emit the same amount of radiation) would be about −18 °C. The surface temperature of this hypothetical planet is 33 °C below Earth's actual surface temperature of approximately 14 °C.

The basic mechanism can be qualified in a number of ways, none of which affect the fundamental process. The atmosphere near the surface is largely opaque to thermal radiation (with important exceptions for "window" bands), and most heat loss from the surface is by sensible heat and latent heat transport. Radiative energy losses become increasingly important higher in the atmosphere, largely because of the decreasing concentration of water vapor, an important greenhouse gas. It is more realistic to think of the greenhouse effect as applying to a "surface" in the mid-troposphere, which is effectively coupled to the surface by a lapse rate. The simple picture also assumes a steady state, but in the real world, there are variations due to the diurnal cycle as well as the seasonal cycle and weather disturbances. Solar heating only applies during daytime. During the night, the atmosphere cools somewhat, but not greatly, because its emissivity is low. Diurnal temperature changes decrease with height in the atmosphere.

Within the region where radiative effects are important, the description given by the idealized greenhouse model becomes realistic. Earth's surface, warmed to a temperature around 255 K, radiates long-wavelength, infrared heat in the range of 4–100 μm. At these wavelengths, greenhouse gases that were largely transparent to incoming solar radiation are more absorbent. Each layer of atmosphere with greenhouses gases absorbs some of the heat being radiated upwards from lower layers. It reradiates in all directions, both upwards and downwards; in equilibrium (by definition) the same amount as it has absorbed. This results in more warmth below. Increasing the concentration of the gases increases the amount of absorption and reradiation, and thereby further warms the layers and ultimately the surface below.

Greenhouse gases—including most diatomic gases with two different atoms (such as carbon monoxide, CO) and all gases with three or more atoms—are able to absorb and emit infrared radiation. Though more than 99% of the dry atmosphere is IR transparent (because the main constituents—N_2, O_2, and Ar—are not able to directly absorb or emit infrared radiation), intermolecular collisions cause the energy absorbed and emitted by the greenhouse gases to be shared with the other, non-IR-active, gases.

GREENHOUSE GASES

Atmospheric gases only absorb some wavelengths of energy but are transparent to others. The absorption patterns of water vapor (blue peaks) and carbon dioxide (pink peaks) overlap in some wavelengths. Carbon dioxide is not as strong a greenhouse gas as water vapor, but it absorbs energy in wavelengths (12-15 micrometers) that water vapor does not, partially closing the "window" through which heat radiated by the surface would normally escape to space. (Illustration NASA, Robert Rohde)

By their percentage contribution to the greenhouse effect on Earth the four major gases are:

- water vapor, 36–70%
- carbon dioxide, 9–26%
- methane, 4–9%
- ozone, 3–7%

It is not physically realistic to assign a specific percentage to each gas because the absorption and emission bands of the gases overlap (hence the ranges given above). The major non-gas contributor to Earth's greenhouse effect, clouds, also absorb and emit infrared radiation and thus have an effect on the radiative properties of the atmosphere.

ROLE IN CLIMATE CHANGE

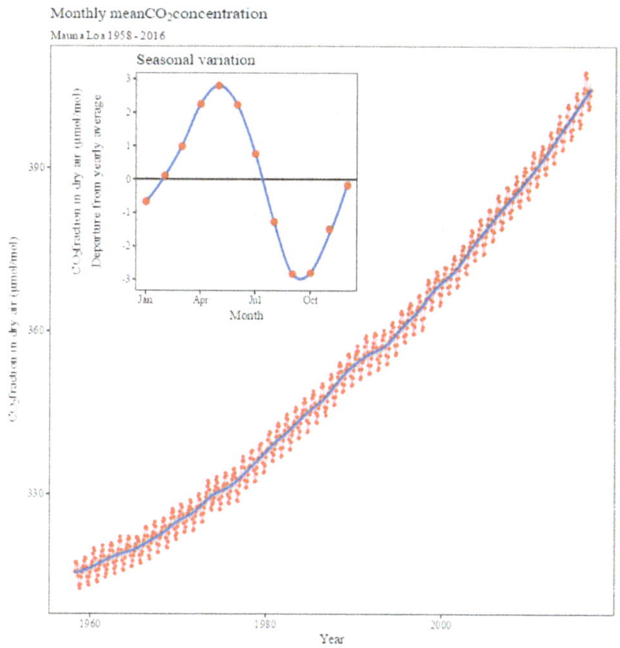

The Keeling Curve of atmospheric CO_2 concentrations measured at Mauna Loa Observatory.

Strengthening of the greenhouse effect through human activities is known as the enhanced (or anthropogenic) greenhouse effect. This increase in radiative forcing from human activity is attributable mainly to increased atmospheric carbon dioxide levels. According to the latest Assessment Report from the Intergovernmental Panel on Climate Change, "*atmospheric concentrations of carbon dioxide, methane and nitrous oxide are unprecedented in at least the last 800,000 years. Their effects, together with those of other anthropogenic drivers, have been detected throughout the climate system and are extremely likely to have been the dominant cause of the observed warming since the mid-20th century*".

CO_2 is produced by fossil fuel burning and other activities such as cement production and tropical deforestation.Measurements of CO_2 from the Mauna Loa observatory show that concentrations have increased from about 313 parts per million (ppm) in 1960 to about 389 ppm in 2010. It reached the 400 ppm milestone on May 9, 2013. The current observed amount of CO_2 exceeds the geological record maxima (~300 ppm) from ice core data. The effect of combustion-produced carbon dioxide on the global climate, a special case of the greenhouse effect first described in 1896 by Svante Arrhenius, has also been called the Callendar effect.

Over the past 800,000 years, ice core data shows that carbon dioxide has varied from values as low as 180 ppm to the pre-industrial level of 270 ppm. Paleoclimatologists consider variations in carbon dioxide concentration to be a fundamental factor influencing climate variations over this time scale.

REAL GREENHOUSES

A modern Greenhouse in RHS Wisley.

The "greenhouse effect" of the atmosphere is named by analogy to greenhouses which become warmer in sunlight. The explanation given in most sources for the warmer temperature in an actual greenhouse is that incident solar radiation in the visible, long-wavelength ultraviolet and short-wavelength infrared range of the spectrum passes through the glass roof and walls and is absorbed by the floor, earth, and contents, which become warmer and re-emit the energy as longer-wavelength infrared radiation. Glass and other materials used for greenhouse walls do not transmit infrared radiation, so the infrared cannot escape via radiative transfer. As the structure is not open to the atmosphere, heat also cannot escape via convection, so the temperature inside the greenhouse rises. The greenhouse effect, due to infrared-opaque "greenhouse gases" including carbon dioxide and methane instead of glass, also affects Earth as a whole; there is no convective cooling because no significant amount of air escapes from Earth.

However, the mechanism by which the atmosphere retains heat—the "greenhouse effect"—is different; a greenhouse is not primarily warmed by the "greenhouse effect". A greenhouse works primarily by allowing sunlight to warm surfaces inside the structure, but then preventing absorbed heat from leaving the structure through convection. The "greenhouse effect" heats Earth because greenhouse gases absorb outgoing radiative energy, heating the atmosphere which then emits radiative energy with some of it going back towards Earth.

A greenhouse is built of any material that passes sunlight, usually glass, or plastic. It mainly warms up because the sun warms the ground and contents inside, which then warms the air in the greenhouse. The air continues to heat up because it is confined within the greenhouse, unlike the environment outside the greenhouse where warm air near the surface rises and mixes with cooler air aloft. This can be demonstrated by opening a small window near the roof of a greenhouse: the temperature will drop considerably. It was demonstrated experimentally (R. W. Wood, 1909) that a "greenhouse" with a cover of rock salt (which is transparent to infrared) heats up an enclosure similarly to one with a glass cover.Thus greenhouses work primarily by preventing convective cooling. More recent quantitative studies suggest that the effect of infrared radiative cooling is not negligibly small, and may have economic implications in a heated greenhouse. Analysis of issues of near-infrared radiation in a greenhouse with screens of a high coefficient of reflection concluded that installation of such screens reduced heat demand by about 8%, and application of dyes to transparent surfaces was suggested. Composite less-reflective glass, or less effective but cheaper anti-reflective coated simple glass, also produced savings.

BODIES OTHER THAN EARTH

In the Solar System, there are also greenhouse effects on Mars, Venus, and Titan. The greenhouse effect on Venus is particularly large because its dense atmosphere consisting mainly of carbon dioxide. Titan has an anti-greenhouse effect, in that its atmosphere absorbs solar radiation but is relatively transparent to infrared radiation. Pluto is also colder than would be expected because evaporation of nitrogen cools it.

A runaway greenhouse effect occurs if positive feedbacks lead to the evaporation of all greenhouse gases into the atmosphere. A runaway greenhouse effect involving carbon dioxide and water vapor is thought to have occurred on Venus.

PALEOCLIMATOLOGY

Paleoclimatology (in British spelling, palaeoclimatology) is the study of changes in climate taken on the scale of the entire history of Earth. It uses a variety of proxy methods from the Earth and life sciences to obtain data previously preserved within things such as rocks, sediments, ice sheets, tree rings, corals, shells and microfossils. It then uses the records to determine the past states of the Earth's various climate regions and its atmospheric system. Studies of past changes in the environment and biodiversity often reflect on the current situation, specifically the impact of climate on mass extinctions and biotic recovery.

The scientific study field of paleoclimate began to form in the early 19th century, when discoveries about glaciations and natural changes in Earth's past climate helped to understand the greenhouse effect.

Reconstructing Ancient Climates

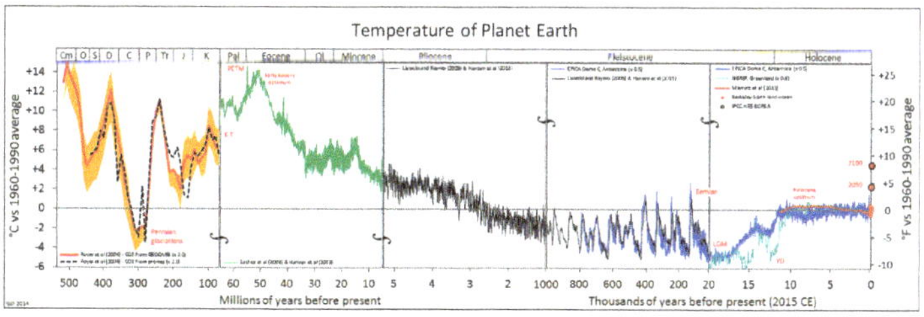

Palaeotemperature graphs compressed together.

Paleoclimatologists employ a wide variety of techniques to deduce ancient climates.

Ice

Mountain glaciers and the polar ice caps/ice sheets provide much data in paleoclimatology. Ice-coring projects in the ice caps of Greenland and Antarctica have yielded data going back several hundred thousand years, over 800,000 years in the case of the EPICA project.

- Air trapped within fallen snow becomes encased in tiny bubbles as the snow is compressed into ice in the glacier under the weight of later years' snow. The trapped air has proven a tremendously valuable source for direct measurement of the composition of air from the time the ice was formed.
- Layering can be observed because of seasonal pauses in ice accumulation and can be used to establish chronology, associating specific depths of the core with ranges of time.
- Changes in the layering thickness can be used to determine changes in precipitation or temperature.
- Oxygen-18 quantity changes ($\delta^{18}O$) in ice layers represent changes in average ocean surface temperature. Water molecules containing the heavier O-18 evaporate at a higher temperature than water molecules containing the normal Oxygen-16 isotope. The ratio of O-18 to O-16 will be higher as temperature increases. It also depends on other factors such as the water's salinity and the volume of water locked up in ice sheets. Various cycles in those isotope ratios have been detected.
- Pollen has been observed in the ice cores and can be used to understand which plants were present as the layer formed. Pollen is produced in abundance and its distribution is typically well understood. A pollen count for a specific layer can be produced by observing the total amount of pollen categorized by type (shape) in a controlled sample of that layer. Changes in plant frequency over time can be plotted through statistical analysis of pollen counts in the core. Knowing which plants were present leads to an understanding of precipitation and temperature, and types of fauna present. Palynology includes the study of pollen for these purposes.
- Volcanic ash is contained in some layers, and can be used to establish the time of the layer's formation. Each volcanic event distributed ash with a unique set of properties (shape and color of particles, chemical signature). Establishing the ash's source will establish a range of time to associate with layer of ice.

Dendroclimatology

Climatic information can be obtained through an understanding of changes in tree growth. Generally, trees respond to changes in climatic variables by speeding up or slowing down growth, which in turn is generally reflected a greater or lesser thickness in growth rings. Different species, however, respond to changes in climatic variables in different ways. A tree-ring record is established by compiling information from many living trees in a specific area. Older intact wood that has escaped decay can extend the time covered by the record by matching the ring depth changes to contemporary specimens. By using that method, some areas have tree-ring records dating back a few thousand years. Older wood not connected to a contemporary record can be dated generally with radiocarbon techniques. A tree-ring record can be used to produce information regarding precipitation, temperature, hydrology, and fire corresponding to a particular area.

Sedimentary Content

On a longer time scale, geologists must refer to the sedimentary record for data.

- Sediments, sometimes lithified to form rock, may contain remnants of preserved vegetation, animals, plankton or pollen, which may be characteristic of certain climatic zones.
- Biomarker molecules such as the alkenones may yield information about their temperature of formation.
- Chemical signatures, particularly Mg/Ca ratio of calcite in Foraminifera tests, can be used to reconstruct past temperature.
- Isotopic ratios can provide further information. Specifically, the $\delta^{18}O$ record responds to changes in temperature and ice volume, and the $\delta^{13}C$ record reflects a range of factors, which are often difficult to disentangle.

Sea floor core sample labelled to identify the exact spot on the sea floor where the sample was taken. Slight variations in location can make a significant difference in the chemical and biological composition of the sediment sample.

Sedimentary facies

On a longer time scale, the rock record may show signs of sea level rise and fall, and features such as "fossilised" sand dunes can be identified. Scientists can get a grasp of long term climate by studying sedimentary rock going back billions of years. The division of earth history into separate periods is largely based on visible changes in sedimentary rock layers that demarcate major changes in conditions. Often, they include major shifts in climate.

Sclerochronology

Corals

Coral "rings" are similar to tree rings except that they respond to different things, such as the water temperature, freshwater influx, pH changes, and wave action. From there, certain equipment can be used to derive the sea surface temperature and water salinity from the past few centuries. The $\delta^{18}O$ of coralline red algae provides a useful proxy of the combined sea surface temperature and sea surface salinity at high latitudes and the tropics, where many traditional techniques are limited.

Limitations

A multinational consortium, the European Project for Ice Coring in Antarctica (EPICA), has drilled an ice core in Dome C on the East Antarctic ice sheet and retrieved ice from roughly 800,000 years ago. The international ice core community has, under the auspices of International Partnerships in Ice Core Sciences (IPICS), defined a priority project to obtain the oldest possible ice core record from Antarctica, an ice core record reaching back to or towards 1.5 million years ago. The deep marine record, the source of most isotopic data, exists only on oceanic plates, which are eventually subducted: the oldest remaining material is 200 million years old. Older sediments are also more prone to corruption by diagenesis. Resolution and confidence in the data decrease over time.

Notable climate events in Earth history

Knowledge of precise climatic events decreases as the record goes back in time, but some notable climate events are known

- Faint young Sun paradox (start)
- Huronian glaciation (~2400 Mya Earth completely covered in ice probably due to Great Oxygenation Event)
- Later Neoproterozoic Snowball Earth (~600 Mya, precursor to the Cambrian Explosion)
- Andean-Saharan glaciation (~450 Mya)
- Carboniferous Rainforest Collapse (~300 Mya)
- Permian–Triassic extinction event (251.4 Mya)
- Oceanic Anoxic Events (~120 Mya, 93 Mya, and others)
- Cretaceous–Paleogene extinction event (66 Mya)
- Paleocene–Eocene Thermal Maximum (Paleocene–Eocene, 55Mya)
- Younger Dryas/The Big Freeze (~11,000 BC)

- Holocene climatic optimum (~7000–3000 BC)
- Climate changes of 535-536 (535–536 AD)
- Medieval warm period (900–1300)
- Little Ice Age (1300–1800)
- Year Without a Summer (1816)

History of the Atmosphere

Earliest Atmosphere

The first atmosphere would have consisted of gases in the solar nebula, primarily hydrogen. In addition, there would probably have been simple hydrides such as those now found in gas giants like Jupiter and Saturn, notably water vapor, methane and ammonia. As the solar nebula dissipated, the gases would have escaped, partly driven off by the solar wind.

Second Atmosphere

The next atmosphere, consisting largely of nitrogen and carbon dioxide and inert gases, was produced by outgassing from volcanism, supplemented by gases produced during the late heavy bombardment of Earth by huge asteroids. A major part of carbon dioxide emissions were soon dissolved in water and built up carbonate sediments.

Water-related sediments have been found dating from as early as 3.8 billion years ago. About 3.4 billion years ago, nitrogen was the major part of the then stable "second atmosphere". An influence of life has to be taken into account rather soon in the history of the atmosphere because hints of early life forms have been dated to as early as 3.5 billion years ago. The fact that it is not perfectly in line with the 30% lower solar radiance (compared to today) of the early Sun has been described as the "faint young Sun paradox".

The geological record, however, shows a continually relatively warm surface during the complete early temperature record of Earth with the exception of one cold glacial phase about 2.4 billion years ago. In the late Archaean eon, an oxygen-containing atmosphere began to develop, apparently from photosynthesizing cyanobacteria (see Great Oxygenation Event) which have been found as stromatolite fossils from 2.7 billion years ago. The early basic carbon isotopy (isotope ratio proportions) was very much in line with what is found today, suggesting that the fundamental features of the carbon cycle were established as early as 4 billion years ago.

Third Atmosphere

The constant rearrangement of continents by plate tectonics influences the long-term evolution of the atmosphere by transferring carbon dioxide to and from large continental carbonate stores. Free oxygen did not exist in the atmosphere until about 2.4 billion years ago, during the Great Oxygenation Event, and its appearance is indicated by the end of the banded iron formations. Until then, any oxygen produced by photosynthesis was consumed by oxidation of reduced materials, notably iron. Molecules of free oxygen did not start to accumulate in the atmosphere until the rate of production of oxygen began to exceed the availability of reducing materials. That point was a shift from a reducing atmosphere to an oxidizing atmosphere. O_2 showed major variations until reaching a steady state of more than 15% by the end of the Precambrian. The following time span was the Phanerozoic eon, during which oxygen-breathing metazoan life forms began to appear.

The amount of oxygen in the atmosphere has fluctuated over the last 600 million years, reaching a peak of 35% during the Carboniferous period, significantly higher than today's 21%. Two main processes govern changes in the atmosphere: plants use carbon dioxide from the atmosphere, releasing oxygen and the breakdown of pyrite and volcanic eruptions release sulfur into the atmosphere, which oxidizes and hence reduces the amount of oxygen in the atmosphere. However, volcanic eruptions also release carbon dioxide, which plants can convert to oxygen. The exact cause of the variation of the amount of oxygen in the atmosphere is not known. Periods with much oxygen in the atmosphere are associated with rapid development of animals. Today's atmosphere contains 21% oxygen, which is high enough for rapid development of animals.

Currently, anthropogenic greenhouse gases are accumulating in the atmosphere, which is the main cause of global warming.

Climate During Geological Ages

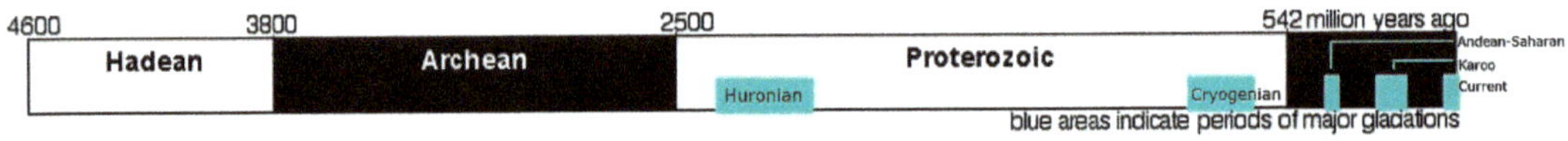

Timeline of glaciations, shown in blue

- The Huronian glaciation, is the first known glaciation in Earth's history, and lasted from 2400-2100 million years ago.
- The Cryogenian glaciation lasted from 850-635 million years ago.
- The Andean-Saharan glaciation lasted from 450–420 million years ago.
- The Karoo glaciation lasted from 360–260 million years ago.
- The Quaternary glaciation is the current glaciation period and begun 2.58 million years ago.

Precambrian Climate

The climate of the late Precambrian showed some major glaciation events spreading over much of the earth. At this time the continents were bunched up in the Rodinia supercontinent. Massive deposits of tillites are found and anomalous isotopic signatures are found, which gave rise to the Snowball Earth hypothesis. As the Proterozoic Eon drew to a close, the Earth started to warm up. By the dawn of the Cambrian and the Phanerozoic, life forms were abundant in the Cambrian explosion with average global temperatures of about 22 °C.

Phanerozoic Climate

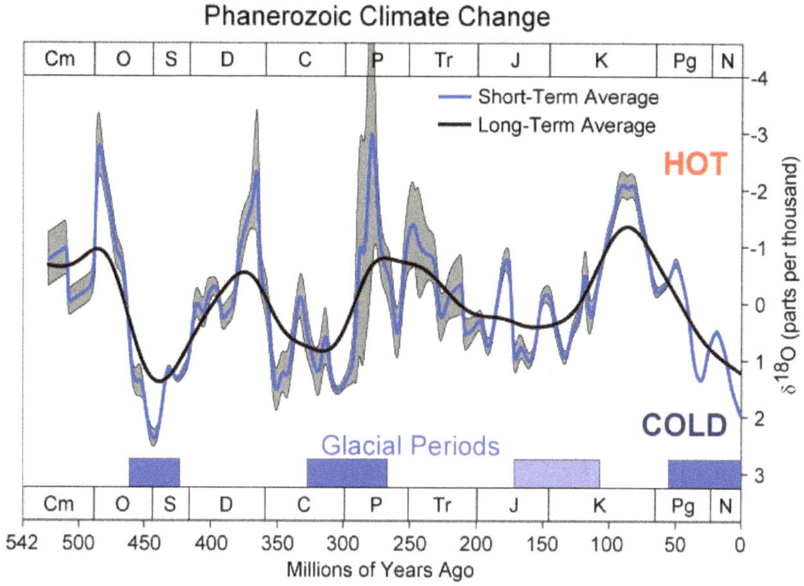

500 million years of climate change.

Major drivers for the preindustrial ages have been variations of the sun, volcanic ashes and exhalations, relative movements of the earth towards the sun and tectonically induced effects as for major sea currents, watersheds and ocean oscillations. In the early Phanerozoic, increased atmospheric carbon dioxide concentrations have been linked to driving or amplifying increased global temperatures.Royer et al. 2004 found a climate sensitivity for the rest of the Phanerozoic which was calculated to be similar to today's modern range of values.

The difference in global mean temperatures between a fully glacial Earth and an ice free Earth is estimated at approximately 10 °C, though far larger changes would be observed at high latitudes and smaller ones at low latitudes . One requirement for the development of large scale ice sheets seems to be the arrangement of continental land masses at or near the poles. The constant rearrangement of continents by plate tectonics can also shape long-term climate evolution. However, the presence or absence of land masses at the poles is not sufficient to guarantee glaciations or exclude polar ice caps. Evidence exists of past warm periods in Earth's climate when polar land masses similar to Antarctica were home to deciduous forests rather than ice sheets.

The relatively warm local minimum between Jurassic and Cretaceous goes along with an increase of subduction and mid-ocean ridge volcanism due to the breakup of the Pangea supercontinent.

Superimposed on the long-term evolution between hot and cold climates have been many short-term fluctuations in climate similar to, and sometimes more severe than, the varying glacial and interglacial states of the present ice age. Some of the most severe fluctuations, such as the Paleocene-Eocene Thermal Maximum, may be related to rapid climate changes due to sudden collapses of natural methane clathrate reservoirs in the oceans.

A similar, single event of induced severe climate change after a meteorite impact has been proposed as reason for the Cretaceous–Paleogene extinction event. Other major thresholds are the Permian-Triassic, and Ordovician-Silurian extinction events with various reasons suggested.

Quaternary Climate

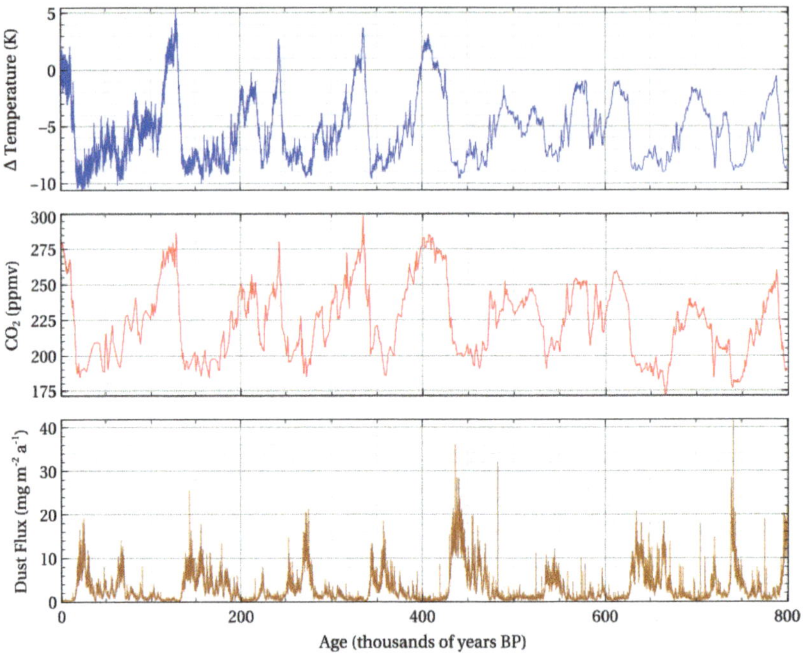

Ice core data for the past 800,000 years . Note length of glacial-interglacial cycles averages ~100,000 years. Blue curve is temperature,red curve is atmospheric CO_2concentrations, and brown curve is dust fluxes. Today's date is on the left side of the graph because the x-axis values represent "age before 1950".

The Quaternary sub-era includes the current climate. There has been a cycle of ice ages for the past 2.2–2.1 million years (starting before the Quaternary in the late Neogene Period).

Note in the graphic on the right the strong 120,000-year periodicity of the cycles, and the striking asymmetry of the curves. This asymmetry is believed to result from complex interactions of feedback mechanisms. It has been observed that ice ages deepen by progressive steps, but the recovery to interglacial conditions occurs in one big step.

The graph below shows the temperature change over the past 12 000 years, from various sources. The thick black curve is an average.

Climate Forcings

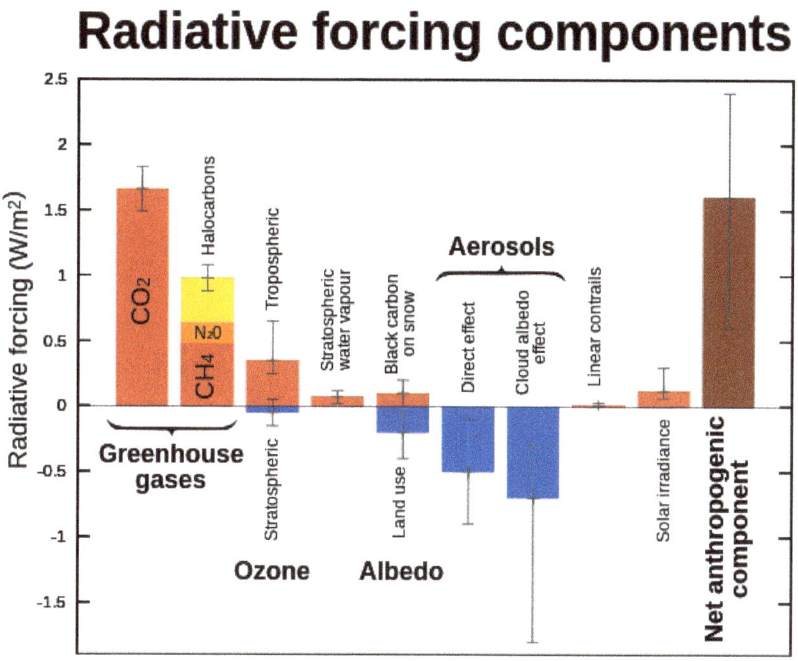

Radiative forcings, IPCC (2007).

The climate forcing is the difference of radiant energy (sunlight) received by the Earth and the outgoing longwave radiation back to space. The radiative forcing is quantified based on the CO_2 amount in the tropopause, in units of watts per square meter to the Earth's surface. Dependent on the radiative balance of incoming and outgoing energy, the Earth either warms up or cools down. Earth radiative balance originates from changes in solar insolation and the concentrations of greenhouse gases and aerosols. Climate change may be due to internal processes in Earth sphere's and/or following external forcings.

Internal Processes and Forcings

The Earth's climate system involves the study of the atmosphere, biosphere, cryosphere, hydrosphere, and lithosphere, and the sum of these processes from Earth sphere's is considered the processes affecting the climate. Greenhouse gases act as the internal forcing of the climate system. Particular interests in climate science and paleoclimatology focuses on the study of Earth climate sensitivity, in response to the sum of forcings.

Examples:

- Thermohaline circulation (Hydrosphere)
- Life (Biosphere)

External forcings

- The Milankovitch cycles determine Earth distance and position to the Sun. The solar insolation, is the total amount of solar radiation received by Earth.
- Volcanic eruptions, are considered an external forcing.
- Human changes of the composition of the atmosphere or land use.

Mechanisms

On timescales of millions of years, the uplift of mountain ranges and subsequently weathering processes of rocks and soils and the subduction of tectonic plates, are an important part of the carbon cycle. The weathering sequesters CO_2, by the reaction of minerals with chemicals (especially silicate weathering with CO_2) and thereby removing CO_2 from the atmosphere and reducing the radiative forcing. The opposite effect is volcanism, responsible for the natural greenhouse effect, by emitting CO_2into the atmosphere, thus affecting glaciation (Ice Age) cycles. James Hansen suggested that humans emit CO_2 10,000 times faster than natural processes have done in the past.

Ice sheet dynamics and continental positions (and linked vegetation changes) have been important factors in the long term evolution of the earth's climate. There is also a close correlation between $CO_2$and temperature, where CO_2 has a strong control over global temperatures in Earth history.

REFERENCES

1. Sahney, S. & Benton, M.J. (2008). "Recovery from the most profound mass extinction of all time" (PDF). Proceedings of the Royal Society: Biological. 275 *(1636): 759–65. doi:10.1098/rspb.2007.1370. PMC 2596898. PMID 18198148.*
2. Halfar, J.; Steneck, R.S.; Joachimski, M.; Kronz, A.; Wanamaker, A.D. (2008). "Coralline red algae as high-resolution climate recorders". Geology. 36 *(6): 463. doi:10.1130/ G24635A.1.*
3. Cobb, K.; Charles, C. D.; Cheng, H; Edwards, R. L. (2003). "El Nino/Southern Oscillation and tropical Pacific climate during the past millennium". Nature. 424 *(6946): 271–6.* doi:10.1038/nature01779. PMID 12867972.
4. Jouzel, Jean; Masson-Delmotte, V.; Cattani, O.; Dreyfus, G.; Falourd, S.; Hoffmann, G.; Minster, B.; Nouet, J.; et al. (10 August 2007). "Orbital and Millennial Antarctic Climate Variability over the Past 800,000 Years". Science. 317 *(5839): 793–796. Bibcode*:2007Sci...317..793J. doi:10.1126/science.1141038. PMID 17615306.
5. Page 1 1 International Partnerships in Ice Core Sciences (IPICS) The oldest ice core: A 1.5 million year record of climate and greenhouse gases from Antarctica" (PDF). Retrieved 22 September 2011.
6. Zahnle, K.; Schaefer, L.; Fegley, B. (2010). "Earth's Earliest Atmospheres". Cold Spring Harbor Perspectives in Biology. 2 *(10): a004895. doi:10.1101/cshperspect.a004895. PMC 2944365. PMID 20573713.*
7. B. Windley: The Evolving Continents. Wiley Press, New York 1984

8. J. Schopf: Earth's Earliest Biosphere: Its Origin and Evolution. Princeton University Press, Princeton, N.J., 1983

9. Christopher R. Scotese, Back to Earth History : Summary Chart for the Precambrian, Paleomar Project

10. Beerling, David (2007). The emerald planet: how plants changed Earth's history. Oxford University press. p. 47. ISBN 9780192806024.

11. Peter Ward: Out of Thin Air: Dinosaurs, Birds, and Earth's Ancient Atmosphere

12. Summary for Policymakers" (PDF). Climate Change 2007: The Physical Science Basis. Contribution of Working Group I to the Fourth Assessment Report of the Intergovernmental Panel on Climate Change. Intergovernmental Panel on Climate Change. 5 February 2007.

13. Rosemarie E. Came, John M. Eiler, Jan Veizer, Karem Azmy, Uwe Brand & Christopher R. Weidman; Eiler; Veizer; Azmy; Brand; Weidman (September 2007). "Coupling of surface temperatures and atmospheric CO_2concentrations during the Palaeozoic era". Nature. 449 *(7159): 198–201. Bibcode*:2007Natur.449..198C. doi:10.1038/nature06085. PMID 17851520.

14. Royer, Dana L. and Robert A. Berner, Isabel P. Montañez, Neil J. Tabor, David J. Beerling *(July 2004). "CO_2 as a primary driver of Phanerozoic climate". GSA Today.* 14 *(3): 4–10.* doi:10.1130/1052-5173(2004)014<4:CAAPDO>2.0.CO;2.

15. Douwe G. Van Der Meer, Richard E. Zeebe, Douwe J. J. van Hinsbergen, Appy Sluijs, Wim Spakman, and Trond H. Torsvik (February 2014). "Plate tectonic controls on atmospheric CO2 levels since the Triassic". PNAS. 111 *(12): 4380–4385. doi:10.1073/pnas.1315657111.*

16. Jouzel, J.; Masson-Delmotte, V.; Cattani, O.; Dreyfus, G.; Falourd, S.; Hoffmann, G.; Minster, B.; Nouet, J.; Barnola, J. M. (2007-08-10). "Orbital and Millennial Antarctic Climate Variability over the Past 800,000 Years". Science. 317 *(5839): 793–796. doi:10.1126/science.1141038. ISSN 0036-8075. PMID 17615306.*

17. Lüthi, Dieter; Le Floch, Martine; Bereiter, Bernhard; Blunier, Thomas; Barnola, Jean-Marc; Siegenthaler, Urs; Raynaud, Dominique; Jouzel, Jean; Fischer, Hubertus (2008-05-15). "High-resolution carbon dioxide concentration record 650,000–800,000 years before present". Nature. 453 *(7193): 379–382.* doi:10.1038/nature06949. ISSN 0028-0836. PMID 18480821.

18. Lambert, F.; Delmonte, B.; Petit, J. R.; Bigler, M.; Kaufmann, P. R.; Hutterli, M. A.; Stocker, T. F.; Ruth, U.; Steffensen, J. P. (2008-04-03). "Dust-climate couplings over the past 800,000 years from the EPICA Dome C ice core". Nature. 452 *(7187): 616–619.* doi:10.1038/nature06763. ISSN 0028-0836.

19. Lambert, F.; Bigler, M.; Steffensen, J. P.; Hutterli, M.; Fischer, H. "Centennial mineral dust variability in high-resolution ice core data from Dome C, Antarctica". Climate of the Past. 8 *(2): 609–623.* doi:10.5194/cp-8-609-2012.

20. "Annex III: Glossary" (PDF). IPCC AR5. Climate change may be due to natural internal processes or external forcings such as modulations of the solar cycles, volcanic eruptions and persistent anthropogenic changes in the composition of the atmosphere or in land use.

21. Caldeira, Ken (June 18, 1992). "Enhanced Cenozoic chemical weathering and the subduction of pelagic carbonate". Nature. 357 *(6379): 578–581.* doi:10.1038/357578a0.

22. Cin-Ty Aeolus Lee, Douglas M. Morton, Mark G. Little, Ronald Kistler, Ulyana N.

Horodyskyj, William P. Leeman, and Arnaud Agranier (January 28, 2008). "Regulating continent growth and composition by chemical weathering". PNAS. 105 *(13): 4981–4986. doi:10.1073/pnas.0711143105.*

23. van der Meer, Douwe (March 25, 2014). "Plate tectonic controls on Atmospheric CO2 since the Triassic". PNAS. 111 *(12): 4380–4385. doi:10.1073/pnas.1315657111.*
24. James Hansen (2009). "The 8 Minute Epoch 65 million Years with James Hansen". University of Oregon.
25. ROYER, D. L.; PAGANI, M.; BEERLING, D. J. (1 July 2012). "Geobiological constraints on Earth system sensitivity to CO2 during the Cretaceous and Cenozoic". Geobiology. 10 *(4): 298–310. doi:10.1111/j.1472-4669.2012.00320.x. PMID 22353368.*
26. Royer, Dana L. (1 December 2006). "CO2-forced climate thresholds during the Phanerozoic". Geochimica et Cosmochimica Acta. 70 *(23): 5665–5675. Bibcode*:2006GeCoA..70.5665R. doi:10.1016/j.gca.2005.11.031.

CHAPTER 4

Carbon Capture and Storage

Chapter Contents

- **Introduction**
- **Capture**
- **Transport**
- **Sequestration**
- **Leakage**
- **Monitoring Geological Sequestration Sites**
- **Carbon Dioxide Recycling - Carbon Capture and Utilization (CCU)**
- **Example CCS Projects**
- **Secarb**
- **Limitations of CCS For Power Stations**
- **Cost**
- **Environmental Effects**
- **Carbon Sequestration**

Carbon capture is a process in which carbon dioxide is extracted from emissions, classically the flue gases produced at large facilities like power plants. By capturing carbon before it can be released, carbon capture reduces the amount of CO2 being released into the atmosphere, thereby reducing some of the damage believed to be caused by carbon emissions. Growing concerns about rising carbon dioxide levels in the early 21st century led to some interest in carbon capture as a technique which could be used to manage carbon emissions.

INTRODUCTION

Carbon capture and storage (CCS) (or carbon capture and sequestration) is the process of capturing waste carbon dioxide (CO_2) from large point sources, such as fossil fuel power plants, transporting it to a storage site, and depositing it where it will not enter the atmosphere, normally an underground geological formation. The aim is to prevent the release of large quantities of CO_2 into the atmosphere (from fossil fuel use in power generation and other industries). It is a potential means of mitigating the contribution of fossil fuel emissions to global warming and ocean acidification. Although CO_2 has been injected into geological formations for several decades for various purposes, including enhanced oil recovery, the long term storage of CO_2 is a relatively new concept. The first commercial example was the Weyburn-Midale Carbon Dioxide Project in 2000. Other examples include SaskPower's Boundary Dam and Mississippi Power's Kemper Project. ‹CCS' can also be used to describe the scrubbing of CO_2 from ambient air as a climate engineering technique.

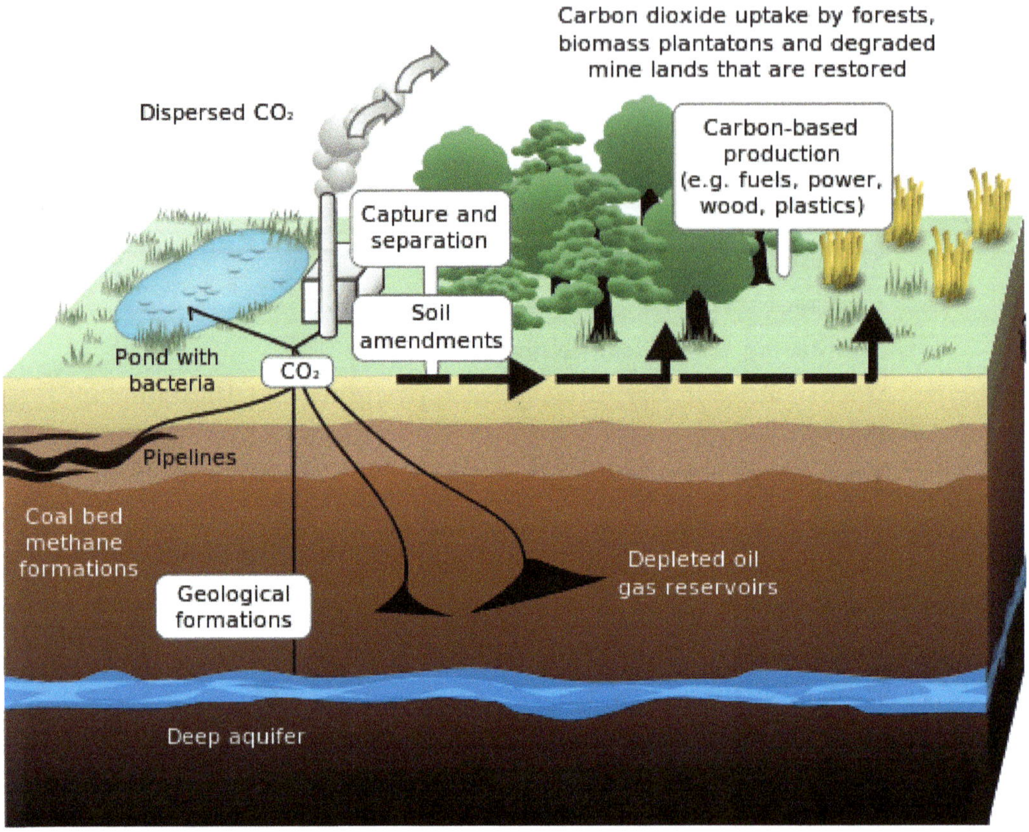

Schematic showing both terrestrial and geological sequestration of carbon dioxide emissions from a coal-fired plant.

An integrated pilot-scale CCS power plant was to begin operating in September 2008 in the eastern German power plant Schwarze Pumpe run by utility Vattenfall, to test the technological feasibility and economic efficiency. CCS applied to a modern conventional power plant could reduce CO_2 emissions to the atmosphere by approximately 80–90% compared to a plant without CCS. The IPCC estimates that the economic potential of CCS could be between 10% and 55% of the total carbon mitigation effort until year 2100.

Carbon dioxide can be captured out of air or fossil fuel power plant flue gas using adsorption (or carbon scrubbing), membrane gas separation, or adsorption technologies. Amines are the leading carbon scrubbing technology. Capturing and compressing CO_2 may increase the energy needs of a coal-fired CCS plant by 25–40%. These and other system costs are estimated to increase the cost per watt energy produced by 21–91% for fossil fuel power plants. Applying the technology to existing plants would be more expensive, especially if they are far from a sequestration site. A 2005 industry report suggests that with successful research, development and deployment (RD&D), sequestered coal-based electricity generation in 2025 may cost less than unsequestered coal-based electricity generation today.

Storage of the CO_2 is envisaged either in deep geological formations, or in the form of mineralcarbonates. Deep ocean storage is not currently considered feasible due to the associated effect of ocean acidification. Geological formations are currently considered the most promising sequestration sites. The National Energy Technology Laboratory (NETL) reported that North America has enough storage capacity for more than 900 years worth of carbon dioxide at current production rates. A general problem is that long term predictions about submarine or underground storage security are very difficult and uncertain, and there is still the risk that CO_2 might leak into the atmosphere.

CAPTURE

Capturing CO_2 is most effective at point sources, such as large fossil fuel or biomass energy facilities, industries with major CO_2 emissions, natural gas processing, synthetic fuel plants and fossil fuel-based hydrogen production plants. Extracting CO_2 from air is also possible, but not very practical because the CO_2 is not concentrated.

Organisms that produce ethanol by fermentation generate cool, essentially pure CO_2 that can be pumped underground. Fermentation produces slightly less CO_2 than ethanol by weight.

Flue gas from the combustion of coal in oxygen has a large concentration of CO2, about 10-15% CO_2whereas natural gas power plant flue gas is about 5-10% CO_2. Therefore, it is more energy and cost efficient to capture CO2 from coal-fired power plants. Impurities in CO_2 streams, like sulfurs and water, could have a significant effect on their phase behaviour and could pose a significant threat of increased corrosion of pipeline and well materials. In instances where CO_2 impurities exist, especially with air capture, a scrubbing separation process would be needed to initially clean the flue gas. According to the Wallula Energy Resource Center in Washington state, by gasifying coal, it is possible to capture approximately 65% of carbon dioxide embedded in it and sequester it in a solid form.

Broadly, three different configurations of technologies for capture exist: post-combustion, pre-combustion, and oxyfuel combustion:

- In *post combustion capture*, the CO_2 is removed after combustion of the fossil fuel — this is the scheme that would be applied to fossil-fuel burning power plants. Here, carbon dioxide is captured from flue gases at power stations or other large point sources. The technology is well understood and is currently used in other industrial applications, although not at the same scale as might be required in a commercial scale power station. Post combustion capture is most popular in research because existing fossil fuel power plants can be retrofitted to include CCS technology in this configuration.

- The technology for *pre-combustion* is widely applied in fertilizer, chemical, gaseous fuel (H_2, CH_4), and power production. In these cases, the fossil fuel is partially oxidized, for instance in a gasifier. The resulting syngas (CO and H_2) is shifted into CO_2 and H_2. The resulting CO_2 can be captured from a relatively pure exhaust stream. The H_2 can now be used as fuel; the carbon dioxide is removed before combustion takes place. There are several advantages and disadvantages when compared to conventional post combustion carbon dioxide capture. The CO_2 is removed after combustion of fossil fuels, but before the flue gas is expanded to atmospheric pressure. This scheme is applied to new fossil fuel burning power plants, or to existing plants where re-powering is an option. The capture before expansion, i.e. from pressurized gas, is standard in almost all industrial CO_2 capture processes, at the same scale as will be required for utility power plants.
- In *oxy-fuel combustion* the fuel is burned in oxygen instead of air. To limit the resulting flame temperatures to levels common during conventional combustion, cooled flue gas is recirculated and injected into the combustion chamber. The flue gas consists of mainly carbon dioxide and water vapour, the latter of which is condensed through cooling. The result is an almost pure carbon dioxide stream that can be transported to the sequestration site and stored. Power plant processes based on oxyfuel combustion are sometimes referred to as "zero emission" cycles, because the CO_2 stored is not a fraction removed from the flue gas stream (as in the cases of pre- and post-combustion capture) but the flue gas stream itself. A certain fraction of the CO_2 generated during combustion will inevitably end up in the condensed water. To warrant the label "zero emission" the water would thus have to be treated or disposed of appropriately. The technique is promising, but the initial air separation step demands a lot of energy.

CO_2 Separation Technologies

Carbon dioxide can be separated out of air or flue gas with absorption, adsorption, or membrane gas separation technologies. Absorption, or carbon scrubbing, with amines is currently the dominant capture technology. Membrane and adsorption technologies are still in the developmental research stages, initiating primary pilot plants in the near future. Metal-Organic Frameworks (MOFs) are a novel class of materials that offer promise for carbon capture using adsorption technologies.

Carbon dioxide adsorbs to a MOF through physisorption or chemisorption based on the porosity and selectivity of the MOF leaving behind a Greenhouse gas poor gas stream that is more environmentally friendly. The carbon dioxide is then stripped off the MOF using temperature swing adsorption (TSA) or pressure swing adsorption (PSA) so the MOF can be reused. Adsorbents and absorbents require regeneration steps where the CO_2 is removed from the sorbent or solution that collected it out of the flue gas in order for the sorbent or solution to be reused. Monoethanolamine (MEA) solutions, the leading amine for capturing CO_2, have a heat capacity between 3-4 J/g K since they are mostly water.Higher heat capacities add to the energy penalty in the solvent regeneration step. Thus, to optimize a MOF for carbon capture, low heat capacities and heats of adsorption are desired. Additionally, high working capacity and high selectivity are desirable in order to capture as much CO_2 as possible from the flue gas. However, there is an energy trade off with selectivity and energy expenditure. As the amount of CO_2 captured increases, the energy, and therefore cost, required to regenerate increases. A large drawback of using MOFs for CCS is the limitations imposed by their chemical and thermal stability.Current research is looking to optimize MOF properties for CCS, but it has proven difficult to find these optimizations that also result in a stable MOF. Metal reservoirs are also a limiting factor to the potential success of MOFs.

Capture is attributed to about two thirds of the total cost of CCS, making it limit the wide-scale deployment of CCS technologies. To optimize a CO_2 capture process would significantly increase the feasibility of CCS since the transport and storage steps of CCS are rather mature technologies.

An alternate method under development is chemical looping combustion (CLC). Chemical looping uses a metal oxide as a solid oxygen carrier. Metal oxide particles react with a solid, liquid or gaseous fuel in a fluidized bed combustor, producing solid metal particles and a mixture of carbon dioxide and water vapor. The water vapor is condensed, leaving pure carbon dioxide, which can then be sequestered. The solid metal particles are circulated to another fluidized bed where they react with air, producing heat and regenerating metal oxide particles that are recirculated to the fluidized bed combustor. A variant of chemical looping is calcium looping, which uses the alternating carbonation and then calcination of a calcium oxide based carrier as a means of capturing CO_2.

A few engineering proposals have been made for the more difficult task of removing CO_2 from the atmosphere – a form of climate engineering – but work in this area is still in its infancy. Capture costs are estimated to be higher than from point sources, but may be feasible for dealing with emissions from diffuse sources such as automobiles and aircraft. The theoretically required energy for air capture is only slightly more than for capture from point sources. The additional costs come from the devices that use the natural air flow. Global Research Technologies demonstrated a pre-prototype of air capture technology in 2007.

TRANSPORT

After capture, the CO_2 would have to be transported to suitable storage sites. This would most likely be done by pipeline, which is generally the cheapest form of transport. In 2008, there were approximately 5,800 km of CO_2 pipelines in the United States, used to transport CO_2 to oil production fields where it is then injected into older fields to extract oil. The injection of CO_2 to produce oil is generally called *Enhanced Oil Recovery* or EOR. In addition, there are several pilot programs in various stages to test the long-term storage of CO_2 in non-oil producing geologic formations.

According to the Congressional Research Service, “There are important unanswered questions about pipeline network requirements, economic regulation, utility cost recovery, regulatory classification of CO_2itself, and pipeline safety. Furthermore, because CO_2 pipelines for enhanced oil recovery are already in use today, policy decisions affecting

CO_2 pipelines take on an urgency that is unrecognized by many. Federal classification of CO_2 as both a commodity (by the Bureau of Land Management) and as a pollutant (by the Environmental Protection Agency) could potentially create an immediate conflict which may need to be addressed not only for the sake of future CCS implementation, but also to ensure consistency of future CCS with CO_2 pipeline operations today."

Ships could also be utilized for transport where pipelines are not feasible. These methods are currently used for transporting CO_2 for other applications.

SEQUESTRATION

Various forms have been conceived for permanent storage of CO_2. These forms include gaseous storage in various deep geological formations (including saline formations and exhausted gas fields), and solid storage by reaction of CO_2 with metal oxides to produce stable carbonates.

Geological Storage

Also known as *geo-sequestration*, this method involves injecting carbon dioxide, generally in supercritical form, directly into underground geological formations. Oil fields, gas fields, saline formations, unmineable coal seams, and saline-filled basalt formations have been suggested as storage sites. Various physical (e.g., highly impermeable caprock) and geochemical trapping mechanisms would prevent the CO_2 from escaping to the surface.

CO_2 is sometimes injected into declining oil fields to increase oil recovery. Approximately 30 to 50 million metric tonnes of CO_2 are injected annually in the United States into declining oil fields. This option is attractive because the geology of hydrocarbon reservoirs is generally well understood and storage costs may be partly offset by the sale of additional oil that is recovered. Disadvantages of old oil fields are their geographic distribution and their limited capacity, as well as the fact that subsequent burning of the additional oil recovered will offset much or all of the reduction in CO_2 emissions.

Unmineable coal seams can be used to store CO_2 because the CO_2 molecules attach to the surface of coal. The technical feasibility, however, depends on the permeability of the coal bed. In the process of absorption the coal releases previously absorbed methane, and the methane can be recovered (enhanced coal bed methane recovery). The sale of the methane can be used to offset a portion of the cost of the CO_2 storage. Burning the resultant methane, however, would negate some of the benefit of sequestering the original CO_2.

Saline formations contain highly mineralized brines, and have so far been considered of no benefit to humans. Saline aquifers have been used for storage of chemical waste in a few cases. The main advantage of saline aquifers is their large potential storage volume and their common occurrence. The major disadvantage of saline aquifers is that relatively little is known about them, especially compared to oil fields. To keep the cost of storage acceptable, the geophysical exploration may be limited, resulting in larger uncertainty about the aquifer structure. Unlike storage in oil fields or coal beds, no side product will offset the storage cost. Leakage of CO_2 back into the atmosphere may be a problem in saline aquifer storage. Current research shows, however, that *trapping mechanisms* such as structural trapping, residual trapping, solubility trapping and mineral trapping could immobilize the CO_2 underground and reduce the risk of leakage.

Enhanced Oil Recovery

Enhanced oil recovery (EOR) is a generic term for techniques used to increase the amount of crude oil that can be extracted from an oil field. In Carbon Capture & Sequestration Enhanced Oil Recovery (CCS EOR), carbon dioxide is injected into an oil field to recover oil that is often never recovered using more traditional methods.

Crude oil development and production in U.S. oil reservoirs can include up to three distinct phases: primary, secondary, and tertiary (or enhanced) recovery. During primary recovery only about 10 percent of a reservoir's original oil in place is typically produced. Secondary recovery techniques extend a field's productive life generally by injecting water or gas to displace oil and drive it to a production wellbore, resulting in the recovery of 20 to 40 percent of the original oil in place. However, with much of the easy-to-produce oil already recovered from U.S. oil fields, producers have attempted several tertiary, or enhanced oil recovery (EOR), techniques that offer prospects for ultimately producing 30 to 60 percent, or more, of the reservoir's original oil in place.

An example of a project that will use CCS EOR is the Kemper Project in Mississippi. Due to the Kemper Project's close proximity to oil fields, the carbon dioxide byproduct from producing electricity will be transported to the neighboring oil fields for EOR.

Ocean Storage

In the past, it was suggested that CO_2 could be stored in the oceans, but this would only exacerbate ocean acidification and has been made illegal under specific regulations. Ocean storage is no longer considered feasible.

Mineral Storage

In this process, CO_2 is exothermically reacted with available metal oxides, which in turn produces stable carbonates (e.g. calcite, magnesite). This process occurs naturally over many years and is responsible for a great amount of surface limestone. The idea of using olivine has been promoted by the geochemist Prof. Schuiling. The reaction rate can be made faster, for example, with a catalyst or by reacting at higher temperatures and/or pressures, or by pre-treatment of the minerals, although this method can require additional energy. The IPCC estimates that a power plant equipped with CCS using mineral storage will need 60–180% more energy than a power plant without CCS.

The economics of mineral carbonation at scale are now being tested in a world-first pilot plant project based in Newcastle, Australia. New techniques for mineral activation and reaction have been developed the GreenMag Group and the University of Newcastle and funded by the New South Wales and Australian Governments to be operational by 2013.

In 2009 it was reported that scientists had mapped 6,000 square miles (16,000 km^2) of rock formations in the U.S. that could be used to store 500 years' worth of U.S. carbon dioxide emissions. A study on mineral sequestration in the US states:

Carbon sequestration by reacting naturally occurring Mg and Ca containing minerals with CO_2 to form carbonates has many unique advantages. Most notabl[e] is the fact that carbonates have a lower energy state than CO_2, which is why mineral carbonation is thermodynamically favorable and occurs naturally (e.g., the weathering of rock over geologic time periods). Secondly, the raw materials such as magnesium based minerals are abundant. Finally, the produced carbonates are unarguably stable and thus re-release of CO_2 into the atmosphere is not an issue. However, conventional carbonation pathways are slow under ambient temperatures and pressures. The significant challenge being addressed by this effort is to identify an industrially and environmentally viable carbonation route that will allow mineral sequestration to be implemented with acceptable economics.

The following table lists principal metal oxides of Earth's Crust. Theoretically, up to 22% of this mineral mass is able to form carbonates.

Earthen Oxide	Percent of Crust	Carbonate	Enthalpy change (kJ/mol)
SiO_2	59.71		
Al_2O_3	15.41		
CaO	4.90	$CaCO_3$	-179
MgO	4.36	$MgCO_3$	-118
Na_2O	3.55	Na_2CO_3	-322
FeO	3.52	$FeCO_3$	-85
K_2O	2.80	K_2CO_3	-393.5
Fe_2O_3	2.63	$FeCO_3$	112
	21.76	All Carbonates	

Ultramafic mine tailings are a readily available source of fine-grained metal oxides that can act as artificial carbon sinks to reduce net greenhouse gas emissions in the mining industry. Accelerating passive CO_2 sequestration via mineral carbonation may be achieved through microbial processes that enhance mineral dissolution and carbonate precipitation.

Energy Requirements

The energy requirements of sequestration processes may be significant. In one paper, sequestration consumed 25 percent of the plant's rated 600-megawatt output capacity.

After adding CO_2 capture and compression, the capacity of the coal-fired power plant is reduced to 457 MW.

LEAKAGE

Lake Nyos as it appeared fewer than two weeks after the eruption; August 29, 1986.

A major concern with CCS is whether leakage of stored CO_2 will compromise CCS as a climate change mitigation option. For well-selected, designed and managed geological storage sites, IPCC estimates that risks are comparable to those associated with current hydrocarbon activity. However, this finding is contested due to a lack of experience in such long term storage. CO_2 could be trapped for millions of years, and although some leakage occurs upwards through the soil, well selected storage sites are likely to retain over 99% of the injected CO_2 over 1000 years. Leakage through the injection pipe is a greater risk.

Although the injection pipe is usually protected with non-return valves to prevent release on a power outage, there is still a risk that the pipe itself could tear and leak due to the pressure. The Berkel en Rodenrijs incident in December 2008 was an example, where a modest release of CO_2 from a pipeline under a bridge resulted in the deaths of some ducks sheltering there. In order to measure accidental carbon releases more accurately and decrease the risk of fatalities through this type of leakage, the implementation of CO_2 alert meters around the project perimeter has been proposed. Malfunction of a carbon dioxide industrial fire suppression system in a large warehouse released CO_2 and 14 citizens collapsed on the nearby public road. A release of CO_2 from a salt mine killed a person at distance of 300 meters.

In 1986 a large leakage of naturally sequestered CO_2 rose from Lake Nyos in Cameroon and asphyxiated 1,700 people. While the carbon had been sequestered naturally, some point to the event as evidence for the potentially catastrophic effects of sequestering carbon artificially. The Lake Nyos disaster resulted from a volcanic event, which very suddenly released as much as a cubic kilometre of CO_2 gas from a pool of naturally occurring CO_2 under the lake in a deep narrow valley. The location of this pool of CO_2 is not a place where man can inject or store CO_2, and this pool was not known about nor monitored until after the occurrence of the natural disaster.

For ocean storage, the retention of CO_2 would depend on the depth. The IPCC estimates 30–85% of the sequestered carbon dioxide would be retained after 500 years for depths 1000–3000 m. Mineral storage is not regarded as having any risks of leakage.

The IPCC recommends that limits be set to the amount of leakage that can take place. This might rule out deep ocean storage as an option.

At the conditions of the deeper oceans, (about 400 bar or 40 MPa, 280 K) water–CO_2(l) mixing is *very* low (where carbonate formation/acidification is the rate limiting step), but the formation of water-CO_2hydrates, a kind of solid water cage that surrounds the CO_2, is favorable.

To further investigate the safety of CO_2 sequestration, Norway's Sleipner gas field can be studied, as it is the oldest plant that stores CO_2 on an industrial scale. According to an environmental assessment of the gas field which was conducted after ten years of operation, the author affirmed that geosequestration of CO_2 was the most definite form of permanent *geological* storage of CO_2:

Available geological information shows absence of major tectonic events after the deposition of the Utsira formation [saline reservoir]. This implies that the geological environment is tectonically stable and a site suitable for carbon dioxide storage. The solubility trapping [is] the most permanent and secure form of geological storage.

In March 2009 StatoilHydro issued a study showing the slow spread of CO_2 in the formation after more than 10 years operation.

Phase I of the Weyburn-Midale Carbon Dioxide Project in Weyburn, Saskatchewan, Canada has determined that the likelihood of stored CO_2 release is less than one percent in 5,000 years. A January 2011 report, however, claimed evidence of leakage in land above that project. This report was strongly refuted by the IEAGHG Weyburn-Midale CO_2 Monitoring and Storage Project, which issued an eight-page analysis of the study, claiming that it showed no evidence of leakage from the reservoir.

The liability of potential leak(s) is one of the largest barriers to large-scale CCS. To assess and reduce such liability, the leakage of stored gasses, particularly carbon dioxide, into the atmosphere may be detected via atmospheric gas monitoring, and can be quantified directly via the eddy covariance flux measurements,

MONITORING GEOLOGICAL SEQUESTRATION SITES

In order to detect carbon dioxide leaks and the effectiveness of geological sequestration sites, different monitoring techniques can be employed to verify that the sequestered carbon stays trapped below the surface in the intended reservoir. Leakage due to injection at improper locations or conditions could result in carbon dioxide being released back into the atmosphere. It is important to be able to detect leaks with enough warning to put a stop to it, and to be able to quantify the amount of carbon that has leaked for purposes such as cap and trade policies, evaluation of environmental impact of leaked carbon, as well as accounting for the total loss and cost of the process. To quantify the amount of carbon dioxide released, should a leak occur, or to closely watch stored CO_2, there are several monitoring methods that can be done at both the surface and subsurface levels.

Subsurface Monitoring

In subsurface monitoring, there are direct and indirect methods to determine the amount

CARBON DIOXIDE RECYCLING - CARBON CAPTURE AND UTILIZATION (CCU)

Recycling CO_2 may offer a response to the global challenge of significantly reducing greenhouse gas emissions from major stationary (industrial) emitters in the near to medium term, but is usually considered a different technological category from CCS. Technologies under development, such as Bio CCS Algal Synthesis, utilises pre-smokestack CO_2 (such as from a coal-fired power station) as a useful feedstock input to the production of oil-rich algae in solar membranes to produce oil for plastics and transport fuel (including aviation fuel), and nutritious stock-feed for farm animal production. The CO_2 and other captured greenhouse gases are injected into the membranes containing waste water and select strains of algae causing, together with sunlight or UV light, an oil rich biomass that doubles in mass every 24 hours.

The Bio CCS Algal Synthesis process is based on earth science photosynthesis: the technology is entirely retrofittable and collocated with the emitter, and the capital outlays may offer a return upon investment due to the high value commodities produced (oil for plastics, fuel and feed).

Bio CCS Algal Synthesis test facilities were being trialed at Australia's three largest coal-fired power stations (Tarong, Queensland; Eraring, NSW; Loy Yang, Victoria) using piped pre-emission smokestack CO_2 (and other greenhouse gases) as feedstock to grow oil-rich algal biomass in enclosed membranes for the production of plastics, transport fuel and nutritious animal feed.

Another potentially useful way of dealing with industrial sources of CO_2 is to convert it into hydrocarbons where it can be stored or reused as fuel or to make plastics. There are a number of projects investigating this possibility.

Carbon dioxide scrubbing variants exist based on potassium carbonate which can be used to create liquid fuels, though this process requires a great deal of energy input. Although the creation of fuel from atmospheric CO_2 is not a climate engineering technique, nor does it actually function as greenhouse gas remediation (because carbon dioxide will be re-released when the fuel is burned) it nevertheless is potentially useful in the creation of a low carbon economy.

Other uses are the production of stable carbonates from silicates (e.g. olivine produces Magnesium carbonate. This process is still in the R&D phase.

Single step Methods: Methanol

A proven process to produce a hydrocarbon is to make methanol. Methanol is easily synthesized from CO_2 and H_2. Based on this fact the idea of a methanol economy was born.

Single step Methods: Hydrocarbons

At the department of Industrial Chemistry and Engineering of Materials at the University of Messina, Italy, there is a project to develop a system which works like a fuel-cell in reverse, whereby a catalyst is used that enables sunlight to split water into hydrogen ions of CO_2 in the reservoir. A direct method would be drilling deep enough to collect fluid sample. This drilling can be difficult and expensive due to the physical properti of the rock. It also only provides data at a specific location. Indirect methods would b to send sound or electromagnetic waves down to the reservoir where it is then reflecte back up to be interpreted. This approach is also expensive but it provides data over much larger region; it does however lack precision. Both direct and indirect monitorin can be done intermittently or continuously.

Seismic Monitoring

Seismic monitoring is a type of indirect subsurface monitoring. It is done by creating vibrational waves either at the surface using a vibroseis truck, or inside a well using spinning eccentric mass. These vibrational waves then propagate through the geological layers and reflect back creating patterns that are read and interpreted by seismometers. It can identify migration pathways of the CO_2 plume.Two examples of monitoring geological sequestration sites using seismic monitoring are the Sleipner sequestration project and the Frio CO_2 Injection test. Although this method can confirm the presence of CO_2 in a given region, it cannot determine the specifics of the environment or concentration of CO_2.

Surface Monitoring

Eddy covariance is a surface monitoring technique that measures the flux of CO_2 from the ground's surface. It involves measuring CO_2 concentrations as well as vertical wind velocities using an anemometer. This provides a measure of the total vertical flux of CO2. Eddy covariance towers could potentially detect leaks, however, the natural carbon cycle, such as photosynthesis and the respiration of plants, would have to be accounted for and a baseline CO_2 cycle would have to be developed for the location of monitoring. An example of Eddy covariance techniques used to monitor carbon sequestration sites is the Shallow Release test. Another similar approach is utilizing accumulation chambers. These chambers are sealed to the ground with an inlet and outlet flow stream connected to a gas analyzer.This also measures the vertical flux of CO_2. The disadvantage of accumulation chambers is its inability to monitor a large region which is necessary in detecting CO_2 leaks over the entire sequestration site.

InSAR Monitoring

InSAR monitoring is another type of surface monitoring. It involves a satellite sending signals down to the Earth's surface where it is reflected back to the satellite's receiver From this, the satellite is able to measure the distance to that point. In CCS, the injectio of CO_2 in deep sublayers of geological sites creates high pressures. These high pressured fluid filled layers affect those above and below it resulting in a change of the surfac landscape. In areas of stored CO_2, the ground's surface often rises due to the hig pressures originating in the deep subsurface layers. These changes in elevation of th Earth's surface corresponds to a change in the distance from the inSAR satellite whic is then detectable and measurable.

and oxygen gas. The ions cross a membrane where they react with the CO_2 to create hydrocarbons.

Two Step Methods

If CO_2 is heated to 2400 °C, it splits into carbon monoxide (CO) and oxygen. The Fischer-Tropsch process can then be used to convert the CO into hydrocarbons. The required temperature can be achieved by using a chamber containing a mirror to focus sunlight on the gas. Rival teams are developing such chambers, at Solarec and at Sandia National Laboratories, both based in New Mexico. According to Sandia these chambers could provide enough fuel to power 100% of domestic vehicles using 5800 km^2; unlike biofuels this would not take fertile land away from crops but would be land that is not being used for anything else. James May, the British TV presenter, visited a demonstration plant in a programme in his Big Ideas series.

EXAMPLE CCS PROJECTS

Industrial-scale projects

As of September 2012, the Global CCS Institute identified 75 large-scale integrated projects in its 2012 Global Status of CCS report which is a net increase of one project since its 2011 Global Status of CCS report. 16 of these projects are in operation or in construction capturing around 36 million tonnes of CO_2 per annum. For more information see Integrated CCS Projects on the Global CCS Institute's website. For information on EU projects see Zero Emissions Platform website. The eight large-scale integrated CCS projects currently in operation are:

In Salah CO_2 Injection — Algeria

In Salah is a fully operational onshore gas field with CO_2 injection. CO_2 is separated from produced gas and reinjected in the producing hydrocarbon reservoir zones. Since 2004, about 1 Mt/a of CO_2 has been captured during natural gas extraction and injected into the Krechba geologic formation at a depth of 1,800m. The Krechba formation is expected to store 17Mt CO_2 over the life of the project.

Injection suspended in 2011 due to concerns about the integrity of the seal.

Sleipner CO_2 Injection — Norway

Sleipner is a fully operational offshore gas field with CO_2 injection initiated in 1996. CO_2 is separated from produced gas and reinjected in the Utsira saline aquifer (800–1000 m below ocean floor) above the hydrocarbon reservoir zones. This aquifer extends much further north from the Sleipner facility at its southern extreme. The large size of the reservoir accounts for why 600 billion tonnes of CO_2 are expected to be stored, long after the Sleipner natural gas project has ended.

Snøhvit CO_2 Injection — Norway

Snøhvit is a fully operational offshore gas field with CO_2 injection. The LNG plant is located onshore. CO_2 is necessarily separated to produce liquefied natural gas (LNG) and then CO_2 is injected in a saline aquifer below the hydrocarbon reservoir zones offshore at a rate of 700,000 t/a into the Tubåen sandstone formation 2,600 m under the seabed for storage. This formation was closed April 2011, and injection started in the Stø-formation where produced gas is taken. Produced CO_2 is increasing, therefore separation capacity may limit production before end 2015 when a new formation will be drilled for CO_2-injection only. (Teknisk Ukeblad nr. 30, 2013, tu.no)

Great Plains Synfuel Plant and Weyburn-Midale Project — Canada

Weyburn-Midale is a coal gasification operation that produces synthetic natural gas and various petrochemicals from coal. This project captures about 2.8 Mt/a of CO_2 from its coal gasification plant located in North Dakota, US, transported by pipeline 320 km across the Canada–US border and injects it into depleting oil fields in Saskatchewan where it is used for enhanced oil recovery (EOR).

Shute Creek Gas Processing Facility — US

Around 7 million tonnes per annum of carbon dioxide are recovered from ExxonMobil's Shute Creek gas processing plant in Wyoming, and transported by pipeline to various oil fields for enhanced oil recovery. This project has been operational since 1986.

Enid Fertilizer — US

The Enid Fertilizer plant sends 675,000 tonnes of CO_2 to be used for EOR. The pipeline and wells are operated separately by Anadarko Petroleum.

Val Verde Natural Gas Plants — US

CO_2 from Mitchell, Gray Ranch, Puckett, and Turrell gas processing plants is transported via the Val Verde and CRC pipelines for EOR (incl. Sharon Ridge EOR field).

Century Plant — US

Occidental Petroleum, along with Sandridge Energy, is operating a West Texas hydrocarbon gas processing plant and related pipeline infrastructure that provides CO_2 for use in EOR. With a total CO_2capture capacity of 8.5 Mt/a expected in 2012, the Century plant would be the largest single industrial source CO_2 capture facility in North America.

Duke Energy East Bend Station — US

Researchers at the Center for Applied Energy Research of the University of Kentucky are currently developing the algae-mediated conversion of coal-fired power plant flue gas to drop-in hydrocarbon fuels. Through their work, these researchers have proven that the carbon dioxide within flue gas from coal-fired power plants can be captured using algae, which can be subsequently harvested and utilized, e.g. as a feedstock for the production of drop-in hydrocarbon fuels.

Canada

The federal government in the 2008 and 2009 budgets has invested approximately $1.4 billion in Carbon Capture and Storage development.

Alberta

Alberta has committed $170 million in 2013/2014 – and a total of $1.3 billion over 15 years – to fund two large-scale CCS projects that will help reduce CO_2 emissions from tar sands refining. In 2010 a grant agreement was signed with the Alberta Carbon Trunk Line. The second is the Quest Project. The Quest project uses amine absorption to capture CO_2 from the Scotford Steam Methane Reformer units. Approximately 1.2 Mtonne CO_2 per year is captured and transported through 64 km of onshore pipeline into a saline aquifer in the Cambrian Basal Sands.

British Columbia

Spectra Energy's Fort Nelson Project is proposed but still needs to secure funding. The total project cost is estimated to be US $12.5 Million. The source of CO_2 will be from the Fort Nelson Natural Gas Processing Plant and will be transported 15 km via an onshore pipeline to middle Devonian carbonate rock that is between 6500 and 7000 feet deep. The process will capture 2.2 Megatonne CO_2 per year using an amine process in its pre-combustion capture. Injections and MVA Operations have already occurred (2014) and is projected to startup in 2018.

Saskatchewan

Led by the province's full-service utility, SaskPower, one of the world's first and largest full production carbon capture facilities is operating at Boundary Dam Power Station. With an initial investment of $1.5 to $1.6 billion, SaskPower will be generating a revenue by selling a portion of the captured CO_2 back into the market to be used for enhanced oil recovery. The project started in May 2011 and became operational in October 2014. The post-combustion full flue gas capture process will capture 1 million tonnes of CO_2 a year.

Pilot Projects

The Alberta Saline Aquifer Project (ASAP), Husky Upgrader and Ethanol Plant pilot, Heartland Area Redwater Project (HARP), Wabamun Area Sequestration Project(WASP), and Aquistore.

Another Canadian initiative is the Integrated CO_2 Network (ICO_2N), a group of industry participants providing a framework for carbon capture and storage development in Canada. Other Canadian organizations related to CCS include CCS 101, Carbon Management Canada, IPAC CO_2, and the Canadian Clean Power Coalition.

Netherlands

In the Netherlands, a 68-megawatt oxyfuel plant ("Zero Emission Power Plant") was being planned to be operational in 2009. This project was later canceled.

ROAD (Rotterdam Capture and Storage Demonstration project) is a joint project by E.ON Benelux and Electrabel Nederland / GDF SUEZ Group. Every year, starting in 2015 ROAD will capture around 1.1 million tonnes of CO_2 at the new power plant on the Maasvlakte. This will be stored in depleted gas reservoirs under the North Sea.

Developed in the Netherlands, an electrocatalysis by a copper complex helps reduce carbon dioxide to oxalic acid.

Norway

In Norway, the CO_2 Technology Centre (TCM) at Mongstad began construction in 2009, and completed in 2012. It includes two capture technology plants (one advanced amine and one chilled ammonia), both capturing fluegas from two sources. This includes a gas-fired power plant and refinery cracker fluegas (similar to coal-fired power plant fluegas).

In addition to this, the Mongstad site was also planned to have a full-scale CCS demonstration plant. The project was delayed to 2014, 2018, and then indefinitely. The project cost rose to USD 985 million.Then in October 2011, Aker Solutions' wrote off its investment in Aker Clean Carbon, declaring the carbon sequestration market to be "dead".

On 1 October 2013 Norway asked Gassnova not to sign any contracts for Carbon capture and storage outside Mongstad.

Poland

In Belchatów, Poland, a lignite-fired energy plant of more than 858 MW is planned to be in operation in 2013.

United States

In November 2008, the DOE awarded a $66.9 million eight-year grant to a research partnership headed by Montana State University to demonstrate that underground geologic formations "can store huge volumes of carbon dioxide economically, safely and permanently".

Researchers under the Big Sky Regional Carbon Sequestration Project plan to inject up to one million tonnes of CO_2 into sandstone beneath southwestern Wyoming.

In the United States, four different synthetic fuel projects are moving forward, which have publicly announced plans to incorporate carbon capture and storage:

- American Clean Coal Fuels, in their Illinois Clean Fuels (ICF) project, is developing a 30,000-barrel (4,800 m^3) per day biomass and coal to liquids project in Oakland, Illinois, which will market the CO_2 created at the plant for enhanced oil recovery applications. By combining sequestration and biomass feedstocks, the ICF project will achieve dramatic reductions in the life-cycle carbon footprint of the fuels they produce. If sufficient biomass is used, the plant should have the capability to go life-cycle carbon negative, meaning that effectively, for each gallon of their fuel that is used, carbon is pulled out of the air, and put into the ground.
- Baard Energy, in their Ohio River Clean Fuels project, is developing a 53,000 bbl/d (8,400 m^3/d) coal and biomass to liquids project, which has announced plans to market the plant's CO_2 for enhanced oil recovery.
- Rentech is developing a 29,600-barrel (4,710 m^3) per day coal and biomass to liquids plant in Natchez, Mississippi, which will market the plant's CO_2 for enhanced oil recovery. The first phase of the project is expected in 2011.
- DKRW is developing a 15,000–20,000-barrel (2,400–3,200 m^3) per day coal to liquids plant in Medicine Bow, Wyoming, which will market its plant's CO_2 for enhanced oil recovery. The project is expected to begin operation in 2013.

In October 2009, the U.S. Department of Energy awarded grants to twelve Industrial Carbon Capture and Storage (ICCS) projects to conduct a Phase 1 feasibility study. The DOE plans to select 3 to 4 of those projects to proceed into Phase 2, design and construction, with operational startup to occur by 2015. Battelle Memorial Institute, Pacific Northwest Division, Boise, Inc., and Fluor Corporation are studying a CCS system for capture and storage of CO_2 emissions associated with the pulp and paper production industry. The site of the study is the Boise White Paper L.L.C. paper mill located near the township of Wallula in Southeastern Washington State. The plant generates approximately 1.2 MMT of CO_2 annually from a set of three recovery boilers that are mainly fired with black liquor, a recycled byproduct formed during the pulping of wood for paper-making. Fluor Corporation will design a customized version of their Econamine Plus carbon capture technology. The Fluor system also will be designed to remove residual quantities of remnant air pollutants from stack gases as part of the CO_2capture process. Battelle is leading preparation of an Environmental Information Volume (EIV) for the entire project, including geologic storage of the captured CO_2 in deep flood basalt formations that exist in the greater region. The EIV will describe the necessary site characterization work, sequestration system infrastructure, and monitoring program to support permanent sequestration of the CO_2 captured at the plant.

In addition to individual carbon capture and sequestration projects, there are a number of U.S. programs designed to research, develop, and deploy CCS technologies on a broad scale. These include the National Energy Technology Laboratory's (NETL) Carbon Sequestration Program, regional carbon sequestration partnerships and the Carbon Sequestration Leadership Forum (CSLF).

SECARB

In October 2007, the Bureau of Economic Geology at the University of Texas at Austin received a 10-year, $38 million subcontract to conduct the first intensively monitored long-term project in the United States studying the feasibility of injecting a large volume of CO_2 for underground storage. The project is a research program of the Southeast Regional Carbon Sequestration Partnership (SECARB), funded by the National Energy Technology Laboratory of the U.S. Department of Energy (DOE).

The SECARB partnership will demonstrate CO_2 injection rate and storage capacity in the Tuscaloosa-Woodbine geologic system that stretches from Texas to Florida. The region has the potential to store more than 200 billion tons of CO_2 from major point sources in the region, equal to about 33 years of overall U.S. emissions at present rates. Beginning in fall 2007, the project will inject CO_2 at the rate of one million tons per year, for up to 1.5 years, into brine up to 10,000 feet (3,000 m) below the land surface near the Cranfield oil field, which lays about 15 miles (24 km) east of Natchez, Mississippi. Experimental equipment will measure the ability of the subsurface to accept and retain CO_2.

Kemper Project

The Kemper Project, is an electrical generating station currently under construction in Kemper County, Mississippi. Mississippi Power, a subsidiary of Southern Company, began construction of the plant in 2010. The project is central to President Obama's Climate Plan. Once operational, the Kemper Project will be a first-of-its-kind electricity plant to employ gasification and carbon capture technologies at this scale.

However, there have been project management problems. The power plant construction has been delayed and is scheduled to open in the third quarter of 2016, more than two years behind schedule, at a cost of $6.6 billion—three times original cost estimate. According to a Sierra Club analysis, Kemper is the most expensive power plant ever built for the watts of electricity it will generate.

Texas Clean Energy Project

The Texas Clean Energy Project (TCEP) is a "NowGen" Integrated Gasification Combined Cycle (IGCC) facility that will incorporate carbon capture, utilization and storage (CCUS) technology in a first-of-its-kind commercial clean coal power plant. This project is expected to be operational in 2018. It will be the first US-based power plant to combine both IGCC and capture 90% of its emissions.

Big Brown Steam Electric Station

Examples of carbon sequestration at an existing US coal plant can be found at utility company Luminant's pilot version at its Big Brown Steam Electric Station in Fairfield, Texas. This system is converting carbon from smokestacks into baking soda. Skyonic plans to circumvent storage problems of liquid CO_2 by storing baking soda in mines, landfills, to be sold as industrial or food grade baking soda, Green Fuel Technologies is piloting and implementing algae based carbon capture, circumventing storage issues by then converting algae into fuel or feed, though this may lead to re-release of the carbon.

United Kingdom

The government of the United Kingdom launched a first tender process for a CCS demonstration project in 2007. The project were to use post-combustion technology on coal-fired power generation at 300–400 megawatts or equivalent. The project aimed to be operational by 2014. The Government announced in June 2008 that four companies had pre-qualified for the competition: BP Alternative Energy International Limited, EON UK Plc, Peel Power Limited and Scottish Power Generation Limited. BP subsequently withdrew from the competition, claiming it could not find a power generator partner, and RWE npower sought a judicial review of the process after it did not qualify. This first CCS tender was cancelled in late 2011 when government could not reach agreement with the ScottishPower/Shell/National Grid consortium on terms and cost, for the project based on retrofitting the existing Longannet coal-fired power station in Scotland.

A second tender process was launched by government in 2012 as part of DECC's CCS Commercialisation Programme and two bidders, namely the Shell Peterhead gas-fired power station CCS project and the White Rose CCS project based upon a new oxy-fuel coal-fired unit at Drax power station were selected in 2013 to proceed to a funded front-end engineering and design phase. This second tender was cancelled in November 2015 following a government spending review at the time of the Chancellor's Autumn Statement.

Doosan Babcock has modified their Clean Combustion Test Facility (CCTF) in Renfrew, Scotland to create the largest Oxyfuel test facility currently in the world. Oxyfuel firing on pulverized coal with recycled flue gas demonstrates the operation of a full scale 40 MW burner for use in coal-fired boilers. Sponsors of the project include the UK Department for Business Enterprise and Regulatory Reform (BERR,) as well as a group of industrial sponsors and university partners comprising Scottish and Southern Energy (Prime Sponsor), E.ON UK PLC, Drax Power Limited, ScottishPower, EDF Energy, Dong Energy Generation, Air Products Plc (Sponsors), and Imperial College and University of Nottingham (University Partners).

In 2009 UK firm 2Co Energy was awarded planning permission for a £5bn power station and carbon-capture-and-storage project at Hatfield, near Doncaster and £164m of EU funding. Technology company Samsung agreed to take a 15% stake in the project. It is planned to construct a 60 km (37 mi) pipeline from Stainforth, near Hatfield in South Yorkshire to Barmston in the East Riding of Yorkshire. CO_2 would then be stored in natural porous rock beneath the North Sea. National Grid believes the project has the potential to reduce CO_2 emissions from power stations across Yorkshire and the Humber by up to 90% and both the proposed White Rose CCS project at Drax Power Station in North Yorkshire and the proposed Don Valley Power Project at Hatfield could benefit from the scheme.

In the Northeast of England, The Northeast of England Process Industry Cluster (NEPIC) of commodity chemical manufacturers are amongst the largest single point producers of carbon dioxide in the United Kingdom and they have created within NEPIC the Process Industry Carbon Capture and Storage Initiative (PICCSI) to study the possibility of a carbon capture and storage (CCS) solution being provided for the chemical and steel manufacturing industry on Teesside, as well as for any carbon based energy production. This CCS technology option is being considered as a result of climate change regulations and the carbon taxation that could become a prohibitive cost for such energy intensive industries.

The Crown Estate is responsible for storage rights on the UK continental shelf and it has facilitated work on offshore carbon dioxide storage technical and commercial issues.

China

In Beijing, as of 2009, one major power plant is capturing and re-selling a small fraction of its CO_2emissions.

Germany

The German industrial area of Schwarze Pumpe, about 4 kilometres (2.5 mi) south of the city of Spremberg, is home to the world's first demonstration CCS coal plant, the Schwarze Pumpe power station. The mini pilot plant is run by an Alstom-built oxy-fuel boiler and is also equipped with a flue gas cleaning facility to remove fly ash and sulphur dioxide. The Swedish company Vattenfall AB invested some €70 million in the two-year project, which began operation September 9, 2008. The power plant, which is rated at 30 megawatts, is a pilot project to serve as a prototype for future full-scale power plants. 240 tonnes a day of CO_2 are being trucked 350 kilometers (220 mi) where it will be injected into an empty gas field. Germany's BUND group called it a "fig leaf". For each tonne of coal burned, 3.6 tonnes of carbon dioxide is produced. The CCS program at Schwarze Pumpe ended in 2014 due to inviable costs and energy use.

German utility RWE operates a pilot-scale CO_2 scrubber at the lignite-fired Niederaußem power station built in cooperation with BASF (supplier of detergent) and Linde engineering.

In Jänschwalde, Germany, a plan is in the works for an Oxyfuel boiler, rated at 650 thermal MW (around 250 electric MW), which is about 20 times more than Vattenfall's 30 MW pilot plant under construction, and compares to today's largest Oxyfuel test rigs of 0.5 MW. Post-combustion capture technology will also be demonstrated at Jänschwalde.

Australia

The Federal Resources and Energy Minister Martin Ferguson opened the first geosequestration project in the southern hemisphere in April 2008. The demonstration plant is near Nirranda South in South Western Victoria. (35.31°S 149.14°E) The plant is owned by the Cooperative Research Centre for Greenhouse Gas Technologies (CO2CRC). CO2CRC is a non profit research collaboration supported by government and industry. The project has stored and monitored over 65,000 tonnes of carbon dioxide-rich gas which was extracted from a natural gas reservoir via a well, compressed and piped 2.25 km to a new well. There the gas has been injected into a depleted natural gas reservoir approximately two kilometers below the surface. The project has moved to a second stage and is investigating carbon dioxide trapping in a saline aquifer 1500 meters below the surface. The Otway Project is a research and demonstration project, focused on comprehensive monitoring and verification.

This plant does not propose to capture CO_2 from coal-fired power generation, though two CO2CRC demonstration projects at a Victorian power station and research gasifier are demonstrating solvent, membrane, and adsorbent capture technologies from coal combustion. Currently, only small-scale projects are storing CO_2 stripped from the products of combustion of coal burnt for electricity generation at coal-fired power stations. Work currently being carried out by the GreenMag Group and the University of Newcastle and funded by the New South Wales and Australian Governments and industry intends to have a working mineral carbonation pilot plant in operation by 2013.

View the full list of Zero Emission Projects for fossil fuel power plant in Europe.

LIMITATIONS OF CCS FOR POWER STATIONS

Critics say large-scale CCS deployment is unproven and decades away from being commercialized. They say that it is risky and expensive and that a better option is renewable energy. Some environmental groups point out that CCS technology leaves behind dangerous waste material that has to be stored, just like nuclear power stations.

Another limitation of CCS is its energy penalty. The technology is expected to use between 10 and 40 percent of the energy produced by a power station. Wide-scale adoption of CCS may erase efficiency gains in coal power plants of the last 50 years, and increase resource consumption by one third. Even taking the fuel penalty into account, however, overall levels of CO_2 abatement would remain high at approximately 80–90%, compared to a plant without CCS. It is possible for CCS, when combined with biomass, to result in net negative emissions. Though, all of the currently (as of Feb 2011) operational BECCS (Bio-energy with carbon capture and storage) plants operate on point emissions other than power stations, such as biofuel refineries.

The use of CCS can reduce CO_2 emissions from the stacks of coal power plants by 85–90% or more, but it has no effect on CO_2 emissions due to the mining and transport of coal. It will actually "increase such emissions and of air pollutants per unit of net delivered power and will increase all ecological, land-use, air-pollution, and water-pollution impacts from coal mining, transport, and processing, because the CCS system requires 25% more energy, thus 25% more coal combustion, than does a system without CCS".

Another concern regards the permanence of storage schemes. Opponents to CCS claim that safe and permanent storage of CO_2 cannot be guaranteed and that even very low leakage rates could undermine any climate mitigation effect. In 1986 a large leakage of naturally sequestered CO_2 rose from Lake Nyos in Cameroon and asphyxiated 1,700 people. While the carbon had been sequestered naturally, some point to the event as evidence for the potentially catastrophic effects of sequestering carbon artificially.

On one hand, Greenpeace claims that CCS could lead to a doubling of coal plant costs. It is also claimed by opponents to CCS that money spent on CCS will divert investments away from other solutions to climate change. On the other hand, CCS is pointed out as economically attractive in comparison to other forms of low carbon electricity generation and seen by the IPCC and others as a critical component for meeting mitigation targets such as 450 ppm and 350 ppm.

COST

Although the processes involved in CCS have been demonstrated in other industrial applications, no commercial scale projects which integrate these processes exist; the costs therefore are somewhat uncertain. Some recent credible estimates indicate that the cost of capturing and storing carbon dioxide is US$60 per ton, corresponding to an increase in electricity prices of about US 6c per kWh (based on typical coal-fired power plant emissions of 0.97 kg (2.13 lb) CO_2 per kWh). This would double the typical US industrial electricity price (now at around 6c per kWh) and increase the typical retail residential electricity price by about 50% (assuming 100% of power is from coal, which may not necessarily be the case, as this varies from state to state). Similar (approximate) price increases would likely be expected in coal dependent countries such as Australia, because the capture technology and chemistry, as well as the transport and injection costs from such power plants would not, in an overall sense, vary significantly from country to country.

The reasons that CCS is expected to cause such power price increases are several. Firstly, the increased energy requirements of capturing and compressing CO_2 significantly raises the operating costs of CCS-equipped power plants. In addition, there are added investment and capital costs. The process would increase the fuel requirement of a plant with CCS by about 25% for a coal-fired plant, and about 15% for a gas-fired plant. The cost of this extra fuel, as well as storage and other system costs, are estimated to increase the costs of energy from a power plant with CCS by 30–60%, depending on the specific circumstances. Pre-commercial CCS demonstration projects are likely to be more expensive than mature CCS technology; the total additional costs of an early large-scale CCS demonstration project are estimated to be €0.5-1.1 billion per project over the project lifetime. Other applications are possible. In the belief that use of sequestered carbon could be harnessed to offset the cost of capture and storage, Walker Architects published the first CO_2 gas CAES application, proposing the use of sequestered CO_2 for Energy Storage on October 24, 2008. To date the feasibility of such potential offsets to the cost have not been examined.

An estimate of costs of energy with and without CCS (2002 US$ per kWh)

	Natural gas combined cycle	**Pulverized coal**	**Integrated gasification combined cycle**
Without capture (reference plant)	0.03–0.05	0.04–0.05	0.04–0.06
With capture and geological storage	0.04–0.08	0.06–0.10	0.06–0.09
(Cost of capture and geological storage)	0.01–0.03	0.02–0.05	0.02–0.03

With capture and Enhanced oil recovery	0.04–0.07	0.05–0.08	0.04–0.08
All costs refer to costs for energy from newly built, large-scale plants. Natural gas combined cycle costs are based on natural gas prices of US$2.80–4.40 per GJ (LHV based). Energy costs for PC and IGCC are based on bituminous coal costs of US$1.00–1.50 per GJ LHV. Note that the costs are very dependent on fuel prices (which change continuously), in addition to other factors such as capital costs. Also note that for EOR, the savings are greater for higher oil prices. Current gas and oil prices are substantially higher than the figures used here. All figures in the table are from Table 8.3a in [IPCC, 2005].			

The cost of CCS depends on the cost of capture and storage, which varies according to the method used. Geological storage in saline formations or depleted oil or gas fields typically cost US$0.50–8.00 per tonne of CO_2 injected, plus an additional US$0.10–0.30 for monitoring costs. When storage is combined with enhanced oil recovery to extract extra oil from an oil field, however, the storage could yield net benefits of US$10–16 per tonne of CO_2 injected (based on 2003 oil prices). This would likely negate some of the effect of the carbon capture when the oil was burnt as fuel. Even taking this into account, as the table above shows, the benefits do not outweigh the extra costs of capture.

Cost of electricity generated by different sources including those incorporating CCS technologies can be found in cost of electricity by source. If CO_2 capture was part of a fuel cycle then the CO_2 would have value rather than be a cost. The proposed Solar Fuel or methane cycle proposed by the Fraunhofer Society amongst others is an example. This "solar fuel" cycle uses the excess electrical renewable energy to create hydrogen via electrolysis of water. The hydrogen is then combined with CO_2 to create synthetic natural gas SNG and stored in the gas network. See the latest Cost Report on the Cost of CO_2 Capture produced by the Zero Emissions Platform

Governments around the world have provided a range of different types of funding support to CCS demonstration projects, including tax credits, allocations and grants. The funding is associated with both a desire to accelerate innovation activities for CCS as a low-carbon technology and the need for economic stimulus activities. As of 2011, approximately US$23.5bn has been made available to support large-scale CCS demonstration projects around the world.

Carbon Capture and Storage and the Kyoto Protocol

One way to finance future CCS projects could be through the Clean Development Mechanism of the Kyoto Protocol. At COP16 in 2010, The Subsidiary Body for Scientific and Technological Advice, at its thirty-third session, issued a draft document recommending the inclusion of Carbon dioxide capture and storage in geological formations in Clean Development Mechanism project activities. At COP17 in Durban, a final agreement was reached enabling CCS projects to receive support through the Clean Development Mechanism.

ENVIRONMENTAL EFFECTS

The theoretical merit of CCS systems is the reduction of CO_2 emissions by up to 90%, depending on plant type. Generally, environmental effects from use of CCS arise during power production, CO_2capture, transport, and storage. Issues relating to storage are discussed in those sections.

Additional energy is required for CO_2 capture, and this means that substantially more fuel has to be used to produce the same amount of power, depending on the plant type. For new super-critical pulverized coal (PC) plants using current technology, the extra energy requirements range from 24 to 40%, while for natural gas combined cycle (NGCC) plants the range is 11–22% and for coal-based gasification combined cycle (IGCC) systems it is 14–25% [IPCC, 2005]. Obviously, fuel use and environmental problems arising from mining and extraction of coal or gas increase accordingly. Plants equipped with flue-gas desulfurization (FGD) systems for sulfur dioxide control require proportionally greater amounts of limestone, and systems equipped with selective catalytic reduction systems for nitrogen oxides produced during combustion require proportionally greater amounts of ammonia.

IPCC has provided estimates of air emissions from various CCS plant designs (see table below). While CO_2 is drastically reduced though never completely captured, emissions of air pollutants increase significantly, generally due to the energy penalty of capture. Hence, the use of CCS entails a reduction in air quality. Type and amount of air pollutants still depends on technology. CO_2 is captured with alkaline solvents catching the acidic CO_2 at low temperatures in the absorber and releasing CO_2 at higher temperatures in a desorber. Chilled Ammonia CCS Plants have inevitable ammonia emissions to air. "Functionalized Ammonia" emit less ammonia, but amines may form secondary amines and these will emit volatile nitrosamines by a side reaction with nitrogendioxide, which is present in any flue gas even after DeNOx. Nevertheless, there are advanced amines in testing with little to no vapor pressure to avoid these amine- and consecutive nitrosamine emissions. Nevertheless, all the capture plants amines have in common, that practically 100% of remaining sulfur dioxide from the plant is washed out of the flue gas, the same applies to dust/ash.

Emissions to air from plants with CCS (kg/MWh)

	Natural gas combined cycle	**Pulverized coal**	**Integrated gasification combined cycle**
CO_2	0.43 (-89%)	1.07 (−87%)	0.97 (−88%)
NO_X	0.11 (+22%)	0.77 (+31%)	0.1 (+11%)
SO_X	—	0.001 (−99.7%)	0.33 (+17.9%)
Ammonia	0.002 (before: 0)	0.23 (+2200%)	—

Based on Table 3.5 in [IPCC, 2005]. Between brackets the increase or decrease compared to a similar plant without CCS.

CARBON SEQUESTRATION

Schematic showing both terrestrial and geological sequestration of carbon dioxide emissions from a coal-fired plant.

Carbon sequestration is the process involved in carbon capture and the long-term storage of atmospheric carbon dioxide. Carbon sequestration involves long-term storage of carbon dioxide or other forms of carbon to either mitigate or defer global warming and avoid dangerous climate change. It has been proposed as a way to slow the atmospheric and marine accumulation of greenhouse gases, which are released by burning fossil fuels.

Carbon dioxide (CO_2) is naturally captured from the atmosphere through biological, chemical, and physical processes. Artificial processes have been devised to produce similar effects,including large-scale, artificial capture and sequestration of industrially produced CO_2 using subsurface saline aquifers, reservoirs, ocean water, aging oil fields, or other carbon sinks.

Description

Carbon sequestration is the process involved in carbon capture and the long-term storage of atmospheric carbon dioxide (CO_2) and may refer specifically to:

- "The process of removing carbon from the atmosphere and depositing it in a reservoir." When carried out deliberately, this may also be referred to as carbon dioxide removal, which is a form of geoengineering.
- Carbon capture and storage, where carbon dioxide is removed from flue gases (e.g., at power stations) before being stored in underground reservoirs.
- Natural biogeochemical cycling of carbon between the atmosphere and reservoirs, such as by chemical weathering of rocks.

Carbon sequestration describes long-term storage of carbon dioxide or other forms of carbon to either mitigate or defer global warming and avoid dangerous climate change. It has been proposed as a way to slow the atmospheric and marine accumulation of greenhouse gases, which are released by burning fossil fuels.

Carbon dioxide is naturally captured from the atmosphere through biological, chemical or physical processes. Some anthropogenic sequestration techniques exploit these natural processes, while some use entirely artificial processes.

Carbon dioxide may be captured as a pure by-product in processes related to petroleum refining or from flue gases from power generation. CO_2 sequestration includes the storage part of carbon capture and storage, which refers to large-scale, artificial capture and sequestration of industrially produced CO_2 using subsurface saline aquifers, reservoirs, ocean water, aging oil fields, or other carbon sinks.

Biological Processes

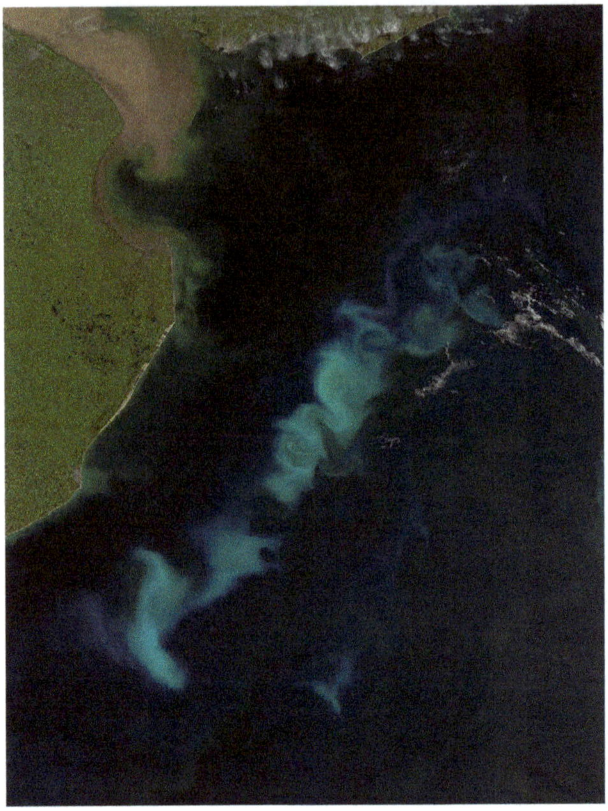

An oceanic phytoplankton bloom in the South Atlantic Ocean, off the coast of Argentina. Encouraging such blooms with iron fertilization could lock up carbon on the seabed.

Biosequestration or carbon sequestration through biological processes affects the global carbon cycle. Examples include major climatic fluctuations, such as the Azolla event, which created the current Arctic climate. Such processes created fossil fuels, as well as clathrate and limestone. By manipulating such processes, geoengineers seek to enhance sequestration.

Peat Production

Peat bogs are a very important carbon store. Peat bogs act as a sink for carbon due to the accumulation of partially decayed biomass that would otherwise continue to decay completely. There is a variance on how much the peatlands act as a carbon sink or carbon source that can be linked to varying climates in different areas of the world and different times of the year. By creating new bogs, or enhancing existing ones, the amount of carbon that is sequestered by bogs would increase.

Forestry

Reforestation is the replanting of trees on marginal crop and pasture lands to incorporate carbon from atmospheric CO_2 into biomass. For this process to succeed the carbon must not return to the atmosphere from mass burning or rotting when the trees die. To this end, land allotted to the trees must not be converted to other uses and management of the frequency of disturbances might be necessary in order to avoid extreme events. Alternatively, the wood from them must itself be sequestered, e.g., via biochar, bio-energy with carbon storage (BECS), landfill or ‹stored' by use in e.g. construction. Short of growth in perpetuity, however, reforestation with long-lived trees (>100 years) will sequester carbon for a more graduated release, minimizing impact during the expected carbon crisis of the 21st century.

Urban Forestry

Urban Forestry increases the amount of carbon taken up in cities by adding new tree sites and the sequestration of carbon occurs over the lifetime of the tree. It is generally practiced and maintained on smaller scales, like in cities. The results of urban forestry can have different results depending on the type of vegetation that is being used, so it can function as a sink but can also function as a source of emissions. Along with sequestration by the plants which is difficult to measure but seems to have little effect on the overall amount of carbon dioxide that is uptaken, the vegetation can have indirect effects on carbon by reducing need for energy consumption.

Wetland Restoration

Wetland soil is an important carbon sink; 14.5% of the world's soil carbon is found in wetlands, while only 6% of the world's land is composed of wetlands.

Agriculture

Globally, soils are estimated to contain approximately 1,500 gigatons of organic carbon to 1 m depth, more than the amount in vegetation and the atmosphere.

Modification of agricultural practices is a recognized method of carbon sequestration as soil can act as an effective carbon sink offsetting as much as 20% of 2010 carbon dioxide emissions annually. (See No-till)

Carbon emission reduction methods in agriculture can be grouped into two categories: reducing and/or displacing emissions and enhancing carbon removal. Some of these reductions involve increasing the efficiency of farm operations (e.g. more fuel-efficient equipment) while some involve interruptions in the natural carbon cycle. Also, some effective techniques (such as the elimination of stubble burning) can negatively impact other environmental concerns (increased herbicide use to control weeds not destroyed by burning).

Reducing Emissions

Increasing yields and efficiency generally reduces emissions as well, since more food results from the same or less effort. Techniques include more accurate use of fertilizers, less soil disturbance, better irrigation, and crop strains bred for locally beneficial traits and increased yields.

Replacing more energy intensive farming operations can also reduce emissions. Reduced or no-till farming requires less machine use and burns correspondingly less fuel per acre. However, no-till usually increases use of weed-control chemicals and the residue now left on the soil surface is more likely to release its CO_2 to the atmosphere as it decays, reducing the net carbon reduction.

In practice, most farming operations that incorporate post-harvest crop residues, wastes and byproducts back into the soil provide a carbon storage benefit. This is particularly the case for practices such as field burning of stubble - rather than releasing almost all of the stored CO_2 to the atmosphere, tillage incorporates the biomass back into the soil.

Enhancing Carbon Removal

All crops absorb CO_2 during growth and release it after harvest. The goal of agricultural carbon removal is to use the crop and its relation to the carbon cycle to permanently sequester carbon within the soil. This is done by selecting farming methods that return biomass to the soil and enhance the conditions in which the carbon within the plants will be reduced to its elemental nature and stored in a stable state. Methods for accomplishing this include:

- Use cover crops such as grasses and weeds as temporary cover between planting seasons
- Concentrate livestock in small paddocks for days at a time so they graze lightly but evenly. This encourages roots to grow deeper into the soil. Stock also till the soil with their hooves, grinding old grass and manures into the soil.
- Cover bare paddocks with hay or dead vegetation. This protects soil from the sun and allows the soil to hold more water and be more attractive to carbon-capturing microbes.
- Restore degraded land, which slows carbon release while returning the land to agriculture or other use.

Agricultural sequestration practices may have positive effects on soil, air, and water quality, be beneficial to wildlife, and expand food production. On degraded croplands, an increase of 1 ton of soil carbon pool may increase crop yield by 20 to 40 kilograms per hectare of wheat, 10 to 20 kg/ ha for maize, and 0.5 to 1 kg/ha for cowpeas.

The effects of soil sequestration can be reversed. If the soil is disrupted or tillage practices are abandoned, the soil becomes a net source of greenhouse gases. Typically after 15 to 30 years of sequestration, soil becomes saturated and ceases to absorb carbon. This implies that there is a global limit to the amount of carbon that soil can hold.

Many factors affect the costs of carbon sequestration including soil quality, transaction costs and various externalities such as leakage and unforeseen environmental damage.

Because reduction of atmospheric CO_2 is a long-term concern, farmers can be reluctant to adopt more expensive agricultural techniques when there is not a clear crop, soil, or economic benefit. Governments such as Australia and New Zealand are considering allowing farmers to sell carbon credits once they document that they have sufficiently increased soil carbon content.

Ocean-Related

Iron Fertilization

Ocean iron fertilization is an example of such a geoengineering technique. Iron fertilizationattempts to encourage phytoplankton growth, which removes carbon from the atmosphere for at least a period of time. This technique is controversial due to limited understanding of its complete effects on the marine ecosystem, including side effects and possibly large deviations from expected behavior. Such effects potentially include release of nitrogen oxides, and disruption of the ocean's nutrient balance.

Natural iron fertilisation events (e.g., deposition of iron-rich dust into ocean waters) can enhance carbon sequestration. Sperm whales act as agents of iron fertilisation when they transport iron from the deep ocean to the surface during prey consumption and defecation. Sperm whales have been shown to increase the levels of primary production and carbon export to the deep ocean by depositing iron rich feces into surface waters of the Southern Ocean. The iron rich feces causes phytoplankton to grow and take up more carbon from the atmosphere. When the phytoplankton dies, some of it sinks to the deep ocean and takes the atmospheric carbon with it. By reducing the abundance of sperm whales in the Southern Ocean, whaling has resulted in an extra 200,000 tonnes of carbon remaining in the atmosphere each year.

Urea Fertilization

Ian Jones proposes fertilizing the ocean with urea, a nitrogen rich substance, to encourage phytoplankton growth.

Australian company Ocean Nourishment Corporation (ONC) plans to sink hundreds of tonnes of urea into the ocean to boost CO_2-absorbing phytoplankton growth as a way to combat climate change. In 2007, Sydney-based ONC completed an experiment involving 1 tonne of nitrogen in the Sulu Sea off the Philippines.

Mixing Layers

Encouraging various ocean layers to mix can move nutrients and dissolved gases around, offering avenues for geoengineering. Mixing may be achieved by placing large vertical pipes in the oceans to pump nutrient rich water to the surface, triggering blooms of algae, which store carbon when they grow and export carbon when they die. This produces results somewhat similar to iron fertilization. One side-effect is a short-term rise in CO 2, which limits its attractiveness.

Seaweed

Seaweed grows very fast and can theoretically be harvested and processed to generate biomethane, via Anaerobic Digestion to generate electricity, via Cogeneration/CHP or as a replacement for natural gas. One study suggested that if seaweed farms covered 9% of the ocean they could produce enough biomethane to supply Earth's equivalent demand for fossil fuel energy, remove 53 gigatonnes of CO_2 per year from the atmosphere and sustainably produce 200 kg per year of fish, per person, for 10 billion people. Ideal species for such farming and conversion include Laminaria digitata, Fucus serratus and Saccharina latissima.

Physical Processes

Biochar can be landfilled, used as a soil improver or burned using carbon capture and storage

Biomass-Related

Bio-energy with carbon capture and storage (BECCS)

BECCS refers to biomass in power stations and boilers that use carbon capture and storage. The carbon sequestered by the biomass would be captured and stored, thus removing carbon dioxide from the atmosphere.

This technology is sometimes referred to as bio-energy with carbon storage, BECS, though this term can also refer to the carbon sequestration potential in other technologies, such as biochar.

Burial

Burying biomass (such as trees) directly, mimics the natural processes that created fossil fuels.Landfills also represent a physical method of sequestration.

Biochar Burial

Biochar is charcoal created by pyrolysis of biomass waste. The resulting material is added to a landfill or used as a soil improver to create terra preta. Addition of pyrogenic organic carbon (biochar) is a novel strategy to increase the soil-C stock for the long-term and to mitigate global-warming by offsetting the atmospheric C (up to 9.5 Pg C annually).

In the soil, the carbon is unavailable for oxidation to CO_2 and consequential atmospheric release. This is one technique advocated by scientist James Lovelock, creator of the Gaia hypothesis. According to Simon Shackley, “people are talking more about something in the range of one to two billion tonnes a year.”

Ocean Storage

If CO_2 were to be injected to the ocean bottom, the pressures would be great enough for CO_2 to be in its liquid phase. The idea behind ocean injection would be to have stable, stationary pools of CO_2 at the ocean floor. The ocean could potentially hold over a thousand billion tons of CO_2. However, this avenue of sequestration isn’t being as actively pursued because of concerns about the impact on ocean life, and concerns about its stability.

River mouths bring large quantities of nutrients and dead material from upriver into the ocean as part of the process that eventually produces fossil fuels. Transporting material such as crop waste out to sea and allowing it to sink exploits this idea to increase carbon storage. International regulations on marine dumping may restrict or prevent use of this technique.

Geological Sequestration

Geological sequestration refers to the storage of CO_2 underground in depleted oil and gas reservoirs, saline formations, or deep, un-minable coal beds.

Once CO_2 is captured from a gas or coal-fired power plant, it would be compressed to ≈100 bar so that it would be a supercritical fluid. In this fluid form, the CO_2 would be easy to transport via pipeline to the place of storage. The CO_2 would then be injected deep underground, typically around 1 km, where it would be stable for hundreds to millions of years. At these storage conditions, the density of supercritical CO_2 is 600 to 800 kg / m^3. For consumers, the cost of electricity from a coal-fired power plant with carbon capture and storage (CCS) is estimated to be 0.01 - 0.05 $ / kWh higher than without CCS. For reference, the average cost of electricity in the US in 2004 was 0.0762 $ / kWh. In other terms, the cost of CCS would be 20 - 70 $/ton of CO_2 captured. The transportation and injection of CO_2 is relatively cheap, with the capture costs accounting for 70 - 80% of CCS costs.

The important parameters in determining a good site for carbon storage are: rock porosity, rock permeability, absence of faults, and geometry of rock layers. The medium in which the CO_2 is to be stored ideally has a high porosity and permeability, such as sandstone or limestone. Sandstone can have a permeability ranging from 1 to 10^{-5} Darcy, and can have a porosity as high as ≈30%. The porous rock must be capped by a layer of low permeability which acts as a seal, or caprock, for the CO_2. Shale is an example of a very good caprock, with a permeability of 10^{-5} to 10^{-9} Darcy. Once injected, the CO_2plume will rise via buoyant forces, since it is less dense than its surroundings. Once it encounters a caprock, it will spread laterally until it encounters a gap. If there are fault planes near the injection zone, there is a possibility the CO_2 could migrate along the fault to the surface, leaking into the atmosphere, which would be potentially dangerous to life in the surrounding area. Another danger related to carbon sequestration is induced seismicity. If the injection of CO_2 creates pressures that are too high underground, the formation will fracture, causing an earthquake.

While trapped in a rock formation, CO_2 can be in the supercritical fluid phase or dissolve in groundwater/brine. It can also react with minerals in the geologic formation to precipitate carbonates. See CarbFix.

Worldwide storage capacity in oil and gas reservoirs is estimated to be 675 - 900 Gt CO_2, and in un-minable coal seams is estimated to be 15 - 200 Gt CO_2. Deep saline formations have the largest capacity, which is estimated to be 1,000 - 10,000 Gt CO_2. In the US, there is an estimated 160 Gt CO_2 storage capacity.

There are a number of large-scale carbon capture and sequestration projects that have demonstrated the viability and safety of this method of carbon storage, which are summarized here by the Global CCS Institute. The dominant monitoring technique is seismic imaging, where vibrations are generated that propagate through the subsurface. The geologic structure can be imaged from the refracted/reflected waves.

The first large-scale CO_2 sequestration project which began in 1996 is called Sleipner, and is located in the North Sea where Norway's StatoilHydro strips carbon dioxide from natural gas with amine solvents and disposed of this carbon dioxide in a deep saline aquifer. In 2000, a coal-fueled synthetic natural gas plant in Beulah, North Dakota, became the world's first coal-using plant to capture and store carbon dioxide, at the Weyburn-Midale Carbon Dioxide Project.

CO^2 has been used extensively in enhanced crude oil recovery operations in the United States beginning in 1972. There are in excess of 10,000 wells that inject CO2 in the state of Texas alone. The gas comes in part from anthropogenic sources, but is principally from large naturally occurring geologic formations of CO_2. It is transported to the oil-producing fields through a large network of over 5,000 kilometres (3,100 mi) of CO2 pipelines. The use of CO2 for enhanced oil recovery (EOR) methods in heavy oil reservoirs in the Western Canadian Sedimentary Basin (WCSB) has also been proposed. However, transport cost remains an important hurdle. An extensive CO_2 pipeline system does not yet exist in the WCSB. Athabasca oil sands mining that produces CO_2 is hundreds of kilometers north of the subsurface Heavy crude oil reservoirs that could most benefit from CO_2 injection.

Chemical Processes

Developed in the Netherlands, an electrocatalysis by a copper complex helps reduce carbon dioxide to oxalic acid; This conversion uses carbon dioxide as a feedstock to generate oxalic acid.

Mineral Carbonation

Carbon, in the form of CO_2 can be removed from the atmosphere by chemical processes, and stored in stable carbonate mineral forms. This process is known as 'carbon sequestration by mineral carbonation' or mineral sequestration. The process involves reacting carbon dioxide with abundantly available metal oxides–either magnesium oxide (MgO) or calcium oxide (CaO)–to form stable carbonates. These reactions are exothermic and occur naturally (e.g., the weathering of rock over geologic time periods).

$$CaO + CO_2 \rightarrow CaCO_3$$

$$MgO + CO_2 \rightarrow MgCO_3$$

Calcium and magnesium are found in nature typically as calcium and magnesium silicates (such as forsterite and serpentinite) and not as binary oxides. For forsterite and serpentine the reactions are:

$$Mg_2SiO_4 + 2\ CO_2 \rightarrow 2\ MgCO_3 + SiO_2$$

$$Mg_3Si_2O_5(OH)_4 + 3\ CO_2 \rightarrow 3\ MgCO_3 + 2\ SiO_2 + 2\ H_2O$$

The following table lists principal metal oxides of Earth's crust. Theoretically up to 22% of this mineral mass is able to form carbonates.

Earthen Oxide	**Percent of Crust**	**Carbonate**	**Enthalpy change (kJ/mol)**
SiO2	59.71		
Al2O3	15.41		
CaO	4.90	CaCO3	-179
MgO	4.36	MgCO3	-117
Na2O	3.55	Na2CO3	
FeO	3.52	FeCO3	
K2O	2.80	K2CO3	
Fe2O3	2.63	FeCO3	
	21.76	All Carbonates	

These reactions are slightly more favorable at low temperatures. This process occurs naturally over geologic time frames and is responsible for much of the Earth's surface limestone. The reaction rate can be made faster however, by reacting at higher temperatures and/or pressures, although this method requires some additional energy. Alternatively, the mineral could be milled to increase its surface area, and exposed to water and constant abrasion to remove the inert Silica as could be achieved naturally by dumping Olivine in the high energy surf of beaches Experiments suggest the weathering process is reasonably quick (one year) given porous basaltic rocks.

CO_2 naturally reacts with peridotite rock in surface exposures of ophiolites, notably in Oman. It has been suggested that this process can be enhanced to carry out natural mineralisation of CO_2.

Industrial Use

Traditional cement manufacture releases large amounts of carbon dioxide, but newly developed cement types from Novacem can absorb CO2 from ambient air during hardening. A similar technique was pioneered by TecEco, which has been producing "EcoCement" since 2002.

In Estonia, oil shale ash, generated by power stations could be used as sorbents for CO_2 mineral sequestration. The amount of CO2 captured averaged 60 to 65% of the carbonaceous CO_2 and 10 to 11% of the total CO_2 emissions.

Chemical Scrubbers

Various carbon dioxide scrubbing processes have been proposed to remove CO_2 from the air, usually using a variant of the Kraft process. Carbon dioxide scrubbing variants exist based on potassium carbonate, which can be used to create liquid fuels, or on sodium hydroxide. These notably include artificial trees proposed by Klaus Lackner to remove carbon dioxide from the atmosphere using chemical scrubbers.

Ocean-related

Basalt storage

Carbon dioxide sequestration in basalt involves the injecting of CO_2 into deep-sea formations. The CO_2 first mixes with seawater and then reacts with the basalt, both of which are alkaline-rich elements. This reaction results in the release of Ca^{2+} and Mg^{2+} ions forming stable carbonate minerals.

Underwater basalt offers a good alternative to other forms of oceanic carbon storage because it has a number of trapping measures to ensure added protection against leakage. These measures include "geothermal, sediment, gravitational and hydrate formation." Because CO_2 hydrate is denser than CO_2 in seawater, the risk of leakage is minimal. Injecting the CO_2 at depths greater than 2,700 meters (8,900 ft) ensures that the CO2 has a greater density than seawater, causing it to sink.

One possible injection site is Juan de Fuca plate. Researchers at the Lamont-Doherty Earth Observatory found that this plate at the western coast of the United States has a possible storage capacity of 208 gigatons. This could cover the entire current U.S. carbon emissions for over 100 years.

This process is undergoing tests as part of the CarbFix project, resulting in 95% of the injected 250 tonnes of CO2 to solidify into calcite in 2 years, using 25 tonnes of water per tonne of CO2.

Acid Neutralisation

Carbon dioxide forms carbonic acid when dissolved in water, so ocean acidification is a significant consequence of elevated carbon dioxide levels, and limits the rate at which it can be absorbed into the ocean (the solubility pump). A variety of different bases have been suggested that could neutralize the acid and thus increase CO 2 absorption. For example, adding crushed limestone to oceans enhances the absorption of carbon dioxide. Another approach is to add sodium hydroxide to oceans which is produced by electrolysis of salt water or brine, while eliminating the waste hydrochloric acid by reaction with a volcanic silicate rock such as enstatite, effectively increasing the rate of natural weathering of these rocks to restore ocean pH.

Obstruction

Danger of leaks

Carbon dioxide may be stored deep underground. At depth, hydrostatic pressure acts to keep it in a liquid state. Reservoir design faults, rock fissures and tectonic processes may act to release the gas stored into the ocean or atmosphere.

Financial costs

The use of the technology would add an additional 1-5 cents of cost per kilowatt hour, according to estimate made by the Intergovernmental Panel on Climate Change. The financial costs of modern coal technology would nearly double if use of CCS technology were to be required by regulation. The cost of CCS technology differs with the different types of capture technologies being used and with the different sites that it is implemented in, but the costs tend to increase with CCS capture implementation. One study conducted predicted that with new technologies these costs could be lowered but would remain slightly higher than prices without CCS technologies.

Energy requirements

The energy requirements of sequestration processes may be significant. In one paper, sequestration consumed 25 percent of the plant's rated 600 megawatt output capacity.

After adding CO_2 capture and compression, the capacity of the coal-fired power plant is reduced to 457 MW.

REFERENCES

1. "IPCC Special Report Carbon Dioxide Capture and Storage Summary for Policymakers" (PDF). Intergovernmental Panel on Climate Change. Retrieved 2011-10-05.
2. Introduction to Carbon Capture and Storage - Carbon storage and ocean

acidification activity". Commonwealth Scientific and Industrial Research Organisation (CSIRO) and the Global CCS Institute. Retrieved 2013-07-03.

3. Burying the problem., Canadian Geographic Magazine
4. [IPCC, 2005] IPCC special report on Carbon Dioxide Capture and Storage. Prepared by working group III of the Intergovernmental Panel on Climate Change. Metz, B., O. Davidson, H. C. de Coninck, M. Loos, and L.A. Meyer (eds.). Cambridge University Press, Cambridge, United Kingdom and New York, NY, USA, 442 pp. Available in full at www.ipcc.ch (PDF - 22.8MB)
5. Coal Utilization Research Council (CURC) Technology Roadmap, 2005
6. Scientific Facts on CO_2 Capture and Storage, 2012
7. NETL 2007 Carbon Sequestration Atlas, 2007
8. Phelps, J; Blackford, J; Holt, J; Polton, J (2015), "Modelling Large-Scale CO2 Leakages in the North Sea", International Journal of Greenhouse Gas Control, doi:10.1016/j.ijggc.2014.10.013
9. Capturing Carbon Dioxide From Air" (PDF). Retrieved 2011-03-29.
10. News for the Business of Energy". Energy Current. Retrieved 2010-04-02.
11. McDonald, Thomas M., Mason, Jarad A., Kong ,Xueqian, et al. "Cooperative insertion of CO_2 in diamine-appended metal-organic frameworks." Nature 519 (19 March 2015): 303-308. doi:10.1038/nature14327
12. Good plant design and operation for onshore carbon capture installations and onshore pipelines - 5 Carbon dioxide plant design". Energy Institute. Retrieved 2012-03-13.
13. Wallula Energy Resource Center". Wallulaenergy.com. 2007-06-14. Retrieved 2010-04-02.
14. Sumida, Kenji; Rogow, David L.; Mason, Jarad A.; McDonald, Thomas M.; Bloch, Eric D.; Herm, Zoey R.; Bae, Tae-Hyun; Long, Jeffrey R. (2012-02-08). "Carbon Dioxide Capture in Metal–Organic Frameworks". Chemical Reviews 112 (2): 724–781. doi:10.1021/cr2003272. ISSN 0009-2665.
15. Gasification Body" (PDF). Retrieved 2010-04-02.
16. integrated gasification combined cycle for carbon capture storage Claverton Energy Group conference 24th October Bath.
17. Energy Futures Laboratory and the Grantham Institute for Climate Change
18. Bryngelsson, Mårten; Westermark, Mats (2005). Feasibility study of CO_2 *removal from pressurized flue gas in a fully fired combined cycle: the Sargas project. Proceedings of the 18th International Conference on Efficiency, Cost, Optimization, Simulation and Environmental Impact of Energy Systems. pp. 703–10.*
19. Bryngelsson, Mårten; Westermark, Mats (2009). "CO_2 *capture pilot test at a pressurized coal fired CHP plant". Energy Procedia.* **1**: 1403–10. doi:10.1016/j.egypro.2009.01.184.
20. Sweet, William (2008). "Winner: Clean Coal - Restoring Coal's Sheen". IEEE

Spectrum. **45**: 57–60. doi:10.1109/MSPEC.2008.4428318.

21. Herm, Zoey R.; Swisher, Joseph A.; Smit, Berend; Krishna, Rajamani; Long, Jeffrey R. (2011-04-20). "Metal–Organic Frameworks as Adsorbents for Hydrogen Purification and Precombustion Carbon Dioxide Capture". Journal of the American Chemical Society 133 (15): 5664–5667. doi:10.1021/ja111411q. ISSN 0002-7863.
22. Kulkarni, Ambarish R.; Sholl, David S. (2012-06-27). "Analysis of Equilibrium-Based TSA Processes for Direct Capture of CO2 from Air". Industrial & Engineering Chemistry Research 51 (25): 8631–8645. doi:10.1021/ie300691c. ISSN 0888-5885.
23. Millward, Andrew R.; Yaghi, Omar M. (2005-12-01). "Metal–Organic Frameworks with Exceptionally High Capacity for Storage of Carbon Dioxide at Room Temperature". Journal of the American Chemical Society 127 (51): 17998–17999. doi:10.1021/ja0570032. ISSN 0002-7863.
24. Plasynski, Sean; Lang, David A..; Willis, Richard; (2007-2010). "Carbon Dioxide Separation with Novel Microporous Metal Organic Frameworks". National Energy Technology Lab 112 (2): 724–781. doi:10.1021/cr2003272. ISSN 0009-2665.
25. Smit, Berend; Reimer, Jeffrey R.; Oldenburg, Curtis M.; Bourg, Ian C. (2014). Introduction to Carbon Capture and Sequestration. Imperial College Press. ISBN 978-1-78326-327-1.
26. The Global Status of CCS: 2011 - Capture". The Global CCS Institute. Retrieved 2012-03-26.
27. Stolaroff, Joshuah K. (August 17, 2006). Capturing CO_2 *from ambient air: a feasibility assessment (PDF)*(PhD thesis).
28. First Successful Demonstration of Carbon Dioxide Air Capture Technology Achieved by Columbia University Scientist and Private Company". Earth. columbia.edu. Retrieved 2010-04-14.
29. Parfomak, Paul W.; Folger, Peter. "CRS Report for Congress: Carbon Dioxide (CO_2) Pipelines for Carbon Sequestration: Emerging Policy Issues", Updated January 17, 2008 (Order Code RL33971)" (PDF). Assets.opencrs.com.
30. Vann, Adam; Parfomak, Paul W. "CRS Report for Congress: Regulation of Carbon Dioxide (CO_2) Sequestration Pipelines: Jurisdictional Issues", Updated April 15, 2008 (Order Code RL34307)" (PDF). Ncseonline.org. (reviewing federal jurisdictional issues related to CO_2 pipelines and reviewing agency jurisdictional determinations under the Interstate Commerce Act and the Natural Gas Act)
31. "Good plant design and operation for onshore carbon capture installations and onshore pipelines - Storage". Energy Institute. Retrieved 2012-12-11.
32. IPCC "Special Report on Carbon Capture and Storage, pp. 181 and 203 (Chapter 5, "Underground Geological Storage")" (PDF). Retrieved 2010-04-14.
33. Good plant design and operation for onshore carbon capture installations and onshore pipelines - Carbon dioxide storage". Energy Institute. Retrieved 2012-04-10.
34. Carbon Capture and Geological Storage - European Commission". Ec.europa.

eu. Retrieved 2013-10-06.

35. Three Phases Of Oil Recovery". chinaoilfieldtech.com. China Oilfield Technology.
36. Enhanced Oil Recovery". energy.gov. US DOE.
37. Carbon Captures Perfect Storm". Kemper Project.
38. Schuiling, Olaf. "Olaf Schuiling proposes olivine rock grinding". Retrieved 2011-12-23.
39. Bhaduri, Gaurav A.; Šiller, Lidija (17 January 2013). "Nickel nanoparticles catalyse reversible hydration of carbon dioxide for mineralization carbon capture and storage". Catalysis Science & Technology. Royal Society of Chemistry. **3** *(5): 1234–1239.* doi:10.1039/C3CY20791A. Retrieved 13 February 2015. (subscription required (help)).
40. "GreenMag-Newcastle Mineral Carbonation Pilot Plant". www.dpi.nsw.gov.au. 2010-06-06. Retrieved 2010-06-06.
41. Rocks Found That Could Store Greenhouse Gas, Live Science, March 9, 2009
42. Goldberg, Chen, O'Connor, Walters, and Ziock (1998). "CO_2 *Mineral Sequestration Studies in US"(PDF). National Energy Technology Laboratory. Retrieved June 7, 2007.*
43. Wilson, Siobhan A.; Dipple, Gregory M.; Power, Ian M.; Thom, James M.; Anderson, Robert G.; Raudsepp, Mati; Gabites, Janet E.; Southam, Gordon (2009). "Carbon Dioxide Fixation within Mine Wastes of Ultramafic-Hosted Ore Deposits: Examples from the Clinton Creek and Cassiar Chrysotile Deposits, Canada". Economic Geology. **104**: 95–112. doi:10.2113/gsecongeo.104.1.95.
44. Power, Ian M.; Dipple, Gregory M.; Southam, Gordon (2010). "Bioleaching of Ultramafic Tailings by Acidithiobacillus spp. For CO_2 *Sequestration". Environmental Science & Technology.* **44**: 456–62. doi:10.1021/es900986n.
45. Power, Ian M; Wilson, Siobhan A; Thom, James M; Dipple, Gregory M; Southam, Gordon (2007). "Biologically induced mineralization of dypingite by cyanobacteria from an alkaline wetland near Atlin, British Columbia, Canada". Geochemical Transactions. **8**: 13. doi:10.1186/1467-4866-8-13. PMC 2213640. PMID 18053262.
46. Power, Ian M.; Wilson, Siobhan A.; Small, Darcy P.; Dipple, Gregory M.; Wan, Wankei; Southam, Gordon (2011). "Microbially Mediated Mineral Carbonation: Roles of Phototrophy and Heterotrophy". Environmental Science & Technology. **45** *(20): 9061–8.* doi:10.1021/es201648g.
47. Spath, Pamela L.; Mann, Margaret K. (2002). "The Net Energy and Global Warming Potential of Biomass Power Compared to Coal-Fired Electricity with CO2 Sequestration - A Life Cycle Approach" (PDF). Bioenergy Conference. Omnipress. Retrieved 8 May 2012.
48. IPCC Special Report: Carbon Dioxide Capture and Storage Technical Summary" (PDF). Intergovernmental Panel on Climate Change. Retrieved 2011-10-05.
49. Comparison of carbon capture and storage with renewable energy technologies regarding structural, economic, and ecological aspects in Germany" (PDF).

March 2013.

50. University of Sydney: Global warming effect of leakage from CO2 storage" (PDF). March 2013.
51. Global Status of BECCS Projects 2010 - Storage Security". Retrieved 2012-04-05.
52. Natuurwetenschap & Techniek; April 2009; CCS leakage risks
53. Eendensterfte door lek in CO2-leiding (Duck deaths from leaking CO2 pipeline)". March 2013. (in Dutch)
54. Hedlund, Frank Huess (2012). "The extreme carbon dioxide outburst at the Menzengraben potash mine 7 July 1953". Safety Science. **50** *(3): 537–53. doi:10.1016/j.ssci.2011.10.004.*
55. Pentland, William (6 October 2008). "The Carbon Conundrum". Forbes.
56. Wagner, Leonard (2007). "Carbon Capture and Storage" (PDF). Moraassociates.com. Archived from the original (PDF) on March 21, 2012.
57. "Norway: StatoilHydro's Sleipner carbon capture and storage project proceeding successfully". Energy-pedia. 8 March 2009. Retrieved 19 December 2009.
58. Allan Casey, Carbon Cemetery, Canadian Geographic Magazine, Jan/Feb 2008, p. 61
59. Lafleur, Paul (August 27, 2010). "Geochemical Soil Gas Suvey, A Site Investigation of SW30-5-13-W2M, Weyburn Field, Saskatchewan". Retrieved 2011-01-12.. PDF file linked to from press release of 2011-01-11.
60. IEA GHG Weyburn-Midale CO2 Monitoring and Storage Project - Response to the Petro-Find Gehchem Ltd. Study (PDF), Petroleum Technology Research Centre, 2011-01-19, archived from the original (PDF)on 2016-03-22
61. US DOE, 2012. Best Practices for Monitoring, Verification and Accounting of CO2 Stored in Deep Geologic Formations - 2012 Update.
62. Holloway, S., A. Karimjee , M. Akai, R. Pipatti, and K. Rypdal , 2006–2011. Carbon Dioxide Transport, Injection and Geological Storage, in Eggleston H.S., Buendia L., Miwa K., Ngara T., and Tanabe K. (Eds.), IPCC Guidelines for National Greenhouse Gas Inventories, IPCC National Greenhouse Gas Inventories Programme, WMO/UNEP
63. Miles, N., K. Davis, and J. Wyngaard, 2005. Detecting Leaks from Belowground CO2 Reservoirs Using Eddy Covariance in S. Benson (Ed.) Carbon Dioxide Capture for Storage in DeepGeologic Formations, Elsevier, Vol. 2: 1031–1043
64. ^ Jump up to:[a] [b] [c] Smit, Berend; Reimer, Jeffery A.; Oldenburg, Curtis M.; Bourg, Ian C. Introduction to Carbon Capture and Sequestration (The Berkeley Lectures on Energy - Vol. 1 ed.). Imperial College Press.
65. Biondi, Biondo; de Ridder, Sjoerd; Chang, Jason (2013). "Continuous passive - seismic monitoring of CO2 geologic sequestration projects" (PDF). Retrieved 6 May 2016.
66. Review of Offshore Monitoring for CCS Projects". IEAGHG. IEA Greenhouse Gas R&D Programme. Retrieved 6 May 2016.

67. Madsen, Rod; Xu, Liukang; Claassen, Brent; McDermitt, Dayle (February 2009). "Surface Monitoring Method for Carbon Capture and Storage Projects". Energy Procedia. **1** *(1): 2161–2168. Retrieved 7 May 2016.*

68. "InSAR—Satellite-based technique captures overall deformation "picture"". USGS Science for a Changing World. US Geological Survey. Retrieved 6 May 2016.

69. Mineral carbonation and industrial uses of carbon dioxide" (PDF). Retrieved 2011-12-23.

70. ^ Jump up to:[a] [b] Accelerating the uptake of CCS: Industrial use of captured carbon dioxide - Appendix E: CO2 for use in algae cultivation". Global CCS Institute and Parsons Brinckerhoff. Retrieved 2012-12-11.

71. New Scientist No2645, 1 March 2008.

72. Chang, Kenneth (2008-02-19). "Scientists Would Turn Greenhouse Gas Into Gasoline". The New York Times. Retrieved 2010-04-03.

73. Carbon Capture and Utilization Using CO2 to manufacture fuel, chemicals, and materials". Retrieved 2011-12-23.

74. David Biello:Chemical Process Makes Fuel from Carbon Dioxide Scientific American September 16, 2006

75. Researchers Build Machine That Turns CO2 Into Fuel | Inhabitat - Sustainable Design Innovation, Eco Architecture, Green Building". Inhabitat. Retrieved 2013-10-06.

76. "The Global Status of CCS: 2012 - Projects". The Global CCS Institute. Retrieved 2012-10-17.

77. The Global Status of CCS: 2011 | Global Carbon Capture and Storage Institute". Cdn.globalccsinstitute.com. 2011-10-04. Retrieved 2013-10-06.

78. Large-scale". Globalccsinstitute.com. Retrieved 2013-10-06.

79. zeroemissionsplatform.eu". zeroemissionsplatform.eu. Retrieved 2013-10-06.

80. https://sequestration.mit.edu/tools/projects/in_salah.html

81. Catalytic deoxygenation of triglycerides and fatty acids to hydrocarbons over Ni–Al layered double hydroxide". ResearchGate. Retrieved 2015-06-08.

82. Algae CO2 Capture Part 1: How it Works | reveal.uky.edu". reveal.uky.edu. Retrieved 2015-06-08.

83. Carbon Capture and Storage in Canada". Deloitte.

84. Alberta Energy: Carbon Capture and Storage". Energy.alberta.ca. Retrieved 2010-04-02.

85. Enhance Energy Inc. | Enhanced Oil Recovery, Carbon Capture and Storage". Enhanceenergy.com. Retrieved 2010-04-02.

86. Quest project Canada". Shell.ca. 2009-10-08. Retrieved 2010-04-02.

87. Alberta Department of Energy: Quest Carbon Capture and Storage Project Annual Summary Report, 2014" (PDF). Retrieved 2016-07-28.

88. "Canadian CCS Projects". ICO2N. Retrieved 2011-03-01.

89. October 2015 |website=Global CCS Institute
90. “Boundary Dam Carbon Capture Project - Capital Cost”. Wikipedia.
91. SaskPower launches world’s first commercial CCS process”. SaskPower. 2 October 2014. Retrieved 10 October 2014.
92. Demonstrating Carbon Capture and Storage in Canada” (PDF). NRCan. 2010. Retrieved 2011-03-01.
93. Canada’s integrated carbon dioxide (CO_2) capture and storage initiative”. ICO2N. Retrieved 2011-03-01.
94. Canadian CCS Organizations”. ICO2N. Retrieved 2011-03-01.
95. Demonstration project The Netherlands: Zero Emission Power Plant” (PDF). Retrieved 2010-04-14.
96. CCS Project Overview”. Zeroemissionsplatform.eu. Retrieved 2013-10-06.
97. Bouwman, Elisabeth; Angamuthu, Raja; Byers, Philip; Lutz, Martin; Spek, Anthony L. (July 15, 2010). “Electrocatalytic CO2 Conversion to Oxalate by a Copper Complex”. Science. **327** *(5393): 313–315. doi:10.1126/science.1177981. Retrieved 26 September 2014.*
98. Webmaster Gassnova. “TCM homepage”. Tcmda.com. Retrieved 2010-04-14.
99. Marianne Stigset (2011-11-06). “Norway Boosts Mongstad Carbon-Storage Site Cost to $985 Million”. Bloomberg.
100. Aker says may pull plug on carbon capture project”. Reuters UK. 2011-11-04.
101. Tord Lien skrinlegger CO2-utredningene
102. CCS project granted funding under the European Energy Programme for Recovery (EEPR)”. Ccsnetwork.eu. Retrieved 2010-07-13.
103. Key facts: Belchatów”. Microsites.ccsnetwork.eu. Retrieved 2013-10-06.
104. CCS Project Overview”. Zeroemissionsplatform.eu. 2010-02-22. Retrieved 2013-10-06.
105. “SU receives $66.9 million carbon sequestration”, Bozeman Daily Chronicle, 2008-11-18. Retrieved on 2008-18-11.
106. American Clean Coal Fuels”. Cleancoalfuels.com. Archived from the original on 2010-04-09. Retrieved 2010-04-02.
107. Baard Energy Home- An Energy Development Company”. Baardenergy.com. Archived from thc original on 2009-02-25. Retrieved 2010-04-02.
108. Natchez Project”. Rentechinc.com. Retrieved 2010-04-14.
109. DKRW Energy. “Medicine Bow — DKRW Energy”. Dkrwadvancedfuels.com. Archived from the originalon 2011-07-10. Retrieved 2010-04-02.
110. DOE — Fossil Energy: Industrial CCS projects from Recovery Act”. Fossil. energy.gov. Retrieved 2010-04-02.
111. NETL Carbon Sequestration NETL Web site. Retrieved on 2008-21-11.
112. Carbon Sequestration Leadership Forum”. Cslforum.org. Retrieved 2010-04-02.

113. Bureau of Economic Geology Receives $38 Million for First Large-Scale U.S. Test Storing Carbon Dioxide Underground". Jsg.utexas.edu. 2007-10-24. Archived from the original on 2010-06-11. Retrieved 2010-04-14.

114. Public Service Commission to consider Mississippi Power Kemper rate increase on Tuesday". gulflive.com. 4 March 2013.

115. Ian Urbina. Piles of dirty secrets behind model "clean coal" project, New York Times, July 5, 2016.

116. Goldenberg, Suzanne (12 March 2014). "Can Kemper become the first US power plant to use ‹clean coal'?". The Guardian. Retrieved 14 July 2014.

117. Amy, Jeff (December 17, 2015). "Kemper Plant May Get More Money From Congress". Clarion-Ledger.

118. Southern Co.'s Kemper Power Plant Costs Rise Yet Again". Atlanta Business Chronicle. April 4, 2016.

119. Drajem, Mark (April 14, 2014). "Coal's Best Hope Rising With Costliest U.S. Power Plant". Bloomberg Business.

120. Texas Clean Energy Project". www.texascleanenergyproject.com. Retrieved 2016-07-28.

121. The Skymine Process". Skyonic.com. Retrieved 2010-04-14.

122. Archived January 16, 2008, at the Wayback Machine.

123. Microsoft Word — Information Memorandum 151107 - final.DOC" (PDF). Retrieved 2010-04-02.

124. The European Archive | UK Government Web Archive". Berr.gov.uk. Retrieved 2010-04-02.

125. Former Govt dept: Department for Business Enterprise and Regulatory Reform". Nds.coi.gov.uk. 2008-06-30. Retrieved 2010-04-14.

126. BP quits carbon capture competition". Rsc.org. Retrieved 2010-04-02.

127. Carrington, Damian (25 November 2015). "UK cancels pioneering £1bn carbon capture and storage competition". the Guardian. Retrieved 2015-11-26.

128. £7.4M Oxycoal 2 project given go-ahead: News from Doosan Babcock Energy Ltd". Pandct.com. Retrieved 2010-04-02.

129. Samsung backs £5bn Hatfield carbon-capture project". bbc.co.uk. Retrieved 2012-06-19.

130. New public exhibitions for Yorkshire carbon dioxide pipe". bbc.co.uk. Retrieved 2012-06-19.

131. Local residents to have their say on CCS project". nationalgrid.com. Retrieved 2012-06-19.

132. Welcome to the website for the National Grid Yorkshire and Humber carbon capture, transportation and storage (CCS) project.". National Grid. Retrieved 2012-06-19.

133. piccsi.co.uk". piccsi.co.uk. Retrieved 2013-10-06.

134. Carbon capture and storage". thecrownestate.co.uk. Retrieved 2016-03-04.
135. Heard on Morning Edition (2009-04-10). "China Puts Fizz In Bid To Reduce Carbon Emissions". Npr.org. Retrieved 2010-04-14.
136. CCS Project Overview". Zeroemissionsplatform.eu. Retrieved 2013-10-06.
137. Germany leads ‹clean coal' pilot". BBC News. 2008-09-03.
138. Access all areas: Schwarze Pumpe". BBC News. 2008-09-03.
139. 'Emissions-free' power plant pilot fires up in Germany
140. http://www.thelocal.se/20140507/vattenfall-abandons-research-on-co2-storage
141. BASF, RWE Power and Linde are developing new processes for CO_2 *capture in coal-fired power plants". Press Release. Basf.com. 2007-09-28. Retrieved 2010-04-14.*
142. CCS project granted funding under the European Energy Programme for Recovery (EEPR)". Ccsnetwork.eu/. 2010-04-28. Retrieved 2010-07-13.
143. Key facts: Jänschwalde". Microsites.ccsnetwork.eu. Retrieved 2013-10-06.
144. http://news.sbs.com.au/worldnewsaustralia//first_carbon_storage_plant_launched_544064
145. Seeking clean coal science ‹only option'". News.theage.com.au. 2008-04-02. Retrieved 2010-04-14.
146. CO2CRC Otway Project overview". Co2crc.com.au. 2010-02-18. Retrieved 2010-04-14.
147. Demonstrating CCS". Co2crc.com.au. 2010-05-06. Retrieved 2010-05-31.
148. Australia's largest carbon capture project gets underway". energyefficiencynews.com. 2009-10-27. Retrieved 2010-05-31.
149. EU Projects". Zeroemissionsplatform.eu. Retrieved 2013-10-06.
150. Simon Robinson (22 January 2012). "Cutting Carbon: Should We Capture and Store It?". *TIME.*
151. Rochon, Emily et al. False Hope: Why carbon capture and storage won›t save the climateGreenpeace, May 2008, p. 5.
152. http://www.ipcc.ch/pdf/special-reports/srccs/srccs_wholereport.pdf
153. Rhodes, J. S.; Keith, D. W. (2008). "Biomass with capture: Negative emissions within social and environmental constraints: An editorial comment". Climatic Change. **87** *(3–4): 321.* doi:10.1007/s10584-007-9387-4.
154. Karlsson. "Global Status of BECCS Projects 2010" (PDF). Global CCS Institute. Retrieved 6 May 2012.
155. Jacobson, Mark Z.; Delucchi, Mark A. (2010). "Providing all Global Energy with Wind, Water, and Solar Power, Part I: Technologies, Energy Resources, Quantities and Areas of Infrastructure, and Materials"(PDF). Energy Policy. p. 4.
156. 20244 DTI Energy Review_AW" (PDF). Retrieved 2010-04-14.
157. A POLICY STRATEGY FOR CARBON CAPTURE AND STORAGE" (PDF). International Energy Agency. Retrieved 6 May 2012.

158. Meyer, Leo. "CCS in the IPCC Fourth Assessment Report" (PDF). Retrieved 6 May 2012.

159. Science, 27 February 2009, Vol 323, p. 1158, "Stimulus Gives DOE Billions for Carbon-Capture Project"

160. CCS — Assessing the Economics" (PDF). Mckinsey.com. 2008.

161. Solar Fuel". Solar Fuel. Retrieved 2013-10-06.

162. Bioenergy and renewable power methane in integrated 100% renewable energy system" (PDF). Retrieved 2013-10-06.

163. scénario *négaWatt 2011 (France)". Negawatt.org. Retrieved 2013-10-06.*

164. Global Status of CCS Report:2011". Global CCS Institute. Retrieved 2011-12-14.

165. SBSTA Presents Global Carbon dioxide Capture and Storage Data at COP16 Archived July 28, 2011, at the Wayback Machine.

166. Bonner, Mark. "CCS enters the CDM at CMP 7". Global CCS Institute. Retrieved 7 May 2012.

167. IPCC Special Report: Carbon Capture and Storage Technical Summary. IPCC. p. 27" (PDF). Retrieved 2013-10-06.

168. http://www.ekopolitan.com/news/ccs-norway-amines-nitrosamines-and-nitramines-released-carbon-capture-proce

CHAPTER 5

Human Impact on the Environment

Chapter Contents

Environmental technology refers to the field of science concerned with reducing the human impact on the environment through technological advances or improvements. Some common applications of environmental technology deal with reducing energy consumption, limiting man-made damage to the physical environment, and reducing waste. Areas of research in the field may involve cleaner energy sources, improved energy efficiency in transportation and buildings, and methods that decrease or prevent pollution. This is a broad field that draws on many sciences, some of which include chemistry, ecology, and biology. Innovation and advances in environmental science may have commercial applications, save money, or be designed to meet government regulations.

INTRODUCTION

Human impact on the environment or **anthropogenic impact on the environment** includes impacts on biophysical environments, biodiversity, and other resources. The term *anthropogenic* designates an effect or object resulting from human activity. The term was first used in the technical sense by Russian geologist Alexey Pavlov, and was first used in English by British ecologist Arthur Tansley in reference to human influences on climax plant communities. The atmospheric scientist Paul Crutzen introduced the term "anthropocene" in the mid-1970s. The term is sometimes used in the context of pollution emissions that are produced as a result of human activities but applies broadly to all major human impacts on the environment.

The ecosystem of public parks often includes humans feeding the wildlife.

CAUSES

Technology

The applications of technology often result in unavoidable and unexpected environmental impacts, which according to the I = PAT equation is measured as resource use or pollution generated per unit GDP. Environmental impacts caused by the application of technology are often perceived as unavoidable for several reasons. First, given that the purpose of many technologies is to exploit, control, or otherwise "improve" upon nature for the perceived benefit of humanity while at the same time the myriad of processes in nature have been optimized and are continually adjusted by evolution, any disturbance of these natural processes by technology is likely to result in negative environmental consequences.Second, the conservation of mass principle and the first law of thermodynamics (i.e., conservation of energy) dictate that whenever material resources or energy are moved around or manipulated by technology, environmental consequences are inescapable. Third, according to the second law of thermodynamics, order can be increased within a system (such as the human economy) only by increasing disorder or entropy outside the system (i.e., the environment). Thus, technologies can create "order" in the human economy (i.e., order as manifested in buildings, factories, transportation networks, communication systems, etc.) only at the expense of increasing "disorder" in the environment. According to a number of studies, increased entropy is likely to be correlated to negative environmental impacts.

Agriculture

The environmental impact of agriculture varies based on the wide variety of agricultural practices employed around the world. Ultimately, the environmental impact depends on the production practices of the system used by farmers. The connection between emissions into the environment and the farming system is indirect, as it also depends on other climate variables such as rainfall and temperature.

There are two types of indicators of environmental impact: "means-based", which is based on the farmer's production methods, and "effect-based", which is the impact that farming methods have on the farming system or on emissions to the environment. An example of a means-based indicator would be the quality of groundwater, that is effected by the amount of nitrogen applied to the soil. An indicator reflecting the loss of nitrate to groundwater would be effect-based.

The environmental impact of agriculture involves a variety of factors from the soil, to water, the air, animal and soil diversity, plants, and the food itself. Some of the environmental issues that are related to agriculture are climate change, deforestation, genetic engineering, irrigation problems, pollutants, soil degradation, and waste.

Fishing

Fishing down the foodweb.

The environmental impact of fishing can be divided into issues that involve the availability of fish to be caught, such as overfishing, sustainable fisheries, and fisheries management; and issues that involve the impact of fishing on other elements of the environment, such as by-catch and destruction of habitat such as coral reefs.

These conservation issues are part of marine conservation, and are addressed in fisheries science programs. There is a growing gap between how many fish are available to be caught and humanity's desire to catch them, a problem that gets worse as the world population grows.

Similar to other environmental issues, there can be conflict between the fishermen who depend on fishing for their livelihoods and fishery scientists who realize that if future fish populations are to be sustainable then some fisheries must reduce or even close.

The journal *Science* published a four-year study in November 2006, which predicted that, at prevailing trends, the world would run out of wild-caught seafood in 2048. The scientists stated that the decline was a result of overfishing, pollution and other environmental factors that were reducing the population of fisheries at the same time as their ecosystems were being degraded. Yet again the analysis has met criticism as being fundamentally flawed, and many fishery management officials, industry representatives and scientists challenge the findings, although the debate continues. Many countries, such as Tonga, the United States, Australia and New Zealand, and international management bodies have taken steps to appropriately manage marine resources.

Irrigation

The environmental impact of irrigation includes the changes in quantity and quality of soil and water as a result of irrigation and the ensuing effects on natural and social conditions at the tail-end and downstream of the irrigation scheme.

The impacts stem from the changed hydrological conditions owing to the installation and operation of the scheme.

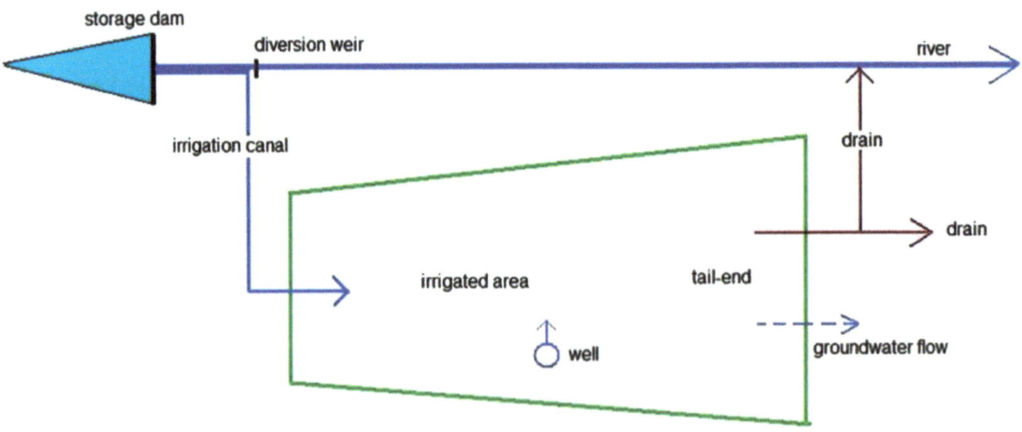

Irrigation scheme hydrology

An irrigation scheme often draws water from the river and distributes it over the irrigated area. As a hydrological result it is found that:

- the downstream river discharge is reduced
- the evaporation in the scheme is increased
- the groundwater recharge in the scheme is increased
- the level of the water table rises
- the drainage flow is increased.

These may be called direct effects.

Effects on soil and water quality are indirect and complex, and subsequent impacts on natural, ecological and socio-economic conditions are intricate. In some, but not all instances, water logging and soil salinization can result. However, irrigation can also be used, together with soil drainage, to overcome soil salinization by leaching excess salts from the vicinity of the root zone.

Irrigation can also be done extracting groundwater by (tube)wells. As a hydrological result it is found that the level of the water descends. The effects may be water mining, land/soil subsidence, and, along the coast, saltwater intrusion.

Irrigation projects can have large benefits, but the negative side effects are often overlooked.Agricultural irrigation technologies such as high powered water pumps, dams, and pipelines are responsible for the large-scale depletion of fresh water resources such as aquifers, lakes, and rivers. As a result of this massive diversion of freshwater, lakes, rivers, and creeks are running dry, severely altering or stressing surrounding ecosystems, and contributing to the extinction of many aquatic species.

Agricultural Land Loss and Soil Erosion

Lal and Stewart estimated global loss of agricultural land by degradation and abandonment at 12 million hectares per year. In contrast, according to Scherr, GLASOD (Global Assessment of Human-Induced Soil Degradation, under the UN Environment Programme) estimated that 6 million hectares of agricultural land per year had been lost to soil degradation since the mid-1940s, and she noted that this magnitude is similar to earlier estimates by Dudal and by Rozanov et al. Such losses are attributable not only to soil erosion, but also to salinization, loss of nutrients and organic matter, acidification, compaction, water logging and subsidence. Human-induced land degradation tends to be particularly serious in dry regions. Focusing on soil properties, Oldeman estimated that about 19 million square kilometers of global land area had been degraded; Dregne and Chou, who included degradation of vegetation cover as well as soil, estimated about 36 million square kilometers degraded in the world's dry regions. Despite estimated losses of agricultural land, the amount of arable land used in crop production globally increased by about 9% from 1961 to 2012, and is estimated to have been 1.396 billion hectares in 2012.

Global average soil erosion rates are thought to be high, and erosion rates on conventional cropland generally exceed estimates of soil production rates, usually by more than an order of magnitude. In the US, sampling for erosion estimates by the US NRCS (Natural Resources Conservation Service) is statistically based, and estimation uses the Universal Soil Loss Equation and Wind Erosion Equation. For 2010, annual average soil loss by sheet, rill and wind erosion on non-federal US land was estimated to be 10.7 t/ha on cropland and 1.9 t/ha on pasture land; the average soil erosion rate on US cropland had been reduced by about 34% since 1982. No-till and low-till practices have become increasingly common on North American cropland used for production of grains such as wheat and barley. On uncultivated cropland, the recent average total soil loss has been 2.2 t/ha per year. In comparison with agriculture using conventional cultivation, it has been suggested that, because no-till agriculture produces erosion rates much closer to soil production rates, it could provide a foundation for sustainable agriculture.

Meat Production

Environmental impacts associated with meat production include use of fossil energy, water and land resources, greenhouse gas emissions, and in some instances, rainforest clearing, water pollution and species endangerment, among other adverse effects. Steinfeld et al. of the FAO estimated that 18% of global anthropogenic GHG (greenhouse gas) emissions (estimated as 100-year carbon dioxide equivalents) are associated in some way with livestock production. A more recent FAO analysis estimated that all agriculture, including the livestock sector, in 2011 accounted for 12% of global anthropogenic GHG emissions expressed as 100-year carbon dioxide equivalents. Similarly, the Intergovernmental Panel on Climate Change has estimated that about 10 to 12% of global anthropogenic GHG emissions (expressed as 100-year carbon dioxide equivalents) were assignable to all of agriculture, including the livestock sector, in 2005 and again in 2010. The percentage assignable to livestock would be some fraction of the percentage for agriculture. The amount assignable to meat production would be some fraction of that assigned to livestock. FAO data indicate that meat accounted for 26% of global livestock product tonnage in 2011. However, many estimates use different sectoral assignment of some emissions. Environmental specialists Jeff Anhang and Robert Goodland with the IFC and World Bank, have put the GHG associated with livestock at 51%, pointing out the FAO report failed to account for the 8,769 metric tons of respiratory CO_2 produced each year, undercounted methane production and land use associated with livestock, and failed to properly categorize emissions related to the slaughtering, processing, packaging, storing and transporting of animals and animal products.

Globally, enteric fermentation (mostly in ruminant livestock) accounts for about 27% of anthropogenic methane emissions, Despite methane's 100-year global warming potential, recently estimated at 28 without and 34 with climate carbon feedbacks, methane emission is currently contributing relatively little to global warming. Over the decade 2000 through 2009, atmospheric methane content increased by an average of only 6 Tg per year (because nearly all natural and anthropogenic methane emission was offset by degradation), while atmospheric carbon dioxide increased by nearly 15,000 Tg per year. At the currently estimated rate of methane degradation, slight reduction of anthropogenic methane emissions, to about 98% of that decade's average, would be expected to result in no further increase of atmospheric methane content. Although reduction of methane emissions would have a rapid effect on warming, the expected effect would be small. Other anthropogenic GHG emissions associated with livestock production include carbon dioxide from fossil fuel consumption (mostly for production, harvesting and transport of feed), and nitrous oxide emissions associated with use of nitrogenous fertilizers, growing of nitrogen-fixing legume vegetation and manure management. Management practices that can mitigate GHG emissions from production of livestock and feed have been identified.

Livestock production, including feed production and grazing, uses about 30% of the earth's ice-free terrestrial surface: about 26% for grazing and about 4% for other feed production. The intensity and duration of grazing use vary greatly and these, together with terrain, vegetation and climate, influence the nature and importance of grazing's environmental impact, which can range from severe to negligible, and in some cases (as noted below) beneficial. Excessive use of vegetation by grazing can be especially conducive to land degradation in dry areas.

Considerable water use is associated with meat production, mostly because of water used in production of vegetation that provides feed. There are several published estimates of water use associated with livestock and meat production, but the amount of water use assignable to such production is seldom estimated. For example, "green water" use is evapotranspirational use of soil water that has been provided directly by precipitation; and "green water" has been estimated to account for 94% of global beef cattle production's "water footprint", and on rangeland, as much as 99.5% of the water use associated with beef production is "green water". However, it would be misleading simply to assign that associated rangeland green water use to beef production, partly because that evapotranspirational use occurs even in the absence of cattle. Even when cattle are present, most of that associated water use can be considered assignable to production of terrestrial environmental values, because it produces root and residue biomass important for erosion control, stabilization of soil structure, nutrient cycling, carbon sequestration, support of numerous primary consumers, many of which support higher trophic levels, etc. Withdrawn water (from surface and groundwater sources) is used for livestock watering, and in some cases is also used for irrigation of forage and feed crops. Whereas all irrigation in the US (including loss in conveyance) is estimated to account for about 38% of US withdrawn freshwater use,irrigation water for production of livestock feed and forage has been estimated to account for about 9%; other withdrawn freshwater use for the livestock sector (for drinking, washdown of facilities, etc.) is estimated at about 0.7%. Because of the preponderance of non-meat products from the livestock sector only some fraction of this water use is assignable to meat production.

Impairment of water quality by manure and other substances in runoff and infiltrating water is a concern, especially where intensive livestock production is carried out. In the US, in a comparison of 32 industries, the livestock industry was found to have a relatively good record of compliance with environmental regulations pursuant to the Clean Water Act and Clean Air Act, but pollution issues from large livestock operations can sometimes be serious where violations occur. Various measures have been suggested by the US Environmental Protection Agency, among others, which can help reduce livestock damage to streamwater quality and riparian environments.

Data of a USDA study indicate that, in 2002, about 0.6% of non-solar energy use in the United States was accounted for by production of meat-producing livestock and poultry. This estimate included embodied energy used in production, such as energy used in manufacture and transport of fertilizer for feed production. (Non-solar energy is specified, because solar energy is used in such processes as photosynthesis and hay-drying.)

Changes in livestock production practices influence the environmental impact of meat production, as illustrated by some beef data. In the US beef production system, practices prevailing in 2007 are estimated to have involved 8.6% less fossil fuel use, 16.3% less greenhouse gas emissions (estimated as 100-year carbon dioxide equivalents), 12.1% less withdrawn water use and 33.0% less land use, per unit mass of beef produced, than in 1977. From 1980 to 2012 in the US, while population increased by 38%, the small ruminant inventory decreased by 42%, the cattle-and-calves inventory decreased by 17%, and methane emissions from livestock decreased by 18%; yet despite the reduction in cattle numbers, US beef production increased over that period.

Some impacts of meat-producing livestock may be considered environmentally beneficial. These include waste reduction by conversion of human-inedible crop residues to food, use of livestock as an alternative to herbicides for control of invasive and noxious weeds and other vegetation management, use of animal manure as fertilizer as a substitute for those synthetic fertilizers that require considerable fossil fuel use for manufacture, grazing use for wildlife habitat enhancement, and carbon sequestration in response to grazing practices, among others.

Palm Oil

A village palm oil press "*malaxeur*" in Bandundu, Democratic Republic of the Congo.

Palm oil, produced from the oil palm, is a basic source of income for many farmers in Southeast Asia, Central and West Africa, and Central America. It is locally used as a cooking oil, exported for use in many commercial food and personal care products and is converted into biofuel. It produces up to 10 times more oil per unit area as soyabeans, rapeseed or sunflowers. Oil palms produce 38% of vegetable oil output on 5% of the world's vegetable-oil farmland. Palm oil is under increasing scrutiny in relation to its effects on the environment.

Introductions and Invasive Species

Introductions of species, particularly plants into new areas, by whatever means and for whatever reasons have brought about major and permanent changes to the environment over large areas. Examples include the introduction of Caulerpa taxifolia into the Mediterranean, the introduction of oat species into the California grasslands, and the introduction of privet, kudzu, and purple loosestrife to North America. Rats, cats, and goats have radically altered biodiversity in many islands. Additionally, introductions have resulted in genetic changes to native fauna where interbreeding has taken place, as with buffalo with domestic cattle, and wolves with domestic dogs.

Energy Industry

The environmental impact of energy harvesting and consumption is diverse. In recent years there has been a trend towards the increased commercialization of various renewable energy sources.

In the real world, consumption of fossil fuel resources leads to global warming and climate change. However, little change is being made in many parts of the world. If the peak oil theory proves true, more explorations of viable alternative energy sources, could be more friendly to the environment.

Rapidly advancing technologies can achieve a transition of energy generation, water and waste management, and food production towards better environmental and energy usage practices using methods of systems ecology and industrial ecology.

Biodiesel

The environmental impact of biodiesel includes energy use, greenhouse gas emissions and some other kinds of pollution. A joint life cycle analysis by the US Department of Agriculture and the US Department of Energy found that substituting 100% biodiesel for petroleum diesel in buses reduced life cycle consumption of petroleum by 95%. Biodiesel reduced net emissions of carbon dioxide by 78.45%, compared with petroleum diesel. In urban buses, biodiesel reduced particulate emissions 32 percent, carbon monoxide emissions 35 percent, and emissions of sulfur oxides 8%, relative to life cycle emissions associated with use of petroleum diesel. Life cycle emissions of hydrocarbons were 35% higher and emission of various nitrogen oxides (NOx) were 13.5% higher with biodiesel. Life cycle analyses by the Argonne National Laboratory have indicated reduced fossil energy use and reduced greenhouse gas emissions with biodiesel, compared with petroleum diesel use. Biodiesel derived from various vegetable oils (e.g. canola or soybean oil), is readily biodegradable in the environment compared with petroleum diesel.

Coal Mining and Burning

The environmental impact of coal mining and -burning is diverse. Legislation passed by the US Congress in 1990 required the United States Environmental Protection Agency (EPA) to issue a plan to alleviate toxic air pollution from coal-fired power plants. After delay and litigation, the EPA now has a court-imposed deadline of March 16, 2011, to issue its report.

Electricity Generation

The environmental impact of electricity generation is significant because modern society uses large amounts of electrical power. This power is normally generated at power plants that convert some other kind of energy into electricity. Each such system has advantages and disadvantages, but many of them pose environmental concerns.

Nuclear Power

The environmental impact of nuclear power results from the nuclear fuel cycle processes including mining, processing, transporting and storing fuel and radioactive fuel waste. Released radioisotopes pose a health danger to human populations, animals and plants as radioactive particles enter organisms through various transmission routes.

Radiation is a carcinogen and causes numerous effects on living organisms and systems. The environmental impacts of nuclear power plant disasters such as the Chernobyl disaster, the Fukushima Daiichi nuclear disaster and the Three Mile Island accident, among others, persist indefinitely, though several other factors contributed to these events including improper management of fail safe systems and natural disasters putting uncommon stress on the generators. The radioactive decay rate of particles varies greatly, dependent upon the nuclear properties of a particular isotope. Radioactive Plutonium-244 has a half-life of 80.8 million years, which indicates the time duration required for half of a given sample to decay, though very little plutonium-244 is produced in the nuclear fuel cycle and lower half-life materials have lower activity thus giving off less dangerous radiation.

Oil Shale Industry

Kiviõli Oil Shale Processing & Chemicals Plant in ida-Virumaa, Estonia.

The environmental impact of the oil shale industry includes the consideration of issues such as land use, waste management, and water and air pollution caused by the extraction and processing of oil shale. Surface mining of oil shale deposits causes the usual environmental impacts of open-pit mining. In addition, the combustion and thermal processing generate waste material, which must be disposed of, and harmful atmospheric emissions, including carbon dioxide, a major greenhouse gas. Experimental in-situ conversion processes and carbon capture and storage technologies may reduce some of these concerns in future, but may raise others, such as the pollution of groundwater.

Petroleum

The environmental impact of petroleum is often negative because it is toxic to almost all forms of life. Climate change exists. Petroleum, commonly referred to as oil, is closely linked to virtually all aspects of present society, especially for transportation and heating for both homes and for commercial activities.

Reservoirs

The Wachusett Dam in Clinton, Massachusetts.

The environmental impact of reservoirs is coming under ever increasing scrutiny as the world demand for water and energy increases and the number and size of reservoirs increases.

Dams and the reservoirs can be used to supply drinking water, generate hydroelectric power, increasing the water supply for irrigation, provide recreational opportunities and flood control. However, adverse environmental and sociological impacts have also been identified during and after many reservoir constructions. Although the impact varies greatly between different dams and reservoirs, common criticisms include preventing sea-run fish from reaching their historical mating grounds, less access to water downstream, and a smaller catch for fishing communities in the area. Advances in technology have provided solutions to many negative impacts of dams but these advances are often not viewed as worth investing in if not required by law or under the threat of fines. Whether reservoir projects are ultimately beneficial or detrimental—to both the environment and surrounding human populations— has been debated since the 1960s and probably long before that. In 1960 the construction of Llyn Celyn and the flooding of Capel Celyn provoked political uproar which continues to this day. More recently, the construction of Three Gorges Dam and other similar projects throughout Asia, Africa and Latin America have generated considerable environmental and political debate.

Wind Power

Wind turbines in an agricultural setting.

Compared to the environmental impact of traditional energy sources, the environmental impact of wind power is relatively minor. Wind powered electricity generation consumes no fuel, and emits no air pollution, unlike fossil fuel power sources. The energy consumed to manufacture and transport the materials used to build a wind power plant is equal to the new energy produced by the plant within a few months. While a wind farm may cover a large area of land, many land uses such as agriculture are compatible, with only small areas of turbine foundations and infrastructure made unavailable for use.

There are reports of bird and bat mortality at wind turbines, as there are around other artificial structures. The scale of the ecological impact may or may not be significant, depending on specific circumstances. Prevention and mitigation of wildlife fatalities, and protection of peat bogs, affect the siting and operation of wind turbines.

There are conflicting reports about the effects of noise on people who live very close to a wind turbine.

Light Pollution

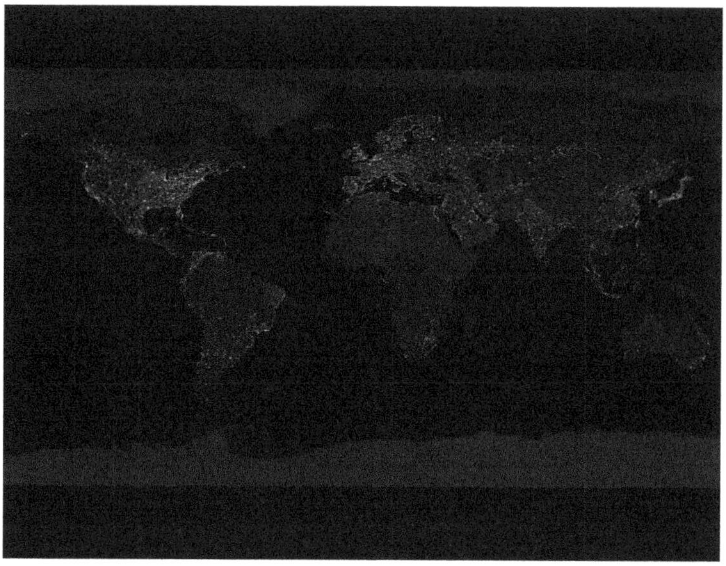

A composite image of artificial light emissions from Earth at night

Artificial light at night is one of the most obvious physical changes that humans have made to the biosphere, and is the easiest form of pollution to observe from space. The main environmental impacts of artificial light are due to light's use as an information source (rather than an energy source). The hunting efficiency of visual predators generally increases under artificial light, changing predator prey interactions. Artificial light also affects dispersal, orientation, migration, and hormone levels, resulting in disrupted circadian rhythms.

Manufactured Products

Cleaning Agents

The environmental impact of cleaning agents is diverse. In recent years, measures have been taken to reduce these effects.

Nanotechnology

Nanotechnology's environmental impact can be split into two aspects: the potential for nanotechnological innovations to help improve the environment, and the possibly novel type of pollution that nanotechnological materials might cause if released into the environment. As nanotechnology is an emerging field, there is great debate regarding to what extent industrial and commercial use of nanomaterials will affect organisms and ecosystems.

Leather

Paint

The environmental impact of paint is diverse. Traditional painting materials and processes can have harmful effects on the environment, including those from the use of lead and other additives. Measures can be taken to reduce environmental impact, including accurately estimating paint quantities so that wastage is minimized, use of paints, coatings, painting accessories and techniques that are environmentally preferred. The United States Environmental Protection Agency guidelines and Green Star ratings are some of the standards that can be applied.

Paper

A pulp and paper mill in New Brunswick, Canada. Although pulp and paper manufacturing requires large amounts of energy, a portion of it comes from burning wood waste.

The environmental impact of paper is significant, which has led to changes in industry and behaviour at both business and personal levels. With the use of modern technology such as the printing press and the highly mechanised harvesting of wood, paper has become a cheap commodity. This has led to a high level of consumption and waste. With the rise in environmental awareness due to the lobbying by environmental organizations and with increased government regulation there is now a trend towards sustainability in the pulp and paper industry.

Pesticides

The environmental impact of pesticides is often greater than what is intended by those who use them. Over 98% of sprayed insecticides and 95% of herbicides reach a destination other than their target species, including nontarget species, air, water, bottom sediments, and food. Pesticide contaminates land and water when it escapes from production sites and storage tanks, when it runs off from fields, when it is discarded, when it is sprayed aerially, and when it is sprayed into water to kill algae.

The amount of pesticide that migrates from the intended application area is influenced by the particular chemical's properties: its propensity for binding to soil, its vapor pressure, its water solubility, and its resistance to being broken down over time. Factors in the soil, such as its texture, its ability to retain water, and the amount of organic matter contained in it, also affect the amount of pesticide that will leave the area. Some pesticides contribute to global warming and the depletion of the ozone layer.

Pharmaceuticals and Personal Care Products

The environmental impact of pharmaceuticals and personal care products (PPCPs) is largely speculative. PPCPs are substances used by individuals for personal health or cosmetic reasons and the products used by agribusiness to boost growth or health of livestock. PPCPs have been detected in water bodies throughout the world. The effects of these chemicals on humans and the environment are not yet known, but to date there is no scientific evidence that they affect human health.

Mining

Acid mine drainage in the Rio Tinto River.

The environmental impact of mining includes erosion, formation of sinkholes, loss of biodiversity, and contamination of soil, groundwater and surface water by chemicals from mining processes. In some cases, additional forest logging is done in the vicinity of mines to increase the available room for the storage of the created debris and soil. Besides creating environmental damage, the contamination resulting from leakage of chemicals also affect the health of the local population. Mining companies in some countries are required to follow environmental and rehabilitation codes, ensuring the area mined is returned to close to its original state. Some mining methods may have significant environmental and public health effects.

Transport

Interstate 10 and Interstate 45 near downtown Houston, Texas in the United States.

The environmental impact of transport is significant because it is a major user of energy, and burns most of the world's petroleum. This creates air pollution, including nitrous oxides and particulates, and is a significant contributor to global warming through emission of carbon dioxide, for which transport is the fastest-growing emission sector. By subsector, road transport is the largest contributor to global warming.

Environmental regulations in developed countries have reduced the individual vehicles emission; however, this has been offset by an increase in the number of vehicles, and more use of each vehicle. Some pathways to reduce the carbon emissions of road vehicles considerably have been studied. Energy use and emissions vary largely between modes, causing environmentalists to call for a transition from air and road to rail and human-powered transport, and increase transport electrification and energy efficiency.

Other environmental impacts of transport systems include traffic congestion and automobile-oriented urban sprawl, which can consume natural habitat and agricultural lands. By reducing transportation emissions globally, it is predicted that there will be significant positive effects on Earth's air quality, acid rain, smog and climate change.

The health impact of transport emissions is also of concern. A recent survey of the studies on the effect of traffic emissions on pregnancy outcomes has linked exposure to emissions to adverse effects on gestational duration and possibly also intrauterine growth.

Aviation

The environmental impact of aviation occurs because aircraft engines emit noise, particulates, and gases which contribute to climate change and global dimming. Despite emission reductions from automobiles and more fuel-efficient and less polluting turbofan and turboprop engines, the rapid growth of air travel in recent years contributes to an increase in total pollution attributable to aviation. In the EU, greenhouse gas emissions from aviation increased by 87% between 1990 and 2006. Among other factors leading to this phenomenon are the increasing number of hypermobile travellers and social factors that are making air travel commonplace, such as frequent flyer programs.

There is an ongoing debate about possible taxation of air travel and the inclusion of aviation in an emissions trading scheme, with a view to ensuring that the total external costs of aviation are taken into account.

Roads

The environmental impact of roads includes the local effects of highways (public roads) such as on noise, light pollution, water pollution, habitat destruction/disturbance and local air quality; and the wider effects including climate change from vehicle emissions. The design, construction and management of roads, parking and other related facilities as well as the design and regulation of vehicles can change the impacts to varying degrees.

Shipping

The environmental impact of shipping includes greenhouse gas emissions and oil pollution. In 2007, carbon dioxide emissions from shipping were estimated at 4 to 5% of the global total, and estimated by the International Maritime Organisation (IMO) to rise by up to 72% by 2020 if no action is taken. There is also a potential for introducing invasive species into new areas through shipping, usually by attaching themselves to the ship's hull.

The First Intersessional Meeting of the IMO Working Group on Greenhouse Gas Emissions from Ships took place in Oslo, Norway on 23–27 June 2008. It was tasked with developing the technical basis for the reduction mechanisms that may form part of a future IMO regime to control greenhouse gas emissions from international shipping, and a draft of the actual reduction mechanisms themselves, for further consideration by IMO's Marine Environment Protection Committee (MEPC).

An Agent Orange spray run, part of Operation Ranch Hand, during the Vietnam War by UC-123B Provider aircraft.

War

As well as the cost to human life and society, there is a significant environmental impact of war. Scorched earth methods during, or after war have been in use for much of recorded history but with modern technology war can cause a far greater devastation on the environment. Unexploded ordnance can render land unusable for further use or make access across it dangerous or fatal.

EFFECTS

Biodiversity

Human impact on biodiversity is significant, humans have caused the extinction of many species, including the dodo and, potentially, large megafaunal species during the last ice age. Though most experts agree that human beings have accelerated the rate of species extinction, the exact degree of this impact is unknown, perhaps 100 to 1000 times the normal background rate of extinction. Some authors have postulated that without human interference the biodiversity of the Earth would continue to grow at an exponential rate.

Coral Reefs

Human impact on coral reefs is significant. Coral reefs are dying around the world. In particular, coral mining, pollution (organic and non-organic), overfishing, blast fishing and the digging of canals and access into islands and bays are serious threats to these ecosystems. Coral reefs also face high dangers from pollution, diseases, destructive fishing practices and warming oceans. In order to find answers for these problems, researchers study the various factors that impact reefs. The list of factors is long, including the ocean's role as a carbon dioxide sink, atmospheric changes, ultraviolet light, ocean acidification, biological virus, impacts of dust storms carrying agents to far flung reefs, pollutants, algal blooms and others. Reefs are threatened well beyond coastal areas.

General estimates show approximately 10% world's coral reefs are already dead. It is estimated that about 60% of the world's reefs are at risk due to destructive, human-related activities. The threat to the health of reefs is particularly strong in Southeast Asia, where 80% of reefs are endangered.

Carbon Cycle

Global warming is the result of increasing atmospheric carbon dioxide concentrations which is caused primarily by the combustion of fossil energy sources such as petroleum, coal, and natural gas, and to an unknown extent by destruction of forests, increased methane, volcanic activity and cement production. Such massive alteration of the global carbon cycle has only been possible because of the availability and deployment of advanced technologies, ranging in application from fossil fuel exploration, extraction, distribution, refining, and combustion in power plants and automobile engines and advanced farming practices. Livestock contributes to climate change both thru the production of greenhouse gases and thru destruction of carbon sinks such as rain-forests. According to the 2006 United Nations/FAO report, 18% of all greenhouse gas emissions found in the atmosphere are due to livestock. The raising of livestock and the land needed to feed them has resulted in the destruction millions of acres of Rainforest and as global demand for meat rises, so too will the demand for land. Ninety-one percent of all rainforest land deforested since 1970 is now used for livestock. Potential negative environmental impacts caused by increasing atmospheric carbon dioxide concentrations are rising global air temperatures, altered hydrogeological cycles resulting in more frequent and severe droughts, storms, and floods, as well as sea level rise and ecosystem disruption.

Nitrogen Cycle

Human impact on the nitrogen cycle is diverse. Agricultural and industrial nitrogen (N) inputs to the environment currently exceed inputs from natural N fixation. As a consequence of anthropogenic inputs, the global nitrogen cycle (Fig. 1) has been significantly altered over the past century. Global atmospheric nitrous oxide (N_2O) mole fractions have increased from a pre-industrial value of ~270 nmol/mol to ~319 nmol/mol in 2005. Human activities account for over one-third of N_2O emissions, most of which are due to the agricultural sector.

REFERENCES

1. Sahney, S., Benton, M.J. and Ferry, P.A. (2010). "Links between global taxonomic diversity, ecological diversity and the expansion of vertebrates on land" (PDF). Biology Letters. 6 (4): 544–547. doi:10.1098/rsbl.2009.1024. PMC 2936204. PMID 20106856.
2. Hawksworth, David L.; Bull, Alan T. (2008). Biodiversity and Conservation in Europe. Springer. p. 3390. ISBN 1402068646.
3. Bampton, M. (1999) "Anthropogenic Transformation" in Encyclopedia of Environmental Science, D. E. Alexander and R. W. Fairbridge (eds.), Kluwer Academic Publishers, Dordrecht, The Netherlands, ISBN 0412740508.
4. Crutzen, Paul and Eugene F. Stoermer. "The ‹Anthropocene'" in International Geosphere-Biosphere Programme Newsletter. 41 (May 2000): 17–18
5. Scott, Michon (2014). "Glossary". NASA Earth Observatory. Retrieved 2008-11-03.
6. Commoner, B. (1971). The closing cycle – Nature, man, and technology, Alfred A. Knopf.
7. Faber, M., Niemes, N. and Stephan, G. (2012). Entropy, environment, and resources, Springer Verlag, Berlin, Germany, ISBN 3642970494.
8. Kümmel, R. (1989). "Energy as a factor of production and entropy as a pollution indicator in macroeconomic modeling". Ecological Economics. 1 (2): 161–180. doi:10.1016/0921-8009(89)90003-7.
9. Ruth, M. (1993). Integrating economics, ecology, and thermodynamics, Kluwer Academic Publishers, ISBN 0792323777.
10. Huesemann, M.H., and J.A. Huesemann (2011). Technofix: Why Technology Won't Save Us or the Environment, Chapter 1, "The inherent unpredictability and unavoidability of unintended consequences", New Society Publishers, ISBN 0865717044.
11. van der Warf, Hayo; Petit, Jean (December 2002). "Evaluation of the environmental impact of agriculture at the farm level: a comparison and analysis of 12 indicator-based methods". Agriculture, Ecosystems and Environment. 93 (1–3): 131–145. doi:10.1016/S0167-8809(01)00354-1. Retrieved 21 April 2015.
12. Oppenlander, Richard (2013). Food Choice and Sustainability. Minneapolis, MN: Langdon Street Press. pp. 120–123. ISBN 978-1-62652-435-4.
13. Myers, R. A.; Worm, B. (2003). "Rapid worldwide depletion of predatory fish communities". Nature. 423 (6937): 280–283. doi:10.1038/nature01610. PMID 12748640.
14. Worm, Boris; Barbier, E. B.; Beaumont, N.; Duffy, J. E.; Folke, C.; Halpern, B. S.; Jackson, J. B. C.; Lotze, H. K.; et al. (2006-11-03). "Impacts of Biodiversity Loss on Ocean Ecosystem Services". Science. 314 (5800): 787–790. doi:10.1126/science.1132294. PMID 17082450.
15. Juliet Eilperin (2009-11-02). "Seafood Population Depleted by 2048, Study

Finds". The Washington Post.

16. FAO. Management of irrigation-induced salt-affected soils. ftp://ftp.fao.org/agl/agll/docs/salinity_brochure_eng.pdf
17. van Hoorn, J. W. and J.G. van Alphen. 2006. Salinity control. In: H.P. Ritzema (ed.), Drainage Principles and Applications. Publication 16, International Institute for Land Reclamation and Improvement (ILRI), Wageningen, The Netherlands. pp. 533–600.
18. Effectiveness and Social/Environmental Impacts of Irrigation Projects: a Review. In: Annual Report 1988, International Institute for Land Reclamation and Improvement (ILRI), Wageningen, The Netherlands, pp. 18–34 . Download from , under nr. 6, or directly as PDF
19. Thakkar, Himanshu (8 November 1999). Assessment of Irrigation in India. World Commission on Dams.
20. Pearce, R. (2006). When the rivers run dry: Water – the defining crisis of the twenty-first century, Beacon Press, ISBN 0807085731.
21. Lal, R. and B. A. Stewart. 1990. Soil degradation. Springer-Verlag, New York.
22. Scherr, S. J. 1999. Soil degradation: a threat to developing country food security by 2020? International Food Policy Research Institute. Washington, D. C.
23. Oldeman, L. R., R. T. A. Hakkeling and W. G. Sambroek. 1990. World map of the status of human-induced soil degradation. An explanatory note. GLASOD, Global Assessment of Soil Degradation. International Soil Reference and Information Centre, Wageningen. http://www.isric.org/sites/default/files/ExplanNote_1.pdfG
24. Eswaran, H., R. Lal and P. F. Reich. 2001. Land degradation: an overview. In. Bridges, E.M. et al. (eds.) Responses to Land Degradation. Proc. 2nd. Int. Conf. Land Degradation and Desertification, Khon Kaen, Thailand. Oxford Press, New Delhi, India.
25. FAOSTAT. [Agricultural statistics database] Food and Agriculture Organization of the United Nations, Rome. http://faostat3.fao.org/
26. Montgomery, D. R. (2007). "Soil erosion and agricultural sustainability". Proc. Natl. Acad. Sci. 104 (33): 13268–13272. doi:10.1073/pnas.0611508104.
27. NRCS. 2013. Summary report 2010 national resources inventory. United States Natural Resources Conservation Service. 163 pp.
28. Steinfeld, H. et al. 2006. Livestock's Long Shadow: Environmental Issues and Options. Livestock, Environment and Development, FAO, Rome. 391 pp.
29. Oppenlander, Richard (2013). Food Choice and Sustainability. Minneapolis, MN: Langdon Street Press. ISBN 978-1-62652-435-4.
30. Tubiello, F. N. et al. (2014). Agriculture, forestry and other land use emissions by sources and removal by sinks 1990–2011. FAO Statistics Division, Working Paper Series ESS/14-02.
31. Intergovernmental Panel on Climate Change. 2007. Climate change 2007, Mitigation of climate change. Fourth Assessment Report.

32. Intergovernmental Panel on Climate Change. (2014). Climate change 2014, Mitigation of climate change. Fifth Assessment Report.
33. Goodland R; Anhang J (Nov–Dec 2009). "Livestock and Climate Change: What if the key actors in climate change were pigs, chickens and cows?". World Watch Magazine. 22 (6): 10–19.
34. Oppenlander, Richard (2013). Food Choice and Sustainability. Minneapolis, MN: Langdon Street Press. pp. 17–25. ISBN 978-1-62652-435-4.
35. Intergovernmental Panel on Climate Change. (2013). Climate change 2013, The physical science basis. Fifth Assessment Report.
36. Dlugokencky, E. J., E. G. Nisbet, R. Fisher and D. Lowry (2011). "Global atmospheric methane: budget, changes and dangers". Philosophical Transactions of the Royal Society A: Mathematical, Physical and Engineering Sciences. 369 (1943): 2058. doi:10.1098/rsta.2010.0341.
37. Boadi, D (2004). "Mitigation strategies to reduce enteric methane emissions from dairy cows: Update review". Can. J. Anim. Sci. 84 (3): 319–335. doi:10.4141/a03-109.
38. Martin, C. et al. 2010. Methane mitigation in ruminants: from microbe to the farm scale. Animal 4 : 351–365.
39. Eckard, R. J. et al. 2010. Options for the abatement of methane and nitrous oxide from ruminant production: A review.. Livestock Science 130: 47–56.
40. Dalal, R.C. et al. 2003. Nitrous oxide emission from Australian agricultural lands and mitigation options: a review. Australian Journal of Soil Research 41: 165–195.
41. Klein, C. A. M.; Ledgard, S. F. (2005). "Nitrous oxide emissions from New Zealand agriculture – key sources and mitigation strategies". Nutrient Cycling in Agroecosystems. 72: 77–85. doi:10.1007/s10705-004-7357-z.
42. Holechek, J. L., R. D. Pieper and C. H. Herbel. (1995). Range management: principles and practices. 2nd Ed. Prentice-Hall.
43. Dregne, H., E., & Chou, N. T. (1992). "Global desertification dimensions and costs", pp. 73–92 in Dregne, H. E. (ed.) Degradation and restoration of arid lands. Texas Technical University.
44. Mekonnen, M. M. and Hoekstra, A. Y. (2010). The green, blue and grey water footprint of farm animals and animal products. Vol. 2: appendices. Value of Water Research Report Series No. 48. UNESCO-IHE Institute for Water Education.
45. Maupin, M. A. et al. (2014). Estimated use of water in the United States 2010. U. S. Geological Survey Circular 1405.
46. Zering, K. D., T. J. Centner, D. Meyer, G. L. Newton, J. M. Sweeten and S. Woodruff. (2012). Water and land issues associated with animal agriculture: a U.S. perspective. CAST Issue Paper No. 50. Council for Agricultural Science and Technology, Ames, Iowa.
47. US EPA. 2000. Profile of the agricultural livestock production industry. U.S. Environmental Protection Agency. Office of Compliance. EPA/310-R-00-002.

156 pp.

48. US EPA. Pasture, Rangeland, and Grazing Operations – Best Management Practices (BMPs) http://www.epa.gov/agriculture/anprgbmp.html

49. Canning, P., A. Charles, S. Huang, K. R. Polenske, and A. Waters. (2010). Energy use in the U. S. food system. USDA Economic Research Service, ERR-94.

50. Capper, J. L. (2011). "The environmental impact of beef production in the United States: 1977 compared with 2007". J. Anim. Sci. 89 (12): 4249–4261. doi:10.2527/jas.2010-3784.

51. US Department of Agriculture Red meat and poultry production.

52. Launchbaugh, K. (ed.) 2006. Targeted Grazing: a natural approach to vegetation management and landscape enhancement. American Sheep Industry. 199 pp.

53. Holechek, J. L.; et al. (1982). "Manipulation of grazing to improve or maintain wildlife habitat". Wildlife Soc. Bull. 10: 204–210.

54. Manley, J. T.; Schuman, G. E.; Reeder, J. D.; Hart, R. H. (1995). "Rangeland soil carbon and nitrogen responses to grazing". J. Soil Water Cons. 50: 294–298.

55. Franzluebbers, A.J.; Stuedemann, J. A. (2010). "Surface soil changes during twelve years of pasture management in the southern Piedmont USA". Soil Sci. Soc. Am. J. 74 (6): 2131–2141. doi:10.2136/sssaj2010.0034.

56. The other oil spill". The Economist. June 24, 2010. Retrieved August 1, 2010.

57. Kay, J. (2002). "On Complexity Theory, Exergy and Industrial Ecology: Some Implications for Construction Ecology", pp. 72–107 in: Kibert C., Sendzimir J., Guy, B. (eds.) Construction Ecology: Nature as the Basis for Green Buildings, London: Spon Press, ISBN 0203166140.

58. Baksh, B.; Fiksel J. (2003). "The Quest for Sustainability: Challenges for Process Systems Engineering"(PDF). AIChE Journal. 49 (6): 1350–1358. doi:10.1002/aic.690490602.

59. USDA-USDoE. (1998). Life cycle inventory of biodiesel and petroleum diesel in an urban bus. NREL/SR-580-24089 UC Category 1503.

60. Huo, H.; Wang, M.; Bloyd, C.; Putsche, V. (2009). "Life-cycle assessment of energy use and greenhouse gas emissions of soybean-derived biodiesel and renewable fuels". Environ. Sci. Technol. 43 (3): 750–756. doi:10.1021/es8011436.

61. Atadashi, I. M; Arou, M. K.; Aziz, A. A. (2010). "High quality biodiesel and its diesel engine application: a review". Renewable and Sustainable Energy Reviews. 14 (7): 1999–2008. doi:10.1016/j.rser.2010.03.020.

62. Environmental impacts of coal power: air pollution. Union of Concerned Scientists

63. Smith, G. (2012). Nuclear roulette: The truth about the most dangerous energy source on earth, Chelsea Green Publishing, ISBN 160358434X.

64. Bartis, Jim (2006-10-26). Unconventional Liquid Fuels Overview (PDF). World Oil Conference. Boston: Association for the Study of Peak Oil and Gas. Retrieved 2007-06-28.

65. Diesendorf, Mark (Summer 2004). "Why Australia needs wind power" (PDF). Dissent. 13: 43–48. Archived from the original (PDF) on July 6, 2011.
66. Eilperin, Juliet; Mufson, Steven (April 16, 2009). "Renewable Energy's Environmental Paradox". The Washington Post. Retrieved 2009-04-17.
67. Wind farms". Royal Society for the Protection of Birds. 2005-09-14. Retrieved 2008-09-07.
68. Lindsay, Richard (October 2004). "WIND FARMS AND BLANKET PEAT The Bog Slide of 16th October 2003 at Derrybrien, Co. Galway, Ireland" (PDF). The Derrybrien Development Cooperatve Ltd. Retrieved 20 May 2009.
69. Kyba, Christopher; Garz, Stefanie; Kuechly, Helga; de Miguel, Alejandro; Zamorano, Jaime; Fischer, Jürgen; Hölker, Franz (23 December 2014). "High-Resolution Imagery of Earth at Night: New Sources, Opportunities and Challenges". Remote Sensing. 7 (1): 1–23. doi:10.3390/rs70100001.
70. Hölker, Franz; Wolter, Christian; Perkin, Elizabeth K.; Tockner, Klement (December 2010). "Light pollution as a biodiversity threat". Trends in Ecology & Evolution. 25 (12): 681–682. doi:10.1016/j.tree.2010.09.007.
71. Miller GT (2004), Sustaining the Earth, 6th edition. Thompson Learning, Inc. Pacific Grove, California. Chapter 9, pp. 211–216, ISBN 0534400876.
72. Part 1. Conditions and provisions for developing a national strategy for biodiversity conservation. Biodiversity Conservation National Strategy and Action Plan of Republic of Uzbekistan. Prepared by the National Biodiversity Strategy Project Steering Committee with the Financial Assistance of The Global Environmental Facility (GEF) and Technical Assistance of United Nations Development Programme (UNDP, 1998). Retrieved on September 17, 2007.
73. Kellogg RL, Nehring R, Grube A, Goss DW, and Plotkin S (February 2000), Environmental indicators of pesticide leaching and runoff from farm fields. United States Department of Agriculture Natural Resources Conservation Service. Retrieved on 2007-10-03.
74. Reynolds, JD (1997), International pesticide trade: Is there any hope for the effective regulation of controlled substances? Florida State University Journal of Land Use & Environmental Law, Volume 131. Retrieved on 2007-10-16.
75. U.S. EPA. Pharmaceuticals and Personal Care Products. Accessed 16 March 2009.
76. Logging of forests and debris dumping. Ngm.nationalgeographic.com (2002-10-17). Retrieved on 2012-05-11.
77. Poisoning by mines. Ngm.nationalgeographic.com (2002-10-17). Retrieved on 2012-05-11.
78. Fuglestvedt, J.; Berntsen, T.; Myhre, G.; Rypdal, K.; Skeie, R. B. (2008). "Climate forcing from the transport sectors". Proceedings of the National Academy of Sciences. 105 (2): 454–458. doi:10.1073/pnas.0702958104.
79. Worldwatch Institute (16 January 2008). "Analysis: Nano Hypocrisy?".
80. Carbon Pathways Analysis – Informing Development of a Carbon Reduction

Strategy for the Transport Sector | Claverton Group. Claverton-energy.com (2009-02-17). Retrieved on 2012-05-11.

81. Environment Canada. "Transportation". Archived from the original on July 13, 2007. Retrieved 30 July 2008.
82. Pereira, G.; et al. (2010). "Residential exposure to traffic emissions and adverse pregnancy outcomes". S.a.p.i.en.s. 3 (1).
83. International Civil Aviation Organization, Air Transport Bureau (ATB). "Aircraft Engine Emissions". Archived from the original on June 1, 2002. Retrieved 2008-03-19.
84. What is the impact of flying?". Enviro.aero. Archived from the original on June 30, 2007. Retrieved 2008-03-19.
85. Carleton, Andrew M.; Lauritsen, Ryan G (2002). "Contrails reduce daily temperature range" (PDF). Nature. 418 (6898): 601. doi:10.1038/418601a. PMID 12167846.
86. Climate change: Commission proposes bringing air transport into EU Emissions Trading Scheme"(Press release). EU press release. 2006-12-20. Retrieved 2008-01-02.
87. Gössling S, Ceron JP, Dubois G, Hall CM, Gössling S, Upham P, Earthscan L (2009). "Hypermobile travellers", pp. 131–151 (Chapter 6) in: Climate Change and Aviation: Issues, Challenges and Solutions, London, ISBN 1844076202.
88. Including Aviation into the EU ETS: Impact on EU allowance prices. ICF Consulting for DEFRA, February 2006.
89. Vidal, John (3 March 2007) CO2 output from shipping twice as much as airlines. The Guardian. Retrieved on 2012-05-11.
90. Greenhouse gas emissions. Imo.org. Retrieved on 2012-05-11.
91. SustainableShipping: (S) News – IMO targets greenhouse gas emissions (17 Jun 2008) – The forum dedicated to marine transportation and the environment. sustainableshipping.com
92. Anthropocene: Have humans created a new geological age?". BBC News. 2011-05-10.
93. May, R.M. (1988). "How many species are there on earth?" (PDF). Science. 241 (4872): 1441–9. doi:10.1126/science.241.4872.1441. PMID 17790039.
94. Coral reefs around the world Guardian.co.uk, 2 September 2009.
95. In The Turf War Against Seaweed, Coral Reefs More Resilient Than Expected". Science Daily. June 3, 2009. Retrieved February 1, 2011.
96. Kleypas, Joan A.; Feely, Richard A.; Fabry, Victoria J.; Langdon, Chris; Sabine, Christopher L.; Robbins, Lisa L. (June 2006). "Impacts of Ocean Acidification on Coral Reefs and Other Marine Calcifiers: A Guide for Future Research" (PDF). Retrieved February 1, 2011.
97. Save Our Seas, 1997 Summer Newsletter, Dr. Cindy Hunter and Dr. Alan Friedlander

98. Tun, K.; Chou, L.M.; Cabanban, A.; Tuan, V.S.; Philreefs; Yeemin, T.; Suharsono; Sour, K.; Lane, D. (2004). C. Wilkinson, ed. "Status of Coral Reefs of the world: 2004": 235–276. Retrieved February 1, 2011.

99. Oppenlander, Richard (2013). Food Choice and Sustainability. Minneapolis, MN: Langdon Street Press. p. 31. ISBN 978-1-62652-435-4.

100. IPCC, 2013: Summary for Policymakers. In: Climate Change 2013: The Physical Science Basis. Contribution of Working Group I to the Fifth Assessment Report of the Intergovernmental Panel on Climate Change. Stocker, T.F., D. Qin, G.-K. Plattner, M. Tignor, S.K. Allen, J. Boschung, A. Nauels, Y. Xia, V. Bex and P.M. Midgley (eds.) Cambridge University Press, Cambridge, United Kingdom and New York, NY, USA. accessed 24 September 2014

101. Galloway, James N.; Aber, John D.; Erisman, JAN Willem; Seitzinger, Sybil P.; Howarth, Robert W.; Cowling, Ellis B.; Cosby, B. Jack (2003). "The Nitrogen Cascade". BioScience. 53 (4): 341. doi:10.1641/0006-3568(2003)053[0341:TNC]2.0.CO;2.

102. Galloway, J. N. (2003). "The Global Nitrogen Cycle", pp. 557–583 in H. D. Holland and K. K. Turekian (eds.) Treatise on Geochemistry. Pergamon Press, Oxford, ISBN 0080437516.

103. Alley, R. et al. (2007). IPCC Climate Change 2007: The Physical Science Basis. Contribution of Working Group I in the Third Assessment Report of Intergovernmental Panel on Climate Change. Report Summary for Policy Makers (SPM).

CHAPTER 6

Climate Change

Chapter Contents

- **Introduction**
- **Climate Change Mitigation**
- **Environmental Migrant**
- **Abrupt Climate Change**

Climate change occurs when the Earth's average temperature changes dramatically over time. As little as one or two degrees can be considered dramatic change because the Earth's ecosystem depends on a very delicate balance, and even small shifts can have a far-reaching impact. A drop in average temperature can also be considered climate change, but in modern times people using the term are usually talking about global warming.

One reason that climate change has become a popular and, at times, controversial topic is that many people believe it is mostly the result of human activity. Burning fossil fuels, such as oil and coal, produces byproducts such as carbon dioxide gas. Since there are not enough plants on Earth to quickly transform all this emitted carbon dioxide into oxygen, the gas remains in the atmosphere. Through a process known as the greenhouse effect, the carbon dioxide traps solar heat, which leads to the warming of the planet.

INTRODUCTION

Climate change is a change in the statistical distribution of weather patterns when that change lasts for an extended period of time (i.e., decades to millions of years). Climate change may refer to a change in average weather conditions, or in the time variation of weather around longer-term average conditions (i.e., more or fewer extreme weather events). Climate change is caused by factors such as biotic processes, variations in solar radiation received by Earth, plate tectonics, and volcanic eruptions. Certain human activities have also been identified as significant causes of recent climate change, often referred to as *global warming*.

Scientists actively work to understand past and future climate by using observations and theoretical models. A climate record—extending deep into the Earth's past—has been assembled, and continues to be built up, based on geological evidence from borehole temperature profiles, cores removed from deep accumulations of ice, floral and faunal records, glacial and periglacial processes, stable-isotope and other analyses of sediment layers, and records of past sea levels. More recent data are provided by the instrumental record. General circulation models, based on the physical sciences, are often used in theoretical approaches to match past climate data, make future projections, and link causes and effects in climate change.

Terminology

The most general definition of *climate change* is a change in the statistical properties (principally its mean and spread) of the climate system when considered over long periods of time, regardless of cause. Accordingly, fluctuations over periods shorter than a few decades, such as El Niño, do not represent climate change.

The term sometimes is used to refer specifically to climate change caused by human activity, as opposed to changes in climate that may have resulted as part of Earth's natural processes. In this sense, especially in the context of environmental policy, the term *climate change* has become synonymous with *anthropogenic global warming*. Within scientific journals, *global warming* refers to surface temperature increases while *climate change* includes global warming and everything else that increasing greenhouse gas levels affect.

A related term is "climatic change". In 1966, the World Meteorological Organization (WMO) proposed the term "climatic change" to encompass all forms of climatic variability on time-scales longer than 10 years, regardless of cause. Change was a given and climatic was used as an adjective to describe this kind of change (as opposed to political or economic change). When it was realized that human activities had a potential to drastically alter the climate, the term climate change replaced climatic change as the dominant term to reflect an anthropogenic cause. Climate change was incorporated in the title of the Intergovernmental Panel on Climate Change (IPCC) and the UN Framework Convention on Climate Change (UNFCCC). Climate change, used as a noun, became an issue rather than the technical description of changing weather.

Causes

On the broadest scale, the rate at which energy is received from the Sun and the rate at which it is lost to space determine the equilibrium temperature and climate of Earth. This energy is distributed around the globe by winds, ocean currents, and other mechanisms to affect the climates of different regions.

Factors that can shape climate are called climate forcings or "forcing mechanisms". These include processes such as variations in solar radiation, variations in the Earth's orbit, variations in the albedo or reflectivity of the continents and oceans, mountain-building and continental drift and changes in greenhouse gas concentrations. There are a variety of climate change feedbacks that can either amplify or diminish the initial forcing. Some parts of the climate system, such as the oceans and ice caps, respond more slowly in reaction to climate forcings, while others respond more quickly. There are also key threshold factors which when exceeded can produce rapid change.

Forcing mechanisms can be either "internal" or "external". Internal forcing mechanisms are natural processes within the climate system itself (e.g., the thermohaline circulation). External forcing mechanisms can be either natural (e.g., changes in solar output) or anthropogenic (e.g., increased emissions of greenhouse gases).

Whether the initial forcing mechanism is internal or external, the response of the climate system might be fast (e.g., a sudden cooling due to airborne volcanic ash reflecting sunlight), slow (e.g. thermal expansion of warming ocean water), or a combination (e.g., sudden loss of albedo in the arctic ocean as sea ice melts, followed by more gradual thermal expansion of the water). Therefore, the climate system can respond abruptly, but the full response to forcing mechanisms might not be fully developed for centuries or even longer.

Internal Forcing Mechanisms

Scientists generally define the five components of earth's climate system to include atmosphere, hydrosphere, cryosphere, lithosphere (restricted to the surface soils, rocks, and sediments), and biosphere. Natural changes in the climate system ("internal forcings") result in internal "climate variability". Examples include the type and distribution of species, and changes in ocean-atmosphere circulations.

Ocean-Atmosphere Variability

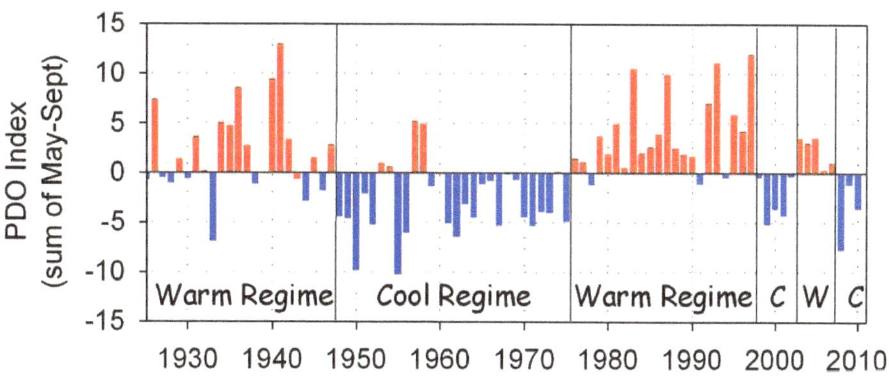

Pacific Decadal Oscillation 1925 to 2010.

The ocean and atmosphere can work together to spontaneously generate internal climate variability that can persist for years to decades at a time. Examples of this type of variability include the El Niño-Southern Oscillation, the Pacific decadal oscillation, and the Atlantic Multidecadal Oscillation. These variations can affect global average surface temperature by redistributing heat between the deep ocean and the atmopshere and/or by altering the cloud/water vapor/sea ice distribution which can affect the total energy budget of the earth.

The oceanic aspects of these circulations can generate variability on centennial timescales due to the ocean having hundreds of times more mass than in the atmosphere, and thus very high thermal inertia. For example, alterations to ocean processes such as thermohaline circulation play a key role in redistributing heat in the world's oceans. Due to the long timescales of this circulation, ocean temperature at depth is still adjusting to effects of the Little Ice Age which occurred between the 1600 and 1800s.

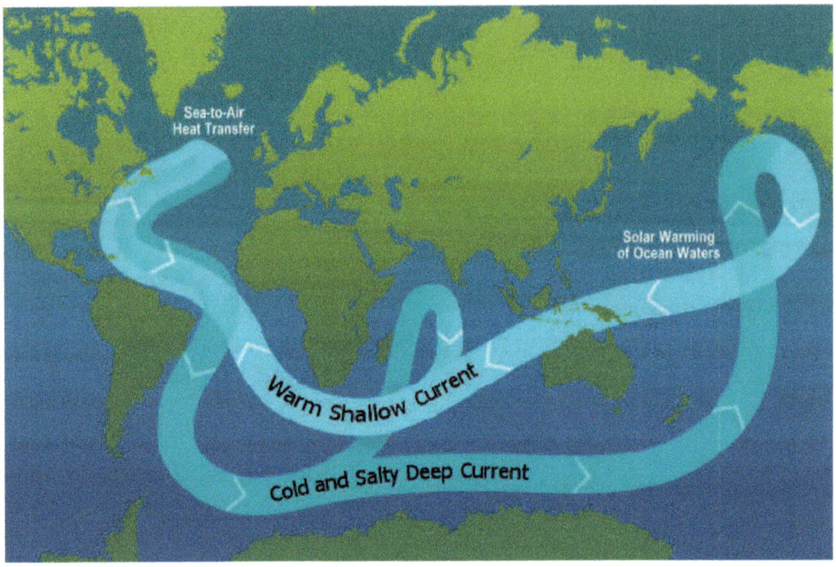

A schematic of modern thermohaline circulation. Tens of millions of years ago, continental-plate movement formed a land-free gap around Antarctica, allowing the formation of the ACC, which keeps warm waters away from Antarctica.

Life

Life affects climate through its role in the carbon and water cycles and through such mechanisms as albedo, evapotranspiration, cloud formation, and weathering.Examples of how life may have affected past climate include:

- glaciation 2.3 billion years ago triggered by the evolution of oxygenic photosynthesis, which depleted the atmosphere of the greenhouse gas carbon dioxide and introduced free oxygen.
- another glaciation 300 million years ago ushered in by long-term burial of decomposition-resistant detritus of vascular land-plants (creating a carbon sink and forming coal)
- termination of the Paleocene-Eocene Thermal Maximum 55 million years ago by flourishing marine phytoplankton

- reversal of global warming 49 million years ago by 800,000 years of arctic azolla blooms
- global cooling over the past 40 million years driven by the expansion of grass-grazer ecosystems

External Forcing Mechanisms

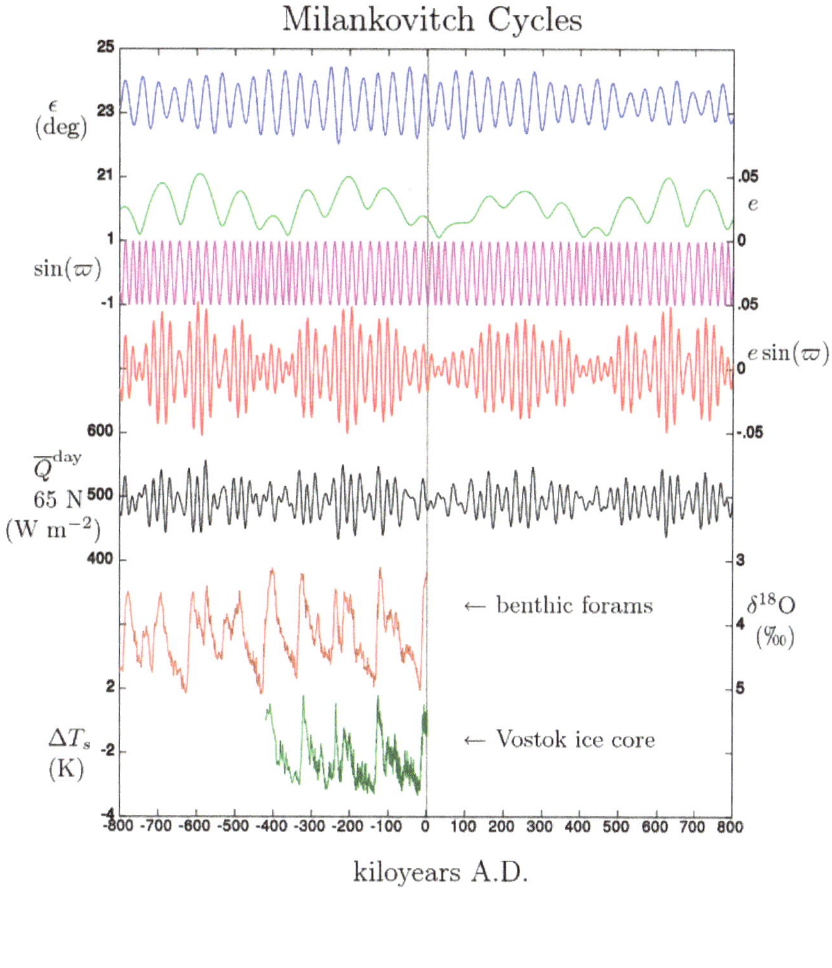

Milankovitch cycles from 800,000 years ago in the past to 800,000 years in the future.

Orbital Variations

Slight variations in Earth's orbit lead to changes in the seasonal distribution of sunlight reaching the Earth's surface and how it is distributed across the globe. There is very little change to the area-averaged annually averaged sunshine; but there can be strong changes in the geographical and seasonal distribution. The three types of orbital variations are variations in Earth's eccentricity, changes in the tilt angle of Earth's axis of rotation, and precession of Earth's axis. Combined together, these produce Milankovitch cycles which have a large impact on climate and are notable for their correlation to glacial and interglacial periods, their correlation with the advance and retreat of the Sahara, and for their appearance in the stratigraphic record.

The IPCC notes that Milankovitch cycles drove the ice age cycles, CO_2 followed temperature change "with a lag of some hundreds of years," and that as a feedback amplified temperature change. The depths of the ocean have a lag time in changing temperature (thermal inertia on such scale). Upon seawater temperature change, the solubility of CO_2 in the oceans changed, as well as other factors impacting air-sea CO_2exchange.

Solar Output

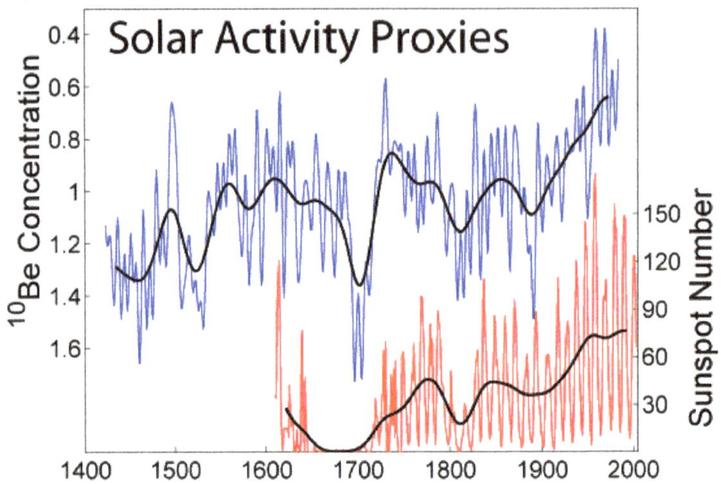

Variations in solar activity during the last several centuries based on observations of sunspots and beryllium isotopes. The period of extraordinarily few sunspots in the late 17th century was the Maunder minimum.

The Sun is the predominant source of energy input to the Earth. Other sources include geothermal energy from the Earth's core, and heat from the decay of radioactive compounds. Both long- and short-term variations in solar intensity are known to affect global climate.

Three to four billion years ago, the Sun emitted only 70% as much power as it does today. If the atmospheric composition had been the same as today, liquid water should not have existed on Earth. However, there is evidence for the presence of water on the early Earth, in the Hadean and Archean eons, leading to what is known as the faint young Sun paradox. Hypothesized solutions to this paradox include a vastly different atmosphere, with much higher concentrations of greenhouse gases than currently exist. Over the following approximately 4 billion years, the energy output of the Sun increased and atmospheric composition changed. The Great Oxygenation Event – oxygenation of the atmosphere around 2.4 billion years ago – was the most notable alteration. Over the next five billion years, the Sun's ultimate death as it becomes a red giant and then a white dwarf will have large effects on climate, with the red giant phase possibly ending any life on Earth that survives until that time.

Solar output also varies on shorter time scales, including the 11-year solar cycle and longer-term modulations. Solar intensity variations possibly as a result of the Wolf, Spörer and Maunder Minimum are considered to have been influential in triggering the Little Ice Age, and some of the warming observed from 1900 to 1950. The cyclical nature of the Sun's energy output is not yet fully understood; it differs from the very slow change that is happening within the Sun as it ages and evolves. Research indicates that solar variability has had effects including the Maunder minimum from 1645 to 1715 A.D., part of the Little Ice Age from 1550 to 1850 A.D. that was marked by relative cooling and greater glacier extent than the centuries before and afterward. Some studies point toward solar radiation increases from cyclical sunspot activity affecting global warming, and climate may be influenced by the sum of all effects (solar variation, anthropogenic radiative forcings, etc.).

Interestingly, a 2010 study *suggests*, "that the effects of solar variability on temperature throughout the atmosphere may be contrary to current expectations."

In an Aug 2011 Press Release, CERN announced the publication in the Nature journal the initial results from its CLOUD experiment. The results indicate that ionisation from cosmic rays significantly enhances aerosol formation in the presence of sulfuric acid and water, but in the lower atmosphere where ammonia is also required, this is insufficient to account for aerosol formation and additional trace vapours must be involved. The next step is to find more about these trace vapours, including whether they are of natural or human origin.

Volcanism

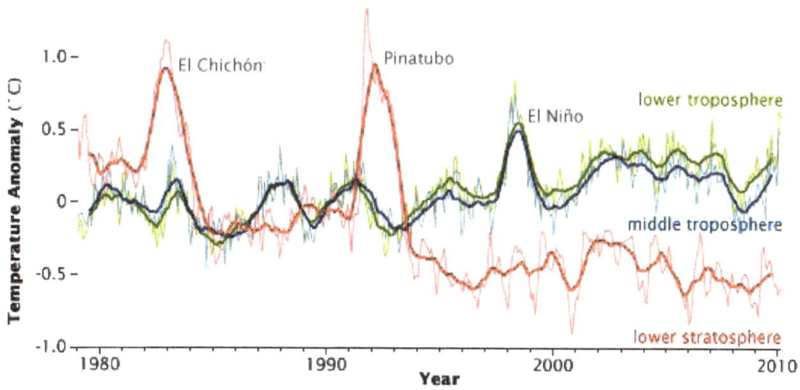

In atmospheric temperature from 1979 to 2010, determined by MSUNASA satellites, effects appear from aerosols released by major volcanic eruptions (El Chichón and Pinatubo). El Niño is a separate event, from ocean variability.

The eruptions considered to be large enough to affect the Earth's climate on a scale of more than 1 year are the ones that inject over 100,000 tons of SO_2 into the stratosphere.

This is due to the optical properties of SO_2 and sulfate aerosols, which strongly absorb or scatter solar radiation, creating a global layer of sulfuric acid haze. On average, such eruptions occur several times per century, and cause cooling (by partially blocking the transmission of solar radiation to the Earth's surface) for a period of a few years.

The eruption of Mount Pinatubo in 1991, the second largest terrestrial eruption of the 20th century, affected the climate substantially, subsequently global temperatures decreased by about 0.5 °C (0.9 °F) for up to three years. Thus, the cooling over large parts of the Earth reduced surface temperatures in 1991-93, the equivalent to a reduction in net radiation of 4 watts per square meter. The Mount Tambora eruption in 1815 caused the Year Without a Summer. Much larger eruptions, known as large igneous provinces, occur only a few times every fifty - hundred million years - through flood basalt, and caused in Earth past global warming and mass extinctions.

Small eruptions, with injections of less than 0.1 Mt of sulfur dioxide into the stratosphere, impact the atmosphere only subtly, as temperature changes are comparable with natural variability. However, because smaller eruptions occur at a much higher frequency, they too have a significant impact on Earth's atmosphere.

Seismic monitoring maps current and future trends in volcanic activities, and tries to develop early warning systems. In climate modelling the aim is to study the physical mechanisms and feedbacks of volcanic forcing.

Volcanoes are also part of the extended carbon cycle. Over very long (geological) time periods, they release carbon dioxide from the Earth's crust and mantle, counteracting the uptake by sedimentary rocks and other geological carbon dioxide sinks. The US Geological Survey estimates are that volcanic emissions are at a much lower level than the effects of current human activities, which generate 100–300 times the amount of carbon dioxide emitted by volcanoes. A review of published studies indicates that annual volcanic emissions of carbon dioxide, including amounts released from mid-ocean ridges, volcanic arcs, and hot spot volcanoes, are only the equivalent of 3 to 5 days of human caused output. The annual amount put out by human activities may be greater than the amount released by supererruptions, the most recent of which was the Toba eruption in Indonesia 74,000 years ago.

Although volcanoes are technically part of the lithosphere, which itself is part of the climate system, the IPCC explicitly defines volcanism as an external forcing agent.

Plate Tectonics

Over the course of millions of years, the motion of tectonic plates reconfigures global land and ocean areas and generates topography. This can affect both global and local patterns of climate and atmosphere-ocean circulation.

The position of the continents determines the geometry of the oceans and therefore influences patterns of ocean circulation. The locations of the seas are important in controlling the transfer of heat and moisture across the globe, and therefore, in determining global climate. A recent example of tectonic control on ocean circulation is the formation of the Isthmus of Panama about 5 million years ago, which shut off direct mixing between the Atlantic and Pacific Oceans. This strongly affected the ocean dynamics of what is now the Gulf Stream and may have led to Northern Hemisphere ice cover.During the Carboniferous period, about 300 to 360 million years ago, plate tectonics may have triggered large-scale storage of carbon and increased glaciation. Geologic evidence points to a "megamonsoonal" circulation pattern during the time of the supercontinent Pangaea, and climate modeling suggests that the existence of the supercontinent was conducive to the establishment of monsoons.

The size of continents is also important. Because of the stabilizing effect of the oceans on temperature, yearly temperature variations are generally lower in coastal areas than they are inland. A larger supercontinent will therefore have more area in which climate is strongly seasonal than will several smaller continents or islands.

Human Influences

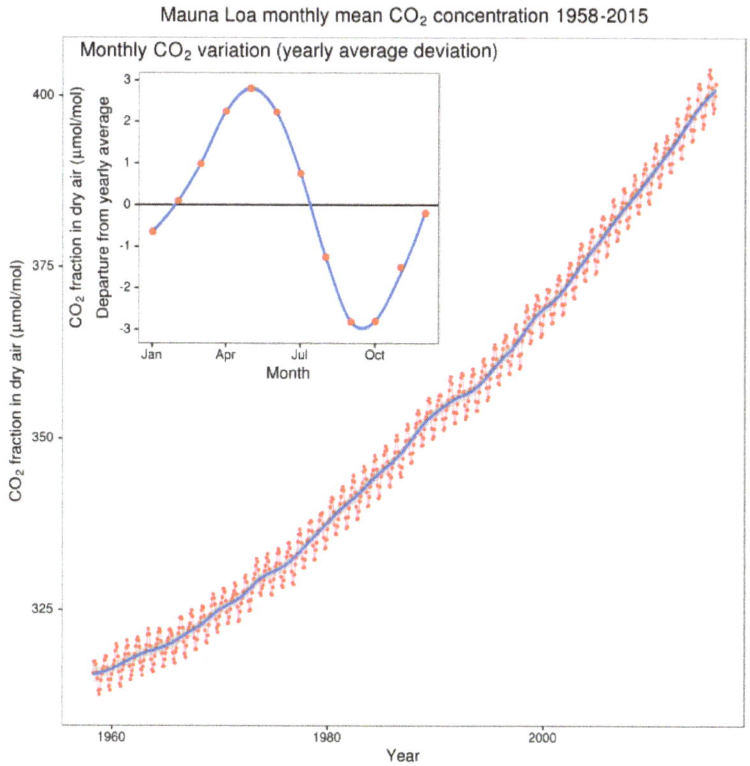

Increase in atmospheric CO_2 levels.

In the context of climate variation, anthropogenic factors are human activities which affect the climate. The scientific consensus on climate change is "that climate is changing and that these changes are in large part caused by human activities," and it "is largely irreversible."

"Science has made enormous inroads in understanding climate change and its causes, and is beginning to help develop a strong understanding of current and potential impacts that will affect people today and in coming decades. This understanding is crucial because it allows decision makers to place climate change in the context of other large challenges facing the nation and the world. There are still some uncertainties, and there always will be in understanding a complex system like Earth's climate. Nevertheless, there is a strong, credible body of evidence, based on multiple lines of research, documenting that climate is changing and that these changes are in large part caused by human activities. While much remains to be learned, the core phenomenon, scientific questions, and hypotheses have been examined thoroughly and have stood firm in the face of serious scientific debate and careful evaluation of alternative explanations."

—United States National Research Council, Advancing the Science of Climate Change

Of most concern in these anthropogenic factors is the increase in CO_2 levels due to emissions from fossil fuel combustion, followed by aerosols (particulate matter in the atmosphere) and the CO_2 released by cement manufacture. Other factors, including land use, ozone depletion, animal agriculture and deforestation, are also of concern in the roles they play – both separately and in conjunction with other factors – in affecting climate, microclimate, and measures of climate variables.

Physical Evidence

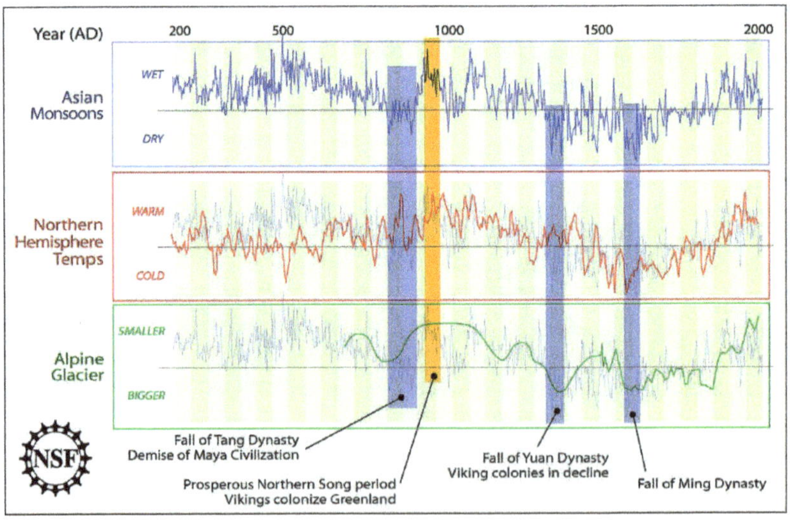

Comparisons between Asian Monsoons from 200 AD to 2000 AD (staying in the background on other plots), Northern Hemisphere temperature, Alpine glacier extent (vertically inverted as marked), and human history as noted by the U.S. NSF.

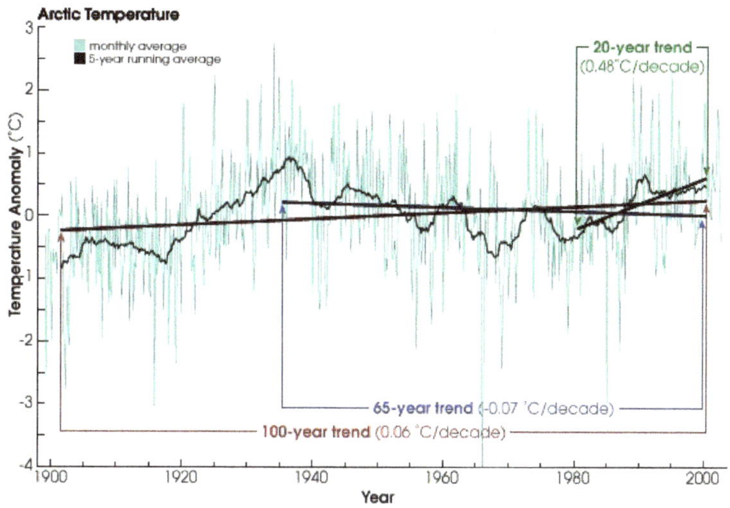

Arctic temperature anomalies over a 100-year period as estimated by NASA. Typical high monthly variance can be seen, while longer-term averages highlight trends.

Evidence for climatic change is taken from a variety of sources that can be used to reconstruct past climates. Reasonably complete global records of surface temperature are available beginning from the mid-late 19th century. For earlier periods, most of the evidence is indirect—climatic changes are inferred from changes in proxies, indicators that reflect climate, such as vegetation, ice cores, dendrochronology, sea level change, and glacial geology.

Temperature Measurements and Proxies

The instrumental temperature record from surface stations was supplemented by radiosonde balloons, extensive atmospheric monitoring by the mid-20th century, and, from the 1970s on, with global satellite data as well. The $^{18}O/^{16}O$ ratio in calcite and ice core samples used to deduce ocean temperature in the distant past is an example of a temperature proxy method, as are other climate metrics noted in subsequent categories.

Historical and Archaeological Evidence

Climate change in the recent past may be detected by corresponding changes in settlement and agricultural patterns. Archaeological evidence, oral history and historical documents can offer insights into past changes in the climate. Climate change effects have been linked to the collapse of various civilizations.

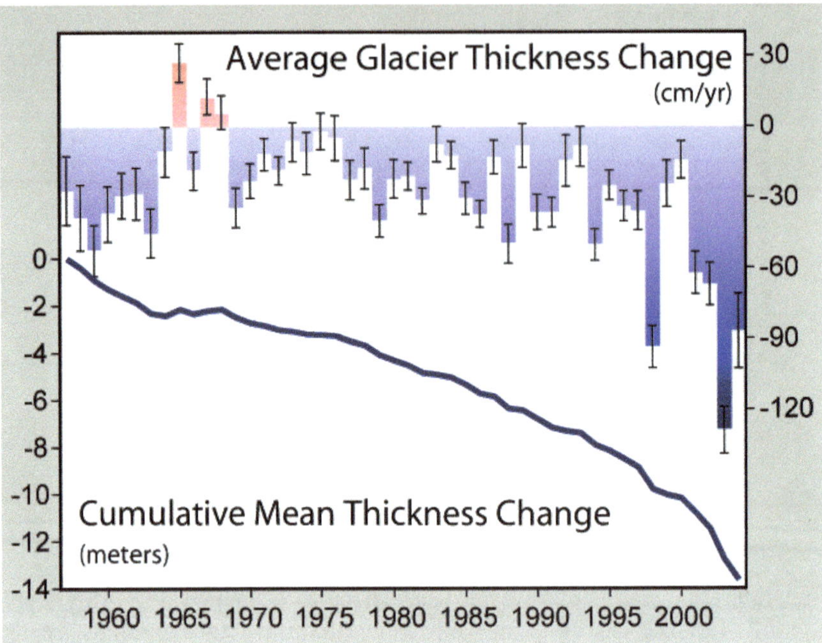

Decline in Thickness of Glaciers Worldwide over the Past Half-Century.

Glaciers

Glaciers are considered among the most sensitive indicators of climate change. Their size is determined by a mass balance between snow input and melt output. As temperatures warm, glaciers retreat unless snow precipitation increases to make up for the additional melt; the converse is also true.

Glaciers grow and shrink due both to natural variability and external forcings. Variability in temperature, precipitation, and englacial and subglacial hydrology can strongly determine the evolution of a glacier in a particular season. Therefore, one must average over a decadal or longer time-scale and/or over many individual glaciers to smooth out the local short-term variability and obtain a glacier history that is related to climate.

A world glacier inventory has been compiled since the 1970s, initially based mainly on aerial photographs and maps but now relying more on satellites. This compilation tracks more than 100,000 glaciers covering a total area of approximately 240,000 km^2, and preliminary estimates indicate that the remaining ice cover is around 445,000 km^2. The World Glacier Monitoring Service collects data annually on glacier retreat and glacier mass balance. From this data, glaciers worldwide have been found to be shrinking significantly, with strong glacier retreats in the 1940s, stable or growing conditions during the 1920s and 1970s, and again retreating from the mid-1980s to present.

The most significant climate processes since the middle to late Pliocene (approximately 3 million years ago) are the glacial and interglacial cycles. The present interglacial period (the Holocene) has lasted about 11,700 years. Shaped by orbital variations, responses such as the rise and fall of continental ice sheets and significant sea-level changes helped create the climate. Other changes, including Heinrich events, Dansgaard–Oeschger events and the Younger Dryas, however, illustrate how glacial variations may also influence climate without the orbital forcing.

Glaciers leave behind moraines that contain a wealth of material—including organic matter, quartz, and potassium that may be dated—recording the periods in which a glacier advanced and retreated. Similarly, by tephrochronological techniques, the lack of glacier cover can be identified by the presence of soil or volcanic tephra horizons whose date of deposit may also be ascertained.

Arctic Sea Ice Loss

Arctic sea ice decline and Climate change in the Arctic

The decline in Arctic sea ice, both in extent and thickness, over the last several decades is further evidence for rapid climate change. Sea ice is frozen seawater that floats on the ocean surface. It covers millions of square miles in the polar regions, varying with the seasons. In the Arctic, some sea ice remains year after year, whereas almost all Southern Ocean or Antarctic sea ice melts away and reforms annually. Satellite observations show that Arctic sea ice is now declining at a rate of 13.3 percent per decade, relative to the 1981 to 2010 average.

Vegetation

A change in the type, distribution and coverage of vegetation may occur given a change in the climate. Some changes in climate may result in increased precipitation and warmth, resulting in improved plant growth and the subsequent sequestration of airborne CO_2. A gradual increase in warmth in a region will lead to earlier flowering and fruiting times, driving a change in the timing of life cycles of dependent organisms. Conversely, cold will cause plant bio-cycles to lag. Larger, faster or more radical changes, however, may result in vegetation stress, rapid plant loss and desertification in certain circumstances. An example of this occurred during the Carboniferous Rainforest Collapse (CRC), an extinction event 300 million years ago. At this time vast rainforests covered the equatorial region of Europe and America. Climate change devastated these tropical rainforests, abruptly fragmenting the habitat into isolated ‹islands' and causing the extinction of many plant and animal species.

Pollen Analysis

Palynology is the study of contemporary and fossil palynomorphs, including pollen. Palynology is used to infer the geographical distribution of plant species, which vary under different climate conditions. Different groups of plants have pollen with distinctive shapes and surface textures, and since the outer surface of pollen is composed of a very resilient material, they resist decay. Changes in the type of pollen found in different layers of sediment in lakes, bogs, or river deltas indicate changes in plant communities. These changes are often a sign of a changing climate. As an example, palynological studies have been used to track changing vegetation patterns throughout the Quaternary glaciationsand especially since the last glacial maximum.

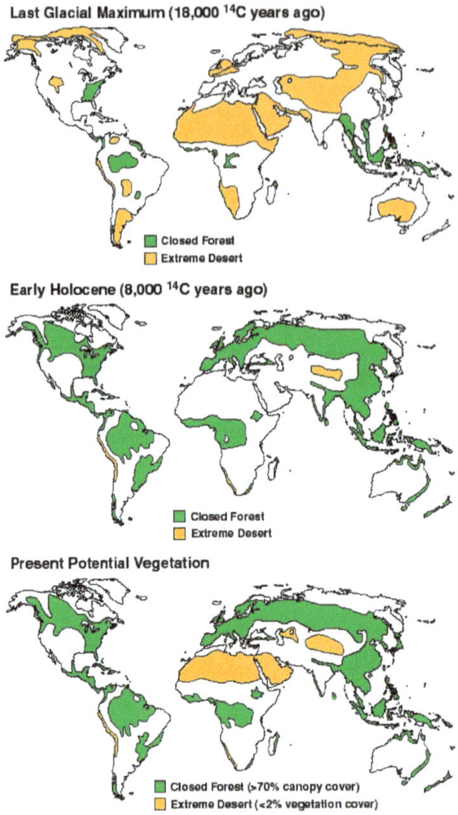

Top: Arid ice age climate

Middle: ***Atlantic Period, warm and wet***

Bottom: Potential vegetation in climate now if not for human effects like agriculture.

Cloud cover and precipitation

Past precipitation can be estimated in the modern era with the global network of precipitation gauges. Surface coverage over oceans and remote areas is relatively sparse, but, reducing reliance on interpolation, satellite clouds and precipitation data has been available since the 1970s. Quantification of climatological variation of precipitation in prior centuries and epochs is less complete but approximated using proxies such as marine sediments, ice cores, cave stalagmites, and tree rings. In July 2016 scientists published evidence of increased cloud cover over polar regions, as predicted by climate models.

Climatological temperatures substantially affect cloud cover and precipitation. For instance, during the Last Glacial Maximum of 18,000 years ago, thermal-driven evaporation from the oceans onto continental landmasses was low, causing large areas of extreme desert, including polar deserts (cold but with low rates of cloud cover and precipitation). In contrast, the world's climate was cloudier and wetter than today near the start of the warm Atlantic Period of 8000 years ago.

Estimated global land precipitation increased by approximately 2% over the course of the 20th century, though the calculated trend varies if different time endpoints are chosen, complicated by ENSO and other oscillations, including greater global land cloud cover precipitation in the 1950s and 1970s than the later 1980s and 1990s despite the positive trend over the century overall. Similar slight overall increase in global river runoff and in average soil moisture has been perceived.

Dendroclimatology

Dendroclimatology is the analysis of tree ring growth patterns to determine past climate variations.Wide and thick rings indicate a fertile, well-watered growing period, whilst thin, narrow rings indicate a time of lower rainfall and less-than-ideal growing conditions.

Ice cores

Analysis of ice in a core drilled from an ice sheet such as the Antarctic ice sheet, can be used to show a link between temperature and global sea level variations. The air trapped in bubbles in the ice can also reveal the CO_2 variations of the atmosphere from the distant past, well before modern environmental influences. The study of these ice cores has been a significant indicator of the changes in CO_2 over many millennia, and continues to provide valuable information about the differences between ancient and modern atmospheric conditions.

Animals

Remains of beetles are common in freshwater and land sediments. Different species of beetles tend to be found under different climatic conditions. Given the extensive lineage of beetles whose genetic makeup has not altered significantly over the millennia, knowledge of the present climatic range of the different species, and the age of the sediments in which remains are found, past climatic conditions may be inferred.

Similarly, the historical abundance of various fish species has been found to have a substantial relationship with observed climatic conditions. Changes in the primary productivity of autotrophs in the oceans can affect marine food webs.

Sea level Change

Global sea level change for much of the last century has generally been estimated using tide gauge measurements collated over long periods of time to give a long-term average.

More recently, altimeter measurements — in combination with accurately determined satellite orbits — have provided an improved measurement of global sea level change. To measure sea levels prior to instrumental measurements, scientists have dated coral reefs that grow near the surface of the ocean, coastal sediments, marine terraces, ooids in limestones, and nearshore archaeological remains. The predominant dating methods used are uranium series and radiocarbon, with cosmogenic radionuclides being sometimes used to date terraces that have experienced relative sea level fall. In the early Pliocene, global temperatures were 1–2˚C warmer than the present temperature, yet sea level was 15–25 meters higher than today.

CLIMATE CHANGE MITIGATION

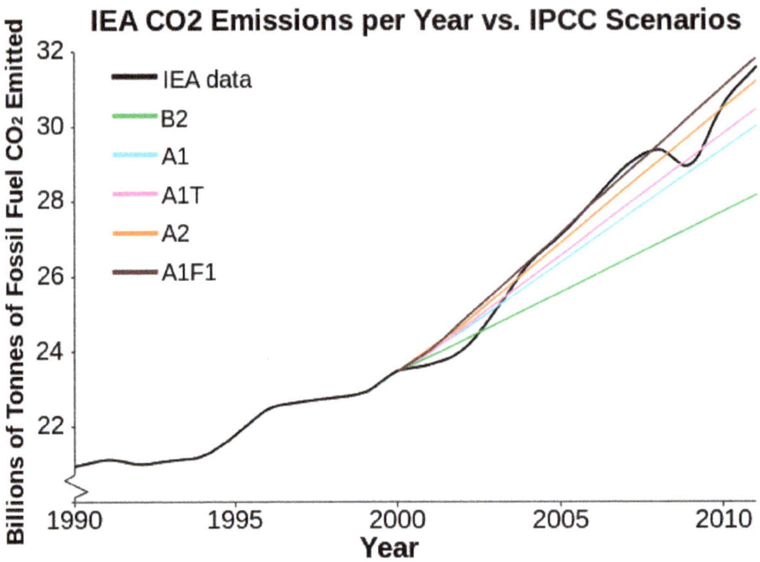

Fossil fuel related CO2 emissions compared to five of IPCC's emissions scenarios. The dips are related to global recessions. Data from IPCC SRES scenarios; Data spreadsheet included with International Energy Agency's "CO2 Emissions from Fuel Combustion 2010 – Highlights"; and Supplemental IEA data. Image source: Skeptical Science.

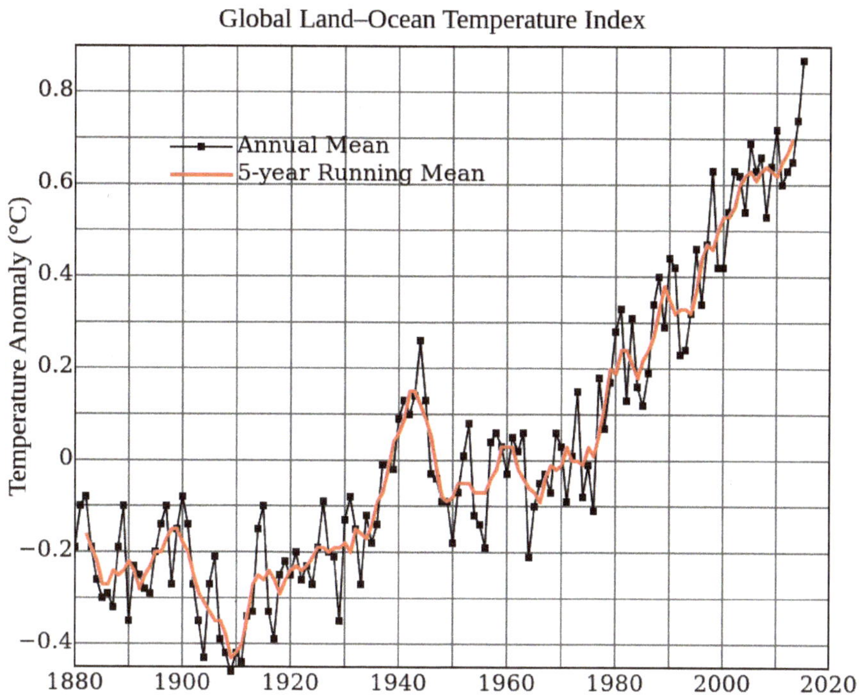

Global mean surface temperature change since 1880, relative to the 1951–1980 mean. The black line is the annual mean and the red line is the 5-year running mean. Source: NASA GISS. Global dimming, from sulfate aerosol air pollution, between 1950 and 1980 is believed to have mitigated global warming somewhat.

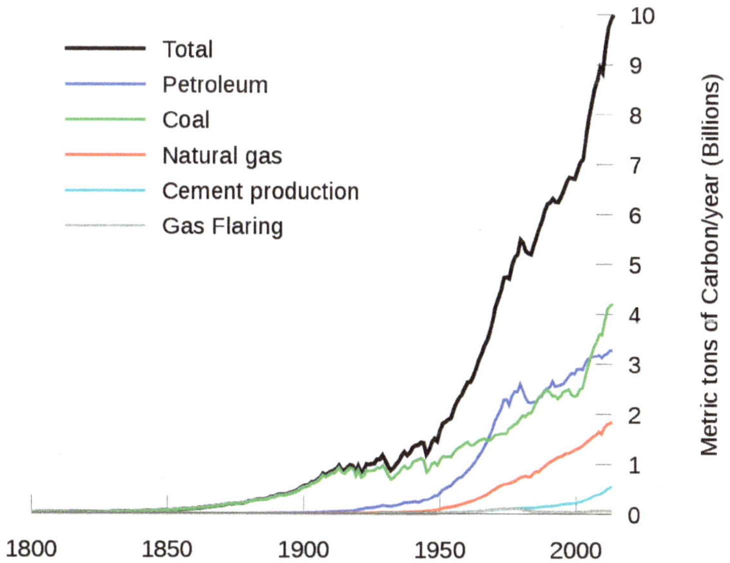

Global carbon dioxide emissions from human activities 1800–2007.

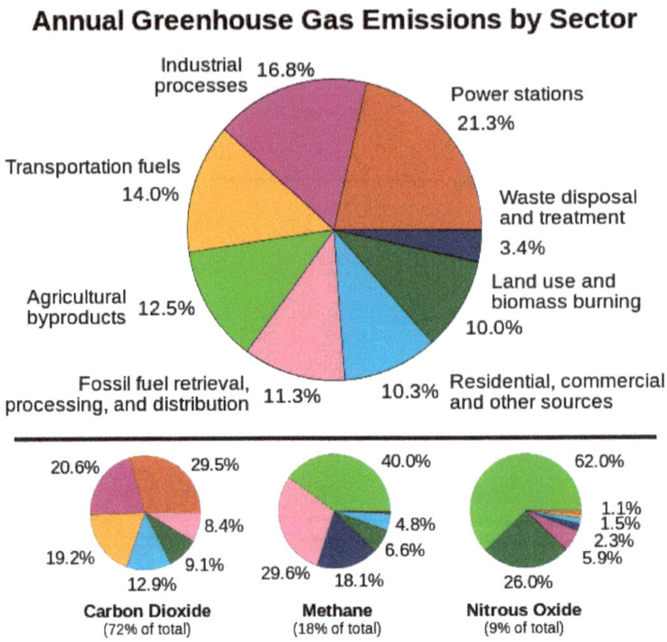

Greenhouse gas emissions by sector. See World Resources Institute for a detailed breakdown.

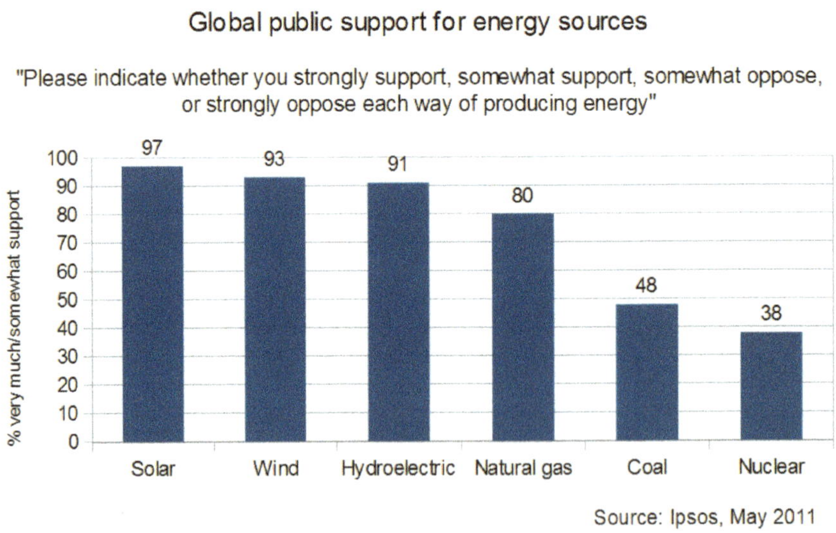

Global public support for energy sources, based on a survey by Ipsos (2011).

Climate change mitigation consists of actions to limit the magnitude or rate of long-term climate change. Climate change mitigation generally involves reductions in human (anthropogenic) emissions of greenhouse gases (GHGs).Mitigation may also be achieved by increasing the capacity of carbon sinks, e.g., through reforestation. Mitigation policies can substantially reduce the risks associated with human-induced global warming.

According to the IPCC's 2014 assessment report, "Mitigation is a public good; climate change is a case of the ‹tragedy of the commons. Effective climate change mitigation will not be achieved if each agent (individual, institution or country) acts independently in its own selfish interest, (See International Cooperation and Emissions Trading) suggesting the need for collective action. Some adaptation actions, on the other hand, have characteristics of a private good as benefits of actions may accrue more directly to the individuals, regions, or countries that undertake them, at least in the short term. Nevertheless, financing such adaptive activities remains an issue, particularly for poor individuals and countries."

Examples of mitigation include phasing out fossil fuels by switching to low-carbon energy sources, such as renewable and nuclear energy, and expanding forests and other "sinks" to remove greater amounts of carbon dioxide from the atmosphere. Energy efficiency may also play a role, for example, through improving the insulation of buildings.Another approach to climate change mitigation is climate engineering.

Most countries are parties to the United Nations Framework Convention on Climate Change (UNFCCC). The ultimate objective of the UNFCCC is to stabilize atmospheric concentrations of GHGs at a level that would prevent dangerous human interference of the climate system. Scientific analysis can provide information on the impacts of climate change, but deciding which impacts are dangerous requires value judgments.

In 2010, Parties to the UNFCCC agreed that future global warming should be limited to below 2.0 °C (3.6 °F) relative to the pre-industrial level. With the Paris Agreement of 2015 this was confirmed, but was revised with a new target laying down "parties will do the best" to achieve warming below 1.5 °C.The current trajectory of global greenhouse gas emissions does not appear to be consistent with limiting global warming to below 1.5 or 2 °C. Other mitigation policies have been proposed, some of which are more stringent or modest than the 2 °C limit.

Greenhouse gas concentrations and Stabilization

Stabilizing CO_2 emissions at their present level would not stabilize its concentration in the atmosphere.

Stabilizing the atmospheric concentration of CO_2 at a constant level would require emissions to be effectively eliminated.

One of the issues often discussed in relation to climate change mitigation is the stabilization of greenhouse gas concentrations in the atmosphere. The United Nations Framework Convention on Climate Change (UNFCCC) has the ultimate objective of preventing "dangerous" anthropogenic (i.e., human) interference of the climate system. As is stated in Article 2 of the Convention, this requires that greenhouse gas (GHG) concentrations are stabilized in the atmosphere at a level where ecosystems can adapt naturally to climate change, food production is not threatened, and economic development can proceed in a sustainable fashion.

There are a number of anthropogenic greenhouse gases. These include carbon dioxide (chemical formula: CO_2), methane (CH_4), nitrous oxide (N_2O), and a group of gases referred to as halocarbons. The emissions reductions necessary to stabilize the atmospheric concentrations of these gases varies. CO_2 is the most important of the anthropogenic greenhouse gases (see radiative forcing).

There is a difference between stabilizing CO_2 emissions and stabilizing atmospheric concentrations of CO_2. Stabilizing emissions of CO_2 at current levels would not lead to a stabilization in the atmospheric concentration of CO_2. In fact, stabilizing emissions at current levels would result in the atmospheric concentration of CO_2 continuing to rise over the 21st century and beyond (see the graphs opposite).

The reason for this is that human activities are adding CO_2 to the atmosphere far faster than natural processes can remove it (see carbon dioxide in Earth's atmosphere for a complete explanation). This is analogous to a flow of water into a bathtub. So long as the tap runs water (analogous to the emission of carbon dioxide) into the tub faster than water escapes through the plughole (the natural removal of carbon dioxide from the atmosphere), then the level of water in the tub (analogous to the concentration of carbon dioxide in the atmosphere) will continue to rise.

According to some studies, stabilizing atmospheric CO_2 concentrations would require anthropogenic CO_2 emissions to be reduced by 80% relative to the peak emissions level. An 80% reduction in emissions would stabilize CO_2 concentrations for around a century, but even greater reductions would be required beyond this. Other research has found that, after leaving room for emissions for food production for 9 billion people and to keep the global temperature rise below 2 °C, emissions from energy production and transport will have to peak almost immediately in the developed world and decline at ca. 10% per annum until zero emissions are reached around 2030. In developing countries energy and transport emissions would have to peak by 2025 and then decline similarly.

Stabilizing the atmospheric concentration of the other greenhouse gasses humans emit also depends on how fast their emissions are added to the atmosphere, and how fast the GHGs are removed. Stabilization for these gases is described in the later section on non-CO_2 GHGs.

Projections

Projections of future greenhouse gas emissions are highly uncertain. In the absence of policies to mitigate climate change, GHG emissions could rise significantly over the 21st century.

Numerous assessments have considered how atmospheric GHG concentrations could be stabilized.The lower the desired stabilization level, the sooner global GHG emissions must peak and decline.GHG concentrations are unlikely to stabilize this century without major policy changes.

Energy consumption by power source

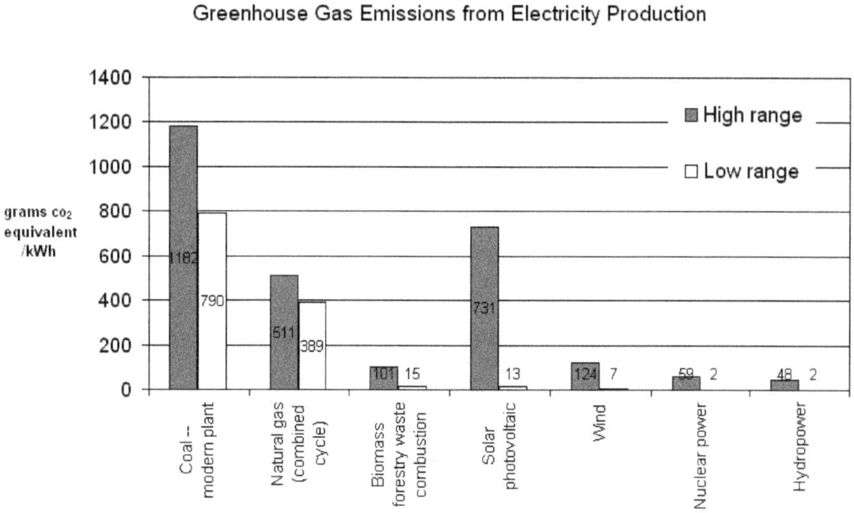

Source: IEA 2000

"Hydropower-Internalised Costs and Externalised Benefits"; Frans H. Koch; International Energy Agency (IEA)-Implementing Agreement for Hydropower Technologies and Programmes; 2000.

To create lasting climate change mitigation, the replacement of high carbon emission intensity power sources, such as conventional fossil fuels—oil, coal and natural gas—with low-carbon power sources is required. Fossil fuels supply humanity with the vast majority of our energy demands, and at a growing rate. In 2012 the IEA noted that coal accounted for half the increased energy use of the prior decade, growing faster than all renewable energy sources. Both hydroelectricity and nuclear power together provide the majority of the generated low-carbon power fraction of global total power consumption.

Fuel type	**Average total global power consumption in TW**		
	1980	**2004**	**2006**
Oil	4.38	5.58	5.74
Gas	1.80	3.45	3.61
Coal	2.34	3.87	4.27
Hydroelectric	0.60	0.93	1.00
Nuclear power	0.25	0.91	0.93
Geothermal, wind, solar energy, wood	0.02	0.13	0.16
Total	**9.48**	**15.0**	**15.8**
Source: The USA Energy Information Administration			

Change and use of energy, by source, in units of (PWh) in that year.				
	Fossil	**Nuclear**	**All renewables**	**Total**
1990	83.374	6.113	13.082	102.569
2000	94.493	7.857	15.337	117.687
2008	117.076	8.283	18.492	143.851
Change 2000–2008	22.583	0.426	3.155	26.164

Methods and Means

This graph shows the projected contribution of various energy sources to world primary electricity consumption (PEC). It is based on a climate change mitigation scenario, in which GHG emissions are substantially reduced over the 21st century. In the scenario, emission reductions are achieved using a portfolio of energy sources, as well as reductions in energy demand.

Assessments often suggest that GHG emissions can be reduced using a portfolio of low-carbon technologies. At the core of most proposals is the reduction of greenhouse gas (GHG) emissions through reducing energy waste and switching to low-carbon power sources of energy. As the cost of reducing GHG emissions in the electricity sector appears to be lower than in other sectors, such as in the transportation sector, the electricity sector may deliver the largest proportional carbon reductions under an economically efficient climate policy.

"Economic tools can be useful in designing climate change mitigation policies." "While the limitations of economics and social welfare analysis, including cost–benefit analysis, are widely documented, economics nevertheless provides useful tools for assessing the pros and cons of taking, or not taking, action on climate change mitigation, as well as of adaptation measures, in achieving competing societal goals. Understanding these pros and cons can help in making policy decisions on climate change mitigation and can influence the actions taken by countries, institutions and individuals."

Other frequently discussed means include energy conservation, increasing fuel economy in automobiles (which includes the use of electric hybrids), charging plug-in hybrids and electric cars by low-carbon electricity, making individual-lifestyle changes (e.g., cycling instead of driving), and changing business practices. Many fossil fuel driven vehicles can be converted to use electricity, the U.S. has an estimated capacity of supporting 73% light duty vehicles (LDV). In terms of transportation, the net result would be a 27% total reduction in emissions of the greenhouse gases carbon dioxide, methane, and nitrous oxide, a 31% total reduction in nitrogen oxides, a slight reduction in nitrous oxide emissions, an increase in particulate matter emissions, the same sulfur dioxide emissions, and the near elimination of carbon monoxide and volatile organic compound emissions (a 98% decrease in carbon monoxide and a 93% decrease in volatile organic compounds). The emissions would be displaced away from street level, where they have "high human-health implications."

A range of energy technologies may contribute to climate change mitigation. These include nuclear power and renewable energy sources such as biomass, hydroelectricity, wind power, solar power, geothermal power, ocean energy, and; the use of carbon sinks, and carbon capture and storage. For example, Pacala and Socolow of Princeton have proposed a 15 part program to reduce CO_2emissions by 1 billion metric tons per year – or 25 billion tons over the 50-year period using today's technologies as a type of Global warming game.

Another consideration is how future socio-economic development proceeds. Development choices (or "pathways") can lead differences in GHG emissions. Political and social attitudes may affect how easy or difficult it is to implement effective policies to reduce emissions.

Alternative Energy Sources

Renewable Energy

The 22,500 MW nameplate capacityThree Gorges Dam in the Peoples Republic of China, the largest hydroelectric power station in the world.

The Shepherds Flat Wind Farm is an 845 megawatt (MW) nameplate capacity, wind farm in the U.S. state of Oregon, each turbine is a nameplate 2 or 2.5 MW electricity generator.

The 150 MW Andasol solar power station is a commercial parabolic trough solar thermal power plant, located in Spain. The Andasol plant uses tanks of molten salt to store solar energy so that it can continue generating electricity for 7.5 hours after the sun has stopped shining.

Renewable energy flows involve natural phenomena such as sunlight, wind, rain, tides, plant growth, and geothermal heat, as the International Energy Agency explains:

Renewable energy is derived from natural processes that are replenished constantly. In its various forms, it derives directly from the sun, or from heat generated deep within the earth. Included in the definition is electricity and heat generated from solar, wind, ocean, hydropower, biomass, geothermal resources, and biofuels and hydrogen derived from renewable resources.

Climate change concerns and the need to reduce carbon emissions are driving increasing growth in the renewable energy industries. Low-carbon renewable energy replaces conventional fossil fuels in three main areas: power generation, hot water/ space heating, and transport fuels. In 2011, the share of renewables in electricity generation worldwide grew for the fourth year in a row to 20.2%. Based on REN21's 2014 report, renewables contributed 19% to supply global energy consumption. This energy consumption is divided as 9% coming from burning biomass, 4.2% as heat energy (non-biomass), 3.8% hydro electricity and 2% as electricity from wind, solar, geothermal, and biomass thermal power plants.

Renewable energy use has grown much faster than anyone anticipated. The Intergovernmental Panel on Climate Change (IPCC) has said that there are few fundamental technological limits to integrating a portfolio of renewable energy technologies to meet most of total global energy demand. At the national level, at least 30 nations around the world already have renewable energy contributing more than 20% of energy supply.

As of 2012, renewable energy accounts for almost half of new electricity capacity installed and costs are continuing to fall.Public policy and political leadership helps to "level the playing field" and drive the wider acceptance of renewable energy technologies. As of 2011, 118 countries have targets for their own renewable energy futures, and have enacted wide-ranging public policies to promote renewables. Leading renewable energy companies include BrightSource Energy, First Solar, Gamesa, GE Energy, Goldwind, Sinovel, Suntech, Trina Solar, Vestas and Yingli.

The incentive to use 100% renewable energy has been created by global warming and other ecological as well as economic concerns. Mark Z. Jacobson says producing all new energy with wind power, solar power, and hydropower by 2030 is feasible and existing energy supply arrangements could be replaced by 2050. Barriers to implementing the renewable energy plan are seen to be "primarily social and political, not technological or economic". Jacobson says that energy costs with a wind, solar, water system should be similar to today's energy costs. According to a 2011 projection by the (IEA)International Energy Agency, solar power generators may produce most of the world's electricity within 50 years, dramatically reducing harmful greenhouse gas emissions. Critics of the "100% renewable energy" approach include Vaclav Smil and James E. Hansen. Smil and Hansen are concerned about the variable output of solar and wind power, NIMBYism, and a lack of infrastructure.

Economic analysts expect market gains for renewable energy (and efficient energy use) following the 2011 Japanese nuclear accidents. In his 2012 State of the Union address, President Barack Obama restated his commitment to renewable energy and mentioned the long-standing Interior Department commitment to permit 10,000 MW of renewable energy projects on public land in 2012.Globally, there are an estimated 3 million direct jobs in renewable energy industries, with about half of them in the biofuels industry.

Some countries, with favorable geography, geology and weather well suited to an economical exploitation of renewable energy sources, already get most of their electricity from renewables, including from geothermal energy in Iceland (100 percent), and Hydroelectric power in Brazil (85 percent), Austria (62 percent), New Zealand (65 percent), and Sweden (54 percent). Renewable power generators are spread across many countries, with wind power providing a significant share of electricity in some regional areas: for example, 14 percent in the U.S. state of Iowa, 40 percent in the northern German state of Schleswig-Holstein, and 20 percent in Denmark. Solar water heating makes an important and growing contribution in many countries, most notably in China, which now has 70 percent of the global total (180 GWth). Worldwide, total installed solar water heating systems meet a portion of the water heating needs of over 70 million households. The use of biomass for heating continues to grow as well. In Sweden, national use of biomass energy has surpassed that of oil. Direct geothermal heating is also growing rapidly. Renewable biofuels for transportation, such as ethanol fuel and biodiesel, have contributed to a significant decline in oil consumption in the United States since 2006. The 93 billion liters of biofuels produced worldwide in 2009 displaced the equivalent of an estimated 68 billion liters of gasoline, equal to about 5 percent of world gasoline production.

Some of the world's largest solar power stations: Ivanpah (CSP) and Topaz (PV)

Nuclear Power

Blue Cherenkov light being produced near the core of the Fission powered Advanced Test Reactor.

Since about 2001 the term "nuclear renaissance" has been used to refer to a possible nuclear power industry revival, driven by rising fossil fuel prices and new concerns about meeting greenhouse gas emission limits. However, in March 2011 the Fukushima nuclear disaster in Japan and associated shutdowns at other nuclear facilities raised questions among some commentators over the future of nuclear power. Platts has reported that "the crisis at Japan's Fukushima nuclear plants has prompted leading energy-consuming countries to review the safety of their existing reactors and cast doubt on the speed and scale of planned expansions around the world".

The World Nuclear Association has reported that nuclear electricity generation in 2012 was at its lowest level since 1999. Several previous international studies and assessments,suggested that as part of the portfolio of other low-carbon energy technologies, nuclear power will continue to play a role in reducing greenhouse gas emissions. Historically, nuclear power usage is estimated to have prevented the atmospheric emission of 64 gigatonnes of CO2-equivalent as of 2013. Public concerns about nuclear power include the fate of spent nuclear fuel, nuclear accidents, security risks, nuclear proliferation, and a concern that nuclear power plants are very expensive. Of these concerns, nuclear accidents and disposal of long-lived radioactive fuel/"waste" have probably had the greatest public impact worldwide. Although generally unaware of it, both of these glaring public concerns are greatly diminished by present passive safety designs, the experimentally proven, "melt-down proof" EBR-II, future molten salt reactors, and the use of conventional and more advanced fuel/"waste" pyroprocessing, with the latter recycling or reprocessing not presently being commonplace as it is often considered to be cheaper to use a once-through nuclear fuel cycle in many countries, depending on the varying levels of intrinsic value given by a society in reducing the long-lived waste in their country, with France doing a considerable amount of reprocessing when compared to the US.

Nuclear power, with a 10.6% share of world electricity production as of 2013, is second only to hydroelectricity as the largest source of low-carbon power. Over 400 reactors generate electricity in 31 countries.

A Yale University review published in the Journal of Industrial Ecology analyzing CO_2 life cycle assessment(LCA) emissions from nuclear power(Light water reactors) determined that: "The collective LCA literature indicates that life cycle GHG emissions from nuclear power are only a fraction of traditional fossil sources and comparable to renewable technologies." While some have raised uncertainty surrounding the future GHG emissions of nuclear power as a result of an extreme potential decline in uranium ore grade without a corresponding increase in the efficiency of enrichment methods. In a scenario analysis of future global nuclear development, as it could be effected by a decreasing global uranium market of average ore grade, the analysis determined that depending on conditions, median life cycle nuclear power GHG emissions could be between 9 and 110 g CO_2-eq/kWh by 2050, with the latter high figure being derived from a "worst-case scenario" that is not "considered very robust" by the authors of the paper, as the "ore grade" in the scenario is lower than the uranium concentration in many lignite coal ashes.

Although this future analyses primarily deals with extrapolations for present Generation II reactor technology, the same paper also summarizes the literature on "FBRs"/Fast Breeder Reactors, of which two are in operation as of 2014 with the newest being the BN-800, for these reactors it states that the "median life cycle GHG emissions ... [are] similar to or lower than [present light water reactors] LWRs and purports to consume little or no uranium ore.

In their 2014 report, the IPCC comparison of energy sources global warming potential per unit of electricity generated, which notably included albedo effects, mirror the median emission value derived from the Warner and Heath Yale meta-analysis for the more common non-breeding Light water reactors, a CO2-equivalent value of 12 g CO2-eq/kWh, which is the lowest global warming forcing of all baseload power sources, with comparable low carbon power baseload sources, such as hydropower and biomass, producing substantially more global warming forcing 24 and 230 g CO2-eq/kWh respectively.

In 2014, Brookings Institution published *The Net Benefits of Low and No-Carbon Electricity Technologies* which states, after performing an energy and emissions cost analysis, that "The net benefits of new nuclear, hydro, and natural gas combined cycle plants far outweigh the net benefits of new wind or solar plants", with the most cost effective low carbon power technology being determined to be nuclear power.

During his presidential campaign, Barack Obama stated, "Nuclear power represents more than 70% of our noncarbon generated electricity. It is unlikely that we can meet our aggressive climate goals if we eliminate nuclear power as an option."

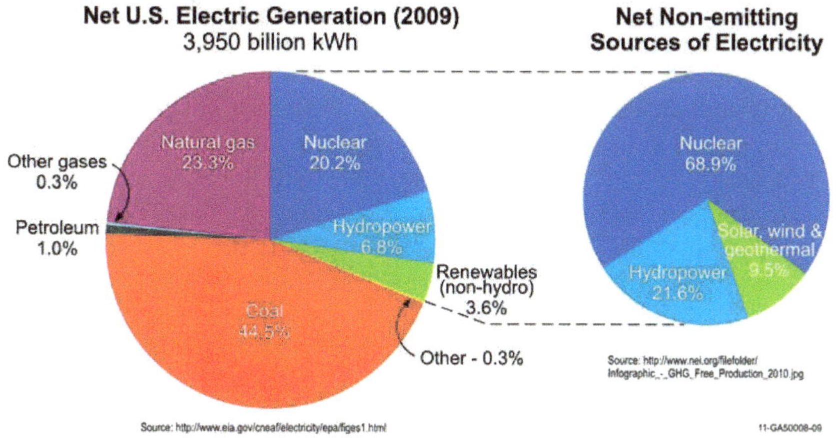

This graph illustrates nuclear power is the United States's largest contributor of non-greenhouse-gas-emitting electric power generation, comprising nearly three-quarters of the non-emitting sources.

Analysis in 2015 by Professor and Chair of Environmental Sustainability Barry W. Brook and his colleagues on the topic of replacing fossil fuels entirely, from the electric grid of the world, has determined that at the historically modest and proven-rate at which nuclear energy was added to and replaced fossil fuels in France and Sweden during each nation's building programs in the 1980s, within 10 years nuclear energy could displace or remove fossil fuels from the electric grid completely, "allow[ing] the world to meet the most stringent greenhouse-gas mitigation targets.". In a similar analysis, Brook had earlier determined that 50% of all global energy, that is not solely electricity, but transportation synfuels etc. could be generated within approximately 30 years, if the global nuclear fission build rate was identical to each of these nation's already proven decadal rates(in units of installed nameplate capacity, GW per year, per unit of global GDP(GW/year/$).

This is in contrast to the completely conceptual paper-studies for a *100% renewable energy* world, which would require an orders of magnitude more costly global investment per year, an investment rate that has no historical precedent, having never been attempted due to its prohibitive cost, and with far greater land area that would be required to be devoted to the wind, wave and solar projects, along with the inherent assumption that humanity will use less, and not more, energy in the future. As Brook notes the "principal limitations on nuclear fission are not technical, economic or fuel-related, but are instead linked to complex issues of societal acceptance, fiscal and political inertia, and inadequate critical evaluation of the real-world constraints facing [the other] low-carbon alternatives."

Nuclear power may be uncompetitive compared with fossil fuel energy sources in countries without a carbon tax program, and in comparison to a fossil fuel plant of the same power output, nuclear power plants take a longer amount of time to construct.

Two new, first of their kind, EPR reactors under construction in Finland and France have been delayed and are running over-budget. However learning from experience, two further EPR reactors under construction in China are on, and ahead, of schedule respectively. As of 2013, according to the IAEA and the European Nuclear Society, worldwide there were 68 civil nuclear power reactors under construction in 15 countries. China has 29 of these nuclear power reactors under construction, as of 2013, with plans to build many more, while in the US the licenses of almost half its reactors have been extended to 60 years, and plans to build another dozen are under serious consideration. There are also a considerable number of new reactors being built in South Korea, India, and Russia. At least 100 older and smaller reactors will "most probably be closed over the next 10–15 years". This is probable only if one does not factor in the ongoing Light Water Reactor Sustainability Program, created to permit the extension of the life span of the USA's 104 nuclear reactors to 60 years. The licenses of almost half of the USA's reactors have been extended to 60 years as of 2008. Two new "passive safety" AP1000 reactors are, as of 2013, being constructed at Vogtle Electric Generating Plant.

Public opinion about nuclear power varies widely between countries. A poll by Gallup International (2011) assessed public opinion in 47 countries. The poll was conducted following a tsunami and earthquake which caused an accident at the Fukushima nuclear power plant in Japan. 49% stated that they held favourable views about nuclear energy, while 43% held an unfavourable view.Another global survey by Ipsos (2011) assessed public opinion in 24 countries. Respondents to this survey showed a clear preference for renewable energy sources over coal and nuclear energy (refer to graph opposite). Ipsos (2012) found that solar and wind were viewed by the public as being more environmentally friendly and more viable long-term energy sources relative to nuclear power and natural gas. However, solar and wind were viewed as being less reliable relative to nuclear power and natural gas. In 2012 a poll done in the UK found that 63% of those surveyed support nuclear power, and with opposition to nuclear power at 11%. In Germany, strong anti-nuclear sentiment led to eight of the seventeen operating reactors being permanently shut down following the March 2011 Fukushima nuclear disaster.

Nuclear fusion research, in the form of the International Thermonuclear Experimental Reactor is underway. Fusion powered electricity generation was initially believed to be readily achievable, as fission power had been. However, the extreme requirements for continuous reactions and plasma containment led to projections being extended by several decades. In 2010, more than 60 years after the first attempts, commercial power production was still believed to be unlikely before 2050. Although rather than an either, or, issue economical fusion-fission hybrid reactors could be built before any attempt at this more demanding commercial "pure-fusion reactor"/DEMO reactor takes place.

Coal to Gas Fuel Switching

Most mitigation proposals imply—rather than directly state—an eventual reduction in global fossil fuel production. Also proposed are direct quotas on global fossil fuel production.

Natural gas emits far fewer greenhouse gases (i.e. CO_2 and methane—CH_4) than coal when burned at power plants, but evidence has been emerging that this benefit could be completely negated by methane leakage at gas drilling fields and other points in the supply chain. A study performed by the Environmental Protection Agency (EPA) and the Gas Research Institute (GRI) in 1997 sought to discover whether the reduction in carbon dioxide emissions from increased natural gas (predominantly methane) use would be offset by a possible increased level of methane emissions from sources such as leaks and emissions. The study concluded that the reduction in emissions from increased natural gas use outweighs the detrimental effects of increased methane emissions. More recent peer-reviewed studies have challenged the findings of this study, with researchers from the National Oceanic and Atmospheric Administration (NOAA) reconfirming findings of high rates of methane (CH4) leakage from natural gas fields.

A 2011 study by noted climate research scientist, Tom Wigley, found that while carbon dioxide (CO_2) emissions from fossil fuel combustion may be reduced by using natural gas rather than coal to produce energy, it also found that additional methane (CH4) from leakage adds to the radiative forcing of the climate system, offsetting the reduction in CO_2 forcing that accompanies the transition from coal to gas. The study looked at methane leakage from coal mining; changes in radiative forcing due to changes in the emissions of sulfur dioxide and carbonaceous aerosols; and differences in the efficiency of electricity production between coal- and gas-fired power generation. On balance, these factors more than offset the reduction in warming due to reduced CO_2 emissions. When gas replaces coal there is additional warming out to 2,050 with an assumed leakage rate of 0%, and out to 2,140 if the leakage rate is as high as 10%. The overall effects on global-mean temperature over the 21st century, however, are small. Petron et al. (2013) and Alvarez et al. (2012) note that estimated that leakage from gas infrastructure is likely to be underestimated. These studies indicate that the exploitation of natural gas as a "cleaner" fuel is questionable. A 2014 meta-study of 20 years of natural gas technical literature shows that methane emissions are consistently underestimated but on a 100-year scale, the climate benefits of coal to gas fuel switching are likely larger than the negative effects of natural gas leakage.

Heat Pump

A heat pump is a device that provides heat energy from a source of heat to a destination called a "heat sink". Heat pumps are designed to move thermal energy opposite to the direction of spontaneous heat flow by absorbing heat from a cold space and releasing it to a warmer one. A heat pump uses some amount of external power to accomplish the work of transferring energy from the heat source to the heat sink.

While air conditioners and freezers are familiar examples of heat pumps, the term "heat pump" is more general and applies to many HVAC (heating, ventilating, and air conditioning) devices used for space heating or space cooling. When a heat pump is used for heating, it employs the same basic refrigeration-type cycle used by an air conditioner or a refrigerator, but in the opposite direction—releasing heat into the conditioned space rather than the surrounding environment. In this use, heat pumps generally draw heat from the cooler external air or from the ground. In heating mode, heat pumps are three to four times more efficient in their use of electric power than simple electrical resistance heaters.

Outside unit of an air-source heat pump.

It has been concluded that heat pumps are the single technology that could reduce the greenhouse gas emissions of households better than every other technology that is available on the market. With a market share of 30% and (potentially) clean electricity, heat pumps could reduce global CO2 emissions by 8% annually. Using ground source heat pumps could reduce around 60% of the primary energy demand and 90% of CO2 emissions in Europe in 2050 and make handling high shares of renewable energy easier. Using surplus renewable energy in heat pumps is regarded as the most effective household means to reduce global warming and fossil fuel depletion.

With significant amounts of fossil fuel used in electricity production, demands on the electrical grid also generate greenhouse gases. Without a high share of low-carbon electricity, a domestic heat pump will produce more carbon emissions than using natural gas.

Fossil fuel Phase-out: Carbon Neutral and Negative Fuels

3,500-4,000 environmental activists blocking a coal mine in Germany to limit climate change (Ende Gelände 2016).

Fossil fuel may be phased-out with carbon neutral and carbon negative pipeline and transportation fuels created with power to gas and gas to liquids technologies. Carbon dioxide from fossil fuel flue gas can be used to produce plastic lumber allowing carbon negative reforestation.

Demand Side Management

Energy efficiency and Conservation

A spiral-type integrated compact fluorescent lamp, use has grown among North American consumers since its introduction in the mid-1990s.

Efficient energy use, sometimes simply called “energy efficiency”, is the goal of efforts to reduce the amount of energy required to provide products and services. For example, insulating a home allows a building to use less heating and cooling energy to achieve and maintain a comfortable temperature. Installing fluorescent lights or natural skylights reduces the amount of energy required to attain the same level of illumination compared to using traditional incandescent light bulbs. Compact fluorescent lights use two-thirds less energy and may last 6 to 10 times longer than incandescent lights.

Energy efficiency has proved to be a cost-effective strategy for building economies without necessarily growing energy consumption. For example, the state of California began implementing energy-efficiency measures in the mid-1970s, including building code and appliance standards with strict efficiency requirements. During the following years, California’s energy consumption has remained approximately flat on a per capita basis while national U.S. consumption doubled. As part of its strategy, California implemented a “loading order” for new energy resources that puts energy efficiency first, renewable electricity supplies second, and new fossil-fired power plants last.

Energy conservation is broader than energy efficiency in that it encompasses using less energy to achieve a lesser energy service, for example through behavioural change, as well as encompassing energy efficiency. Examples of conservation without efficiency improvements would be heating a room less in winter, driving less, or working in a less brightly lit room. As with other definitions, the boundary between efficient energy use and energy conservation can be fuzzy, but both are important in environmental and economic terms. This is especially the case when actions are directed at the saving of fossil fuels.

Reducing energy use is seen as a key solution to the problem of reducing greenhouse gas emissions. According to the International Energy Agency, improved energy efficiency in buildings, industrial processes and transportation could reduce the world’s energy needs in 2050 by one third, and help control global emissions of greenhouse gases.

Demand Side Switching Sources

Fuel switching on the demand side refers to changing the type of fuel used to satisfy a need for an energy service. To meet deep decarbonization goals, like the 80% reduction by 2050 goal being discussed in California and the European Union, many primary energy changes are needed. Energy efficiency alone may not be sufficient to meet these goals, switching fuels used on the demand side will help lower carbon emissions. Progressively coal, oil and eventually natural gas for space and water heating in buildings will need to be reduced. For an equivalent amount of heat, burning natural gas produces about 45 per cent less carbon dioxide than burning coal. There are various ways in which this could happen, and different strategies will likely make sense in different locations. While the system efficiency of a gas furnace may be higher than the combination of natural gas power plant and electric heat, the combination of the same natural gas power plant and an electric heat pump has lower emissions per unit of heat delivered in all but the coldest climates. This is possible because of the very efficient coefficient of performance of heat pumps.

At the beginning of this century 70% of all electricity was generated by fossil fuels, and as carbon free sources eventually make up half of the generation mix, replacing gas or oil furnaces and water heaters with electric ones will have a climate benefit. In areas like Norway, Brazil and Quebec that have abundant hydroelectricity, electric heat and hot water is common.

The economics of switching the demand side from fossil fuels to electricity for heating, will depend on the price of fuels vs electricity and the relative prices of the equipment. The EIA Annual Energy Outlook 2014 suggests that domestic gas prices will rise faster than electricity prices which will encourage electrification in the coming decades. Electrifying heating loads may also provide a flexible resource that can participate in demand response. Since thermostatically controlled loads have inherent energy storage, electrification of heating could provide a valuable resource to integrate variable renewable resources into the grid.

Alternatives to electrification, include decarbonizing pipeline gas through power to gas, biogas, or other carbon neutral fuels. A 2015 study by Energy+Environmental Economics shows that a hybrid approach of decarbonizing pipeline gas, electrification, and energy efficiency can meet carbon reduction goals at a similar cost as only electrification and energy efficiency in Southern California.

Demand Side Grid Management

Expanding intermittent electrical sources such as wind power, creates a growing problem balancing grid fluctuations. Some of the plans include building pumped storage or continental super grids costing billions of dollars. However instead of building for more power,there are a variety of ways to affect the size and timing of electricity demand on the consumer side. Designing for reduced demands on a smaller power grid is more efficient and economic than having extra generation and transmission for intermittentcy, power failures and peak demands. Having these abilities is one of the chief aims of a smart grid.

Time of use metering is a common way to motivate electricity users to reduce their peak load consumption. For instance, running dishwashers and laundry at night after the peak has passed, reduces electricity costs.

Dynamic demand plans have devices passively shut off when stress is sensed on the electrical grid. This method may work very well with thermostats, when power on the grid sags a small amount, a low power temperature setting is automatically selected reducing the load on the grid. For instance millions of refrigerators reduce their consumption when clouds pass over solar installations. Consumers would need to have a smart meter in order for the utility to calculate credits.

Demand response devices could receive all sorts of messages from the grid. The message could be a request to use a low power mode similar to dynamic demand, to shut off entirely during a sudden failure on the grid, or notifications about the current and expected prices for power. This would allow electric cars to recharge at the least expensive rates independent of the time of day. The vehicle-to-grid suggestion would use a car's battery or fuel cell to supply the grid temporarily.

Lifestyle and Behavior

The IPCC Fifth Assessment Report emphasises that behaviour, lifestyle and cultural change have a high mitigation potential in some sectors, particularly when complementing technological and structural change. In general, higher consumption lifestyles have a greater environmental impact. Overall, food accounts for the largest share of consumption-based GHG emissions with nearly 20% of the global carbon footprint, followed by housing, mobility, services, manufactured products, and construction. Food and services are more significant in poor countries, while mobility and manufactured goods are more significant in rich countries.

Dietary Change

A 2014 study into the real-life diets of British people estimates their greenhouse gas contributions (CO_2eq) to be: 7.19 kg/day for high meat-eaters through to 3.81 kg/day for vegetarians and 2.89 kg/day for vegans. The widespread adoption of a vegetarian diet could cut food-related greenhouse gas emissions by 63% by 2050. China introduced new dietary guidelines in 2016 which aim to cut meat consumption by 50% and thereby reduce greenhouse gas emissions by 1 billion tonnes by 2030. A 2016 study concluded that taxes on meat and milk could simultaneously result in reduced greenhouse gas emissions and healthier diets. The study analyzed surcharges of 40% on beef and 20% on milk and suggests that an optimum plan would reduce emissions by 1 billion tonnes per year.

Sinks and Negative Emissions

A carbon sink is a natural or artificial reservoir that accumulates and stores some carbon-containing chemical compound for an indefinite period, such as a growing forest. A negative carbon dioxide emission on the other hand is a permanent removal of carbon dioxide out of the atmosphere, such as directly capturing carbon dioxide in the atmosphere and storing it in geologic formations underground.

The Antarctic Climate and Ecosystems Cooperative Research Centre (ACE-CRC) notes that one third of humankind's annual emissions of CO_2 are absorbed by the oceans. However, this also leads to ocean acidification, with potentially significant impacts on marine life. Acidification lowers the level of carbonate ions available for calcifying organisms to form their shells. These organisms include plankton species that contribute to the foundation of the Southern Ocean food web. However acidification may impact on a broad range of other physiological and ecological processes, such as fish respiration, larval development and changes in the solubility of both nutrients and toxins.

Reforestation and Afforestation

Almost 20 percent (8 $GtCO_2$/year) of total greenhouse-gas emissions were from deforestation in 2007. It is estimated that avoided deforestation reduces CO_2 emissions at a rate of 1 tonne of CO_2 per \$1–5 in opportunity costs from lost agriculture. Reforestation and afforestation, where there was previously no forest, could save at least another 1 $GtCO_2$/year, at an estimated cost of \$5–15/tCO_2.

Transferring land rights to indigenous inhabitants is argued to efficiently conserve forests. Regrowth of forests on abandoned farmland restores more forest than that lost to deforestation.

Transferring rights over land from public domain to its indigenous inhabitants is argued to be a cost effective strategy to conserve forests. This includes the protection of such rights entitled in existing laws, such as India's Forest Rights Act. The transferring of such rights in China, perhaps the largest land reform in modern times, has been argued to have increased forest cover. In Brazil, forested areas given tenure to indigenous groups have even lower rates of clearing than national parks. A 2016 report concludes that modest investments in indigenous land rights will generate economic, social, and environmental returns for the communities involved and for climate protection. The report quantifies the economic value of securing such rights, with a focus on the Amazon region.

With increased intensive agriculture and urbanization, there is an increase in the amount of abandoned farmland. By some estimates, for every half a hectare of original old-growth forest cut down, more than 20 hectares of new secondary forests are growing, even though they do not have the same biodiversity as the original forests and original forests store 60% more carbon than these new secondary forests. According to a study in *Science*, promoting regrowth on abandoned farmland could offset years of carbon emissions.

Avoided Desertification

Restoring grasslands store CO_2 from the air into plant material. Grazing livestock, usually not left to wander, would eat the grass and would minimize any grass growth. However, grass left alone would eventually grow to cover its own growing buds, preventing them from photosynthesizing and the dying plant would stay in place. A method proposed to restore grasslands uses fences with many small paddocks and moving herds from one paddock to another after a day a two in order to mimick natural grazers and allowing the grass to grow optimally. Additionally, when part of leaf matter is consumed by a herding animal, a corresponding amount of root matter is sloughed off too as it would not be able to sustain the previous amount of root matter and while most of the lost root matter would rot and enter the atmosphere, part of the carbon is sequestered into the soil. It is estimated that increasing the carbon content of the soils in the world's 3.5 billion hectares of agricultural grassland by 1% would offset nearly 12 years of CO_2emissions. Allan Savory, as part of holistic management, claims that while large herds are often blamed for desertification, prehistoric lands supported large or larger herds and areas where herds were removed in the United States are still desertifying.

Managed grazing methods are argued to be able to restore grasslands, thereby significantly decreasing atmospheric CO_2levels.

Carbon Capture and Storage

Carbon capture and storage (CCS) is a method to mitigate climate change by capturing carbon dioxide (CO_2) from large point sources such as power plants and subsequently storing it away safely instead of releasing it into the atmosphere. The Intergovernmental Panel on Climate Change says CCS could contribute between 10% and 55% of the cumulative worldwide carbon-mitigation effort over the next 90 years. The International Energy Agency says CCS is "the most important single new technology for CO_2 savings" in power generation and industry. Though it requires up to 40% more energy to run a CCS coal power plant than a regular coal plant, CCS could potentially capture about 90% of all the carbon emitted by the plant. Norway, which first began storing CO_2, has cut its emissions by almost a million tons a year, or about 3% of the country's 1990 levels. As of late 2011, the total CO_2 storage capacity of all 14 projects in operation or under construction is over 33 million tonnes a year. This is broadly equivalent to preventing the emissions from more than six million cars from entering the atmosphere each year.

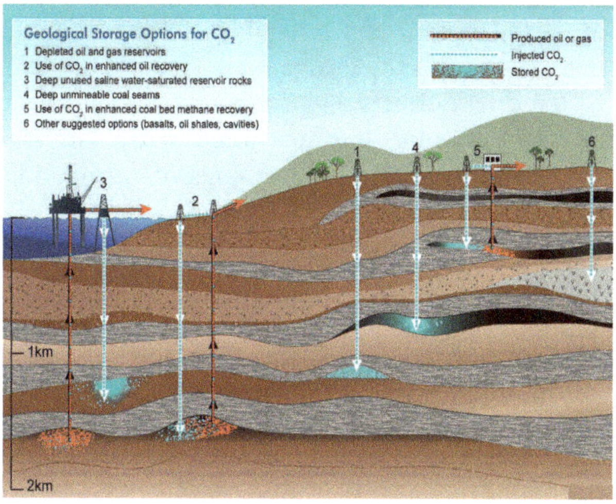

Schematic showing both terrestrial and geological sequestration of carbon dioxide emissions from a coal-fired plant.

Negative Carbon Dioxide Emissions

Creating negative carbon dioxide emissions literally removes carbon from the atmosphere. Examples are direct air capture, biochar, bio-energy with carbon capture and storage and enhanced weathering technologies. These processes are sometimes considered as variations of sinks or mitigation,and sometimes as geoengineering.

In combination with other mitigation measures, sinks in combination with negative carbon emissions are considered crucial for meeting the 350 ppm target, and even the less conservative 450 ppm target.

Geoengineering

Geoengineering is seen by some as an alternative to mitigation and adaptation, but by others as an entirely separate response to climate change. In a literature assessment, Barker *et al.* (2007) described geoengineering as a type of mitigation policy. IPCC (2007) concluded that geoengineering options, such as ocean fertilization to remove CO_2 from the atmosphere, remained largely unproven.It was judged that reliable cost estimates for geoengineering had not yet been published.

Chapter 28 of the National Academy of Sciences report *Policy Implications of Greenhouse Warming: Mitigation, Adaptation, and the Science Base* (1992) defined geoengineering as "options that would involve large-scale engineering of our environment in order to combat or counteract the effects of changes in atmospheric chemistry." They evaluated a range of options to try to give preliminary answers to two questions: can these options work and could they be carried out with a reasonable cost. They also sought to encourage discussion of a third question — what adverse side effects might there be. The following types of option were examined: reforestation, increasing ocean absorption of carbon dioxide (carbon sequestration) and screening out some sunlight. NAS also argued "Engineered countermeasures need to be evaluated but should not be implemented without broad understanding of the direct effects and the potential side effects, the ethical issues, and the risks.». In July 2011 a report by the United States Government Accountability Office on geoengineering found that "[c]limate engineering technologies do not now offer a viable response to global climate change."

Carbon Dioxide Removal

Carbon dioxide removal has been proposed as a method of reducing the amount of radiative forcing. A variety of means of artificially capturing and storing carbon, as well as of enhancing natural sequestration processes, are being explored. The main natural process is photosynthesis by plants and single-celled organisms (see biosequestration). Artificial processes vary, and concerns have been expressed about the long-term effects of some of these processes.

It is notable that the availability of cheap energy and appropriate sites for geological storage of carbon may make carbon dioxide air capture viable commercially. It is, however, generally expected that carbon dioxide air capture may be uneconomic when compared to carbon capture and storage from major sources — in particular, fossil fuel powered power stations, refineries, etc. In such cases, costs of energy produced will grow significantly. However, captured CO_2 can be used to force more crude oil out of oil fields, as Statoil and Shell have made plans to do. CO_2 can also be used in commercial greenhouses, giving an opportunity to kick-start the technology. Some attempts have been made to use algae to capture smokestack emissions, notably the GreenFuel Technologies Corporation, who have now shut down operations.

Solar Radiation Management

The main purpose of solar radiation management seek to reflect sunlight and thus reduce global warming. The ability of stratospheric sulfate aerosols to create a global dimming effect has made them a possible candidate for use in climate engineering projects.

Non-CO_2 Greenhouse Gases

CO_2 is not the only GHG relevant to mitigation, and governments have acted to regulate the emissions of other GHGs emitted by human activities (anthropogenic GHGs). The emissions caps agreed to by most developed countries under the Kyoto Protocol regulate the emissions of almost all the anthropogenic GHGs. These gases are CO_2, methane (CH_4), nitrous oxide (N_2O), the hydrofluorocarbons (HFC), perfluorocarbons (PFC), and sulfur hexafluoride (SF_6).

Stabilizing the atmospheric concentrations of the different anthropogenic GHGs requires an understanding of their different physical properties. Stabilization depends both on how quickly GHGs are added to the atmosphere and how fast they are removed. The rate of removal is measured by the atmospheric lifetime of the GHG in question (see the main GHG article for a list). Here, the lifetime is defined as the time required for a given perturbation of the GHG in the atmosphere to be reduced to 37% of its initial amount. Methane has a relatively short atmospheric lifetime of about 12 years, while N_2O's lifetime is about 110 years. For methane, a reduction of about 30% below current emission levels would lead to a stabilization in its atmospheric concentration, while for N_2O, an emissions reduction of more than 50% would be required.

Methane is a significantly more potent greenhouse gas than carbon dioxide in the amount of heat it can trap, especially in the short term. Burning one molecule of methane generates one molecule of carbon dioxide, indicating there may be no net benefit in using gas as a fuel source. Reducing the amount of waste methane produced in the first place and moving away from use of gas as a fuel source will have a greater beneficial impact, as might other approaches to productive use of otherwise-wasted methane. In terms of prevention, vaccines are being developed in Australia to reduce the significant global warming contributions from methane released by livestock via flatulence and eructation.

Another physical property of the anthropogenic GHGs relevant to mitigation is the different abilities of the gases to trap heat (in the form of infrared radiation). Some gases are more effective at trapping heat than others, e.g., SF_6 is 22,200 times more effective a GHG than CO_2 on a per-kilogram basis. A measure for this physical property is the global warming potential (GWP), and is used in the Kyoto Protocol.

Although not designed for this purpose, the Montreal Protocol has probably benefited climate change mitigation efforts. The Montreal Protocol is an international treaty that has successfully reduced emissions of ozone-depleting substances (for example, CFCs), which are also greenhouse gases.

By Sector

Transport

Bicycles have almost no carbon footprint compared to cars, and canal transport may represent a positive option for certain types of freight in the 21st century

Modern energy-efficient technologies, such as plug-in hybrid electric vehicles, and development of new technologies, such as carbon-neutral synthetic gasoline & Jet fuel, may reduce the consumption of petroleum, land use changes and emissions of carbon dioxide. A shift from air transport and truck transport to electric rail transport would reduce emissions significantly. For electric vehicles, the reduction of carbon emissions will improve further if the way the required electricity is generated is low-carbon power in origin.

Urban Planning

Effective urban planning to reduce sprawl would decrease Vehicle Miles Travelled (VMT), lowering emissions from transportation. Increased use of public transport can also reduce greenhouse gas emissions per passenger kilometer. Between 1982 and 1997, the amount of land consumed for urban development in the United States increased by 47 percent while the nation's population grew by only 17 percent. Inefficient land use development practices have increased infrastructure costs as well as the amount of energy needed for transportation, community services, and buildings.

At the same time, a growing number of citizens and government officials have begun advocating a smarter approach to land use planning. These smart growth practices include compact community development, multiple transportation choices, mixed land uses, and practices to conserve green space. These programs offer environmental, economic, and quality-of-life benefits; and they also serve to reduce energy usage and greenhouse gas emissions.

Approaches such as New Urbanism and Transit-oriented development seek to reduce distances travelled, especially by private vehicles, encourage public transit and make walking and cycling more attractive options. This is achieved through "medium-density", mixed-use planning and the concentration of housing within walking distance of town centers and transport nodes.

Smarter growth land use policies have both a direct and indirect effect on energy consuming behavior. For example, transportation energy usage, the number one user of petroleum fuels, could be significantly reduced through more compact and mixed use land development patterns, which in turn could be served by a greater variety of non-automotive based transportation choices.

Building Design

Emissions from housing are substantial, and government-supported energy efficiency programmes can make a difference.

For institutions of higher learning in the United States, greenhouse gas emissions depend primarily on total area of buildings and secondarily on climate. If climate is not taken into account, annual greenhouse gas emissions due to energy consumed on campuses plus purchased electricity can be estimated with the formula, $E=aS^b$, where a =0.001621 metric tonnes of CO_2 equivalent/square foot or 0.0241 metric tonnes of CO_2 equivalent/ square meter and b = 1.1354.

New buildings can be constructed using passive solar building design, low-energy building, or zero-energy building techniques, using renewable heat sources. Existing buildings can be made more efficient through the use of insulation, high-efficiency appliances (particularly hot water heaters and furnaces), double- or triple-glazed gas-filled windows, external window shades, and building orientation and siting. Renewable heat sources such as shallow geothermal and passive solar energy reduce the amount of greenhouse gasses emitted. In addition to designing buildings which are more energy-efficient to heat, it is possible to design buildings that are more energy-efficient to cool by using lighter-coloured, more reflective materials in the development of urban areas (e.g. by painting roofs white) and planting trees. This saves energy because it cools buildings and reduces the urban heat island effect thus reducing the use of air conditioning.

Agriculture

According to the EPA, agricultural soil management practices can lead to production and emission of nitrous oxide (N2O), a major greenhouse gas and air pollutant. Activities that can contribute to N2O emissions include fertilizer usage, irrigation and tillage. The management of soils accounts for over half of the emissions from the Agriculture sector. Cattle livestocks account for one third of emissions, through methane emissions. Manure management and rice cultivation also produce gaseous emissions.

Methods that significantly enhance carbon sequestration in soil include no-till farming, residue mulching, cover cropping, and crop rotation, all of which are more widely used in organic farming than in conventional farming. Because only 5% of US farmland currently uses no-till and residue mulching, there is a large potential for carbon sequestration.

A 2015 study found that farming can deplete soil carbon and render soil incapable of supporting life. Instead the study showed that conservation farming can protect carbon in soils, and repair damage over time.

The farming practise of cover crops has been recognized as climate-smart agriculture by the White House.

Societal Controls

Another method being examined is to make carbon a new currency by introducing tradeable "personal carbon credits". The idea being it will encourage and motivate individuals to reduce their 'carbon footprint' by the way they live. Each citizen will receive a free annual quota of carbon that they can use to travel, buy food, and go about their business. It has been suggested that by using this concept it could actually solve two problems; pollution and poverty, old age pensioners will actually be better off because they fly less often, so they can cash in their quota at the end of the year to pay heating bills and so forth.

Population

Various organizations promote population control as a means for mitigating global warming. Proposed measures include improving access to family planning and reproductive health care and information, reducing natalistic politics, public education about the consequences of continued population growth, and improving access of women to education and economic opportunities.

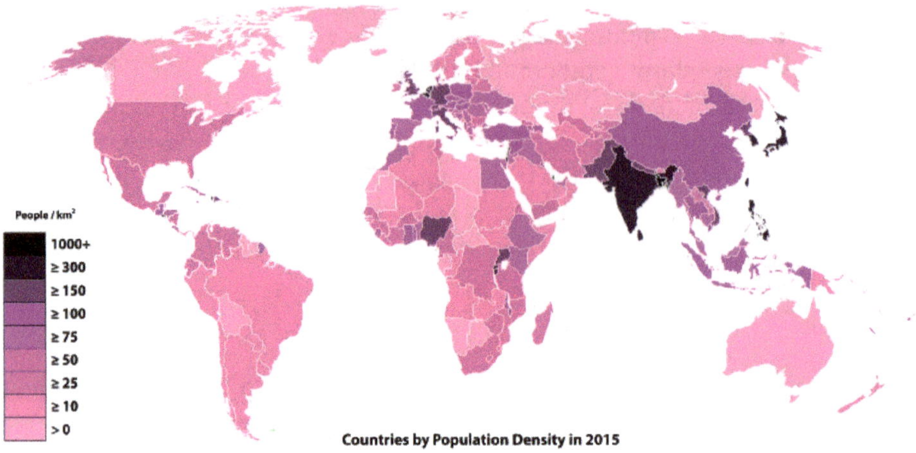

c

Population density by country.

Population control efforts are impeded by there being somewhat of a taboo in some countries against considering any such efforts. Also, various religions discourage or prohibit some or all forms of birth control.

Population size has a different per capita effect on global warming in different countries, since the per capita production of anthropogenic greenhouse gases varies greatly by country.

Costs and Benefits

Costs

The Stern Review proposes stabilising the concentration of greenhouse-gas emissions in the atmosphere at a maximum of 550ppm CO_2e by 2050. The Review estimates that this would mean cutting total greenhouse-gas emissions to three quarters of 2007 levels. The Review further estimates that the cost of these cuts would be in the range −1.0 to +3.5% of World GDP, (i.e. GWP), with an average estimate of approximately 1%. Stern has since revised his estimate to 2% of GWP. For comparison, the Gross World Product (GWP) at PPP was estimated at $74.5 trillion in 2010, thus 2% is approximately $1.5 trillion. The Review emphasises that these costs are contingent on steady reductions in the cost of low-carbon technologies. Mitigation costs will also vary according to how and when emissions are cut: early, well-planned action will minimise the costs.

One way of estimating the cost of reducing emissions is by considering the likely costs of potential technological and output changes. Policy makers can compare the marginal abatement costs of different methods to assess the cost and amount of possible abatement over time. The marginal abatement costs of the various measures will differ by country, by sector, and over time.

Benefits

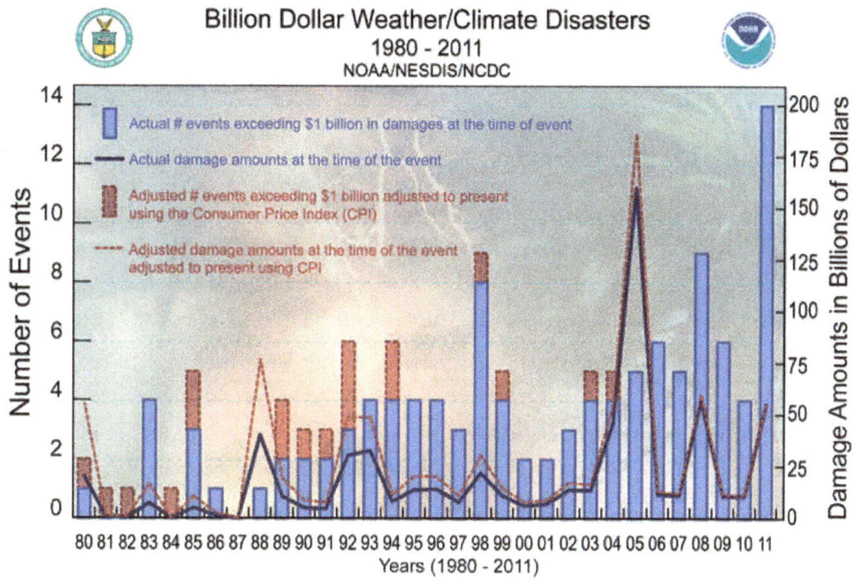

Total extreme weather cost and number of events costing more than $1 billion in the United States from 1980 to 2011.

Yohe *et al.* (2007) assessed the literature on sustainability and climate change.With high confidence, they suggested that up to the year 2050, an effort to cap greenhouse gas (GHG) emissions at 550 ppm would benefit developing countries significantly. This was judged to be especially the case when combined with enhanced adaptation. By 2100, however, it was still judged likely that there would be significant effects of global warming. This was judged to be the case even with aggressive mitigation and significantly enhanced adaptive capacity.

Sharing

One of the aspects of mitigation is how to share the costs and benefits of mitigation policies. There is no scientific consensus over how to share these costs and benefits (Toth *et al.*, 2001). In terms of the politics of mitigation, the UNFCCC's ultimate objective is to stabilize concentrations of GHG in the atmosphere at a level that would prevent "dangerous" climate change (Rogner *et al.*, 2007).

GHG emissions are an important correlate of wealth, at least at present (Banuri *et al.*, 1996, pp. 91–92). Wealth, as measured by per capita income (i.e., income per head of population), varies widely between different countries. Activities of the poor that involve emissions of GHGs are often associated with basic needs, such as heating to stay tolerably warm. In richer countries, emissions tend to be associated with things like cars, central heating, etc. The impacts of cutting emissions could therefore have different impacts on human welfare according to wealth.

Distributing Emissions Abatement Costs

There have been different proposals on how to allocate responsibility for cutting emissions (Banuri *et al.*, 1996, pp. 103–105):

- **Egalitarianism**: this system interprets the problem as one where each person has equal rights to a global resource, i.e., polluting the atmosphere.
- **Basic needs**: this system would have emissions allocated according to basic needs, as defined according to a minimum level of consumption. Consumption above basic needs would require countries to buy more emission rights. From this viewpoint, developing countries would need to be at least as well off under an emissions control regime as they would be outside the regime.
- **Proportionality and polluter-pays principle**: Proportionality reflects the ancient Aristotelian principle that people should receive in proportion to what they put in, and pay in proportion to the damages they cause. This has a potential relationship with the "polluter-pays principle", which can be interpreted in a number of ways:
- *Historical responsibilities*: this asserts that allocation of emission rights should be based on patterns of past emissions. Two-thirds of the stock of GHGs in the atmosphere at present is due to the past actions of developed countries (Goldemberg *et al.*, 1996, p. 29).
- *Comparable burdens and ability to pay*: with this approach, countries would reduce emissions based on comparable burdens and their ability to take on the costs of reduction. Ways to assess burdens include monetary costs per head of population, as well as other, more complex measures, like the UNDP's Human Development Index.
- *Willingness to pay*: with this approach, countries take on emission reductions based on their ability to pay along with how much they benefit from reducing their emissions.

Specific Proposals

- **Ad hoc**: Lashof (1992) and Cline (1992) (referred to by Banuri *et al.*, 1996, p. 106), for example, suggested that allocations based partly on GNP could be a way of sharing the burdens of emission reductions. This is because GNP and economic activity are partially tied to carbon emissions.
- **Equal per capita entitlements**: this is the most widely cited method of distributing abatement costs, and is derived from egalitarianism (Banuri *et al.*, 1996, pp. 106–107). This approach can be divided into two categories. In the first category, emissions are allocated according to national population. In the

second category, emissions are allocated in a way that attempts to account for historical (cumulative) emissions.

- **Status quo**: with this approach, historical emissions are ignored, and current emission levels are taken as a status quo right to emit (Banuri *et al.*, 1996, p. 107). An analogy for this approach can be made with fisheries, which is a common, limited resource. The analogy would be with the atmosphere, which can be viewed as an exhaustible natural resource (Goldemberg *et al.*, 1996, p. 27). In international law, one state recognized the long-established use of another state's use of the fisheries resource. It was also recognized by the state that part of the other state's economy was dependent on that resource.

Governmental and Intergovernmental Action

Many countries, both developing and developed, are aiming to use cleaner technologies (World Bank, 2010, p. 192). Use of these technologies aids mitigation and could result in substantial reductions in CO_2 emissions. Policies include targets for emissions reductions, increased use of renewable energy, and increased energy efficiency. It is often argued that the results of climate change are more damaging in poor nations, where infrastructures are weak and few social services exist. The Commitment to Development Index is one attempt to analyze rich country policies taken to reduce their disproportionate use of the global commons. Countries do well if their greenhouse gas emissions are falling, if their gas taxes are high, if they do not subsidize the fishing industry, if they have a low fossil fuel rate per capita, and if they control imports of illegally cut tropical timber.

Kyoto Protocol

The main current international agreement on combating climate change is the Kyoto Protocol, which came into force on 16 February 2005. The Kyoto Protocol is an amendment to the United Nations Framework Convention on Climate Change (UNFCCC). Countries that have ratified this protocol have committed to reduce their emissions of carbon dioxide and five other greenhouse gases, or engage in emissions trading if they maintain or increase emissions of these gases.

Temperature Targets

The graph on the right shows three "pathways" to meet the UNFCCC's 2 °C target, labelled "global technology", "decentralised solutions", and "consumption change". Each pathway shows how various measures (e.g., improved energy efficiency, increased use of renewable energy) could contribute to emissions reductions. Image credit: PBL Netherlands Environmental Assessment Agency.

Actions to mitigate climate change are sometimes based on the goal of achieving a particular temperature target. One of the targets that has been suggested is to limit the future increase in global mean temperature (global warming) to below 2 °C, relative to the pre-industrial level. The 2 °C target was adopted in 2010 by Parties to the United Nations Framework Convention on Climate Change. Most countries of the world are Parties to the UNFCCC. The target had been adopted in 1996 by the European Union Council.

Feasibility of 2 °C

Temperatures have increased by 0.8 °C compared to the pre-industrial level, and another 0.5–0.7 °C is already committed.The 2 °C rise is typically associated in climate models with a carbon dioxide equivalent concentration of 400–500 ppm by volume; the current (January 2015) level of carbon dioxide alone is 400 ppm by volume, and rising at 1–3 ppm annually. Hence, to avoid a very likely breach of the 2 °C target, CO_2levels would have to be stabilised very soon; this is generally regarded as unlikely, based on current programs in place to date. The importance of change is illustrated by the fact that world economic energy efficiency is improving at only half the rate of world economic growth.

Views in the Literature

There is disagreement among experts over whether or not the 2 °C target can be met. For example, according to Anderson and Bows (2011), "there is little to no chance" of meeting the target. On the other hand, according to Alcamo *et al.* (2013):

- Policies adopted by parties to the UNFCCC are too weak to meet a 2 or 1.5 °C target. However, these targets might still be achievable if more stringent mitigation policies are adopted immediately.
- Cost-effective 2 °C scenarios project annual global greenhouse gas emissions to peak before the year 2020, with deep cuts in emissions thereafter, leading to a reduction in 2050 of 41% compared to 1990 levels.

Discussion on Other Targets

Scientific analysis can provide information on the impacts of climate change and associated policies, such as reducing GHG emissions. However, deciding what policies are best requires value judgements. For example, limiting global warming to 1 °C relative to pre-industrial levels may help to reduce climate change damages more than a 2 °C limit. However, a 1 °C limit may be more costly to achieve than a 2 °C limit.

According to some analysts, the 2 °C "guardrail" is inadequate for the needed degree and timeliness of mitigation. On the other hand, some economic studies suggest more modest mitigation policies.For example, the emissions reductions proposed by Nordhaus (2010) might lead to global warming (in the year 2100) of around 3 °C, relative to pre-industrial levels.

Official Long-Term Target of 1.5 °C

In 2015, two official UNFCCC scientific expert bodies came to the conclusion that, "in some regions and vulnerable ecosystems, high risks are projected even for warming above 1.5°C". This expert position was, together with the strong diplomatic voice of the poorest countries and the island nations in the Pacific, the driving force leading to the decision of the Paris Conference 2015, to lay down this 1.5 °C long-term target on top of the existing 2 °C goal.

Encouraging use Changes

Emissions Tax

An emissions tax on greenhouse gas emissions requires individual emitters to pay a fee, charge or tax for every tonne of greenhouse gas released into the atmosphere. Most environmentally related taxes with implications for greenhouse gas emissions in OECD countries are levied on energy products and motor vehicles, rather than on CO_2 emissions directly.

Emission taxes can be both cost-effective and environmentally effective. Difficulties with emission taxes include their potential unpopularity, and the fact that they cannot guarantee a particular level of emissions reduction. Emissions or energy taxes also often fall disproportionately on lower income classes. In developing countries, institutions may be insufficiently developed for the collection of emissions fees from a wide variety of sources.

Subsidies

According to Mark Z. Jacobson, a program of subsidization balanced against expected flood costs could pay for conversion to 100% renewable power by 2030. Jacobson, and his colleague Mark Delucchi, suggest that the cost to generate and transmit power in 2020 will be less than 4 cents per kilowatt hour (in 2007 dollars) for wind, about 4 cents for wave and hydroelectric, from 4 to 7 cents for geothermal, and 8 cents per kWh for solar, fossil, and nuclear power.

Investment

Another indirect method of encouraging uses of renewable energy, and pursue sustainability and environmental protection, is that of prompting investment in this area through legal means, something that is already being done at national level as well as in the field of international investment.

Carbon emissions trading

With the creation of a market for trading carbon dioxide emissions within the Kyoto Protocol, it is likely that London financial markets will be the centre for this potentially highly lucrative business; the New York and Chicago stock markets may have a lower trade volume than expected as long as the US maintains its rejection of the Kyoto.

However, emissions trading may delay the phase-out of fossil fuels.

In the north-east United States, a successful cap and trade program has shown potential for this solution.

The European Union Emission Trading Scheme (EU ETS) is the largest multi-national, greenhouse gas emissions trading scheme in the world. It commenced operation on 1 January 2005, and all 28 member states of the European Union participate in the scheme which has created a new market in carbon dioxide allowances estimated at 35 billion Euros (US$43 billion) per year. The Chicago Climate Exchange was the first (voluntary) emissions market, and is soon to be followed by Asia's first market (Asia Carbon Exchange). A total of 107 million metric tonnes of carbon dioxide equivalent have been exchanged through projects in 2004, a 38% increase relative to 2003 (78 Mt CO_2e).

Twenty three multinational corporations have come together in the G8 Climate Change Roundtable, a business group formed at the January 2005 World Economic Forum. The group includes Ford, Toyota, British Airways and BP. On 9 June 2005 the Group published a statement stating that there was a need to act on climate change and claiming that market-based solutions can help. It called on governments to establish "clear, transparent, and consistent price signals" through "creation of a long-term policy framework" that would include all major producers of greenhouse gases.

The Regional Greenhouse Gas Initiative is a proposed carbon trading scheme being created by nine North-eastern and Mid-Atlantic American states; Connecticut, Delaware, Maine, Massachusetts, New Hampshire, New Jersey, New York, Rhode Island and Vermont. The scheme was due to be developed by April 2005 but has not yet been completed.

Implementation

Implementation puts into effect climate change mitigation strategies and targets. These can be targets set by international bodies or voluntary action by individuals or institutions. This is the most important, expensive and least appealing aspect of environmental governance.

Funding

Implementation requires funding sources but is often beset by disputes over who should provide funds and under what conditions. A lack of funding can be a barrier to successful strategies as there are no formal arrangements to finance climate change development and implementation. Funding is often provided by nations, groups of nations and increasingly NGO and private sources. These funds are often channelled through the Global Environmental Facility (GEF). This is an environmental funding mechanism in the World Bank which is designed to deal with global environmental issues. The GEF was originally designed to tackle four main areas: biological diversity, climate change, international waters and ozone layer depletion, to which land degradation and persistent organic pollutant were added. The GEF funds projects that are agreed to achieve global environmental benefits that are endorsed by governments and screened by one of the GEF's implementing agencies.

Problems

There are numerous issues which result in a current perceived lack of implementation. It has been suggested that the main barriers to implementation are, Uncertainty, Fragmentation,

Institutional void, Short time horizon of policies and politicians and Missing motives and willingness to start adapting. The relationships between many climatic processes can cause large levels of uncertainty as they are not fully understood and can be a barrier to implementation. When information on climate change is held between the large numbers of actors involved it can be highly dispersed, context specific or difficult to access causing fragmentation to be a barrier. Institutional void is the lack of commonly accepted rules and norms for policy processes to take place, calling into question the legitimacy and efficacy of policy processes. The Short time horizon of policies and politicians often means that climate change policies are not implemented in favour of socially favoured societal issues. Statements are often posed to keep the illusion of political action to prevent or postpone decisions being made. Missing motives and willingness to start adapting is a large barrier as it prevents any implementation.

The issues that arise with a system which involves international government cooperation, such as Cap and Trade, could potentially be improved with a polycentric approach where the rules are enforced by many small sections of authority as apposed to one overall enforcement agency.

Occurrence

Despite a perceived lack of occurrence, evidence of implementation is emerging internationally. Some examples of this are the initiation of NAPA's and of joint implementation. Many developing nations have made National Adaptation Programs of Action (NAPAs) which are frameworks to prioritize adaption needs. The implementation of many of these is supported by GEF agencies. Many developed countries are implementing 'first generation' institutional adaption plans particularly at the state and local government scale. There has also been a push towards joint implementation between countries by the UNFCC as this has been suggested as a cost-effective way for objectives to be achieved.

Territorial Policies

United States

Efforts to reduce greenhouse gas emissions by the United States include energy policies which encourage efficiency through programs like Energy Star, Commercial Building Integration, and the Industrial Technologies Program. On 12 November 1998, Vice President Al Gore symbolically signed the Kyoto Protocol, but he indicated participation by the developing nations was necessary prior its being submitted for ratification by the United States Senate.

In 2007, Transportation Secretary Mary Peters, with White House approval, urged governors and dozens of members of the House of Representatives to block California's first-in-the-nation limits on greenhouse gases from cars and trucks, according to e-mails obtained by Congress. The U.S. Climate Change Science Program is a group of about twenty federal agencies and US Cabinet Departments, all working together to address global warming. The Bush administration pressured American scientists to suppress discussion of global warming, according to the testimony of the Union of Concerned Scientists to the Oversight and Government Reform Committee of the U.S. House of Representatives. "High-quality science" was "struggling to get out," as the Bush administration pressured scientists to tailor their writings on global warming to fit the Bush administration's skepticism, in some cases at the behest of an ex-oil industry lobbyist. "Nearly half of all respondents perceived or personally experienced pressure to eliminate the words 'climate change,' 'global warming' or other similar terms from a variety of communications." Similarly, according to the testimony of senior officers of the Government Accountability Project, the White House attempted to bury the report "National Assessment of the Potential Consequences of Climate Variability and Change," produced by U.S. scientists pursuant to U.S. law. Some U.S. scientists resigned their jobs rather than give in to White House pressure to underreport global warming.

In the absence of substantial federal action, state governments have adopted emissions-control laws such as the Regional Greenhouse Gas Initiative in the Northeast and the Global Warming Solutions Act of 2006 in California.

Developing Countries

In order to reconcile economic development with mitigating carbon emissions, developing countries need particular support, both financial and technical. One of the means of achieving this is the Kyoto Protocol's Clean Development Mechanism (CDM). The World Bank's Prototype Carbon Fund is a public private partnership that operates within the CDM.

An important point of contention, however, is how overseas development assistance not directly related to climate change mitigation is affected by funds provided to climate change mitigation. One of the outcomes of the UNFCC Copenhagen Climate Conference was the Copenhagen Accord, in which developed countries promised to provide US $30 million between 2010 and 2012 of new and additional resources. Yet it remains unclear what exactly the definition of additional is and the European Commission has requested its member states to define what they understand to be additional, and researchers at the Overseas Development Institute have found four main understandings:

- Climate finance classified as aid, but additional to (over and above) the '0.7%' ODA target;
- Increase on previous year's Official Development Assistance (ODA) spent on climate change mitigation;
- Rising ODA levels that include climate change finance but where it is limited to a specified percentage; and
- Increase in climate finance not connected to ODA.

The main point being that there is a conflict between the OECD states budget deficit cuts, the need to help developing countries adapt to develop sustainably and the need to ensure that funding does not come from cutting aid to other important Millennium Development Goals.

However, none of these initiatives suggest a quantitative cap on the emissions from developing countries. This is considered as a particularly difficult policy proposal as the economic growth of developing countries are proportionally reflected in the growth of greenhouse emissions. Critics of mitigation often argue that, the developing countries' drive to attain a comparable living standard to the developed countries would doom the attempt at mitigation of global warming. Critics also argue that holding down emissions would shift the human cost of global warming from a general one to one that was borne most heavily by the poorest populations on the planet.

In an attempt to provide more opportunities for developing countries to adapt clean technologies, UNEP and WTO urged the international community to reduce trade barriers and to conclude the Doha trade round "which includes opening trade in environmental goods and services".

Non-governmental Approaches

While many of the proposed methods of mitigating global warming require governmental funding, legislation and regulatory action, individuals and businesses can also play a part in the mitigation effort.

Choices in Personal Actions and Business Operations

Environmental groups encourage individual action against global warming, often aimed at the consumer. Common recommendations include lowering home heating and cooling usage, burning less gasoline, supporting renewable energy sources, buying local products to reduce transportation, turning off unused devices, and various others.

A geophysicist at Utrecht University has urged similar institutions to hold the vanguard in voluntary mitigation, suggesting the use of communications technologies such as videoconferencing to reduce their dependence on long-haul flights.

Air travel and Shipment

In 2008, climate scientist Kevin Anderson raised concern about the growing effect of rapidly increasing global air transport on the climate in a paper, and a presentation, suggesting that reversing this trend is necessary to reduce emissions.

Part of the difficulty is that when aviation emissions are made at high altitude, the climate impacts are much greater than otherwise. Others have been raising the related concerns of the increasing hypermobility of individuals, whether traveling for business or pleasure, involving frequent and often long distance air travel, as well as air shipment of goods.

Business Opportunities and Risks

On 9 May 2005 Jeff Immelt, the chief executive of General Electric (GE), announced plans to reduce GE's global warming related emissions by one percent by 2012. "GE said that given its projected growth, those emissions would have risen by 40 percent without such action."

On 21 June 2005 a group of leading airlines, airports and aerospace manufacturers pledged to work together to reduce the negative environmental impact of aviation, including limiting the impact of air travel on climate change by improving fuel efficiency and reducing carbon dioxide emissions of new aircraft by fifty percent per seat kilometre by 2020 from 2000 levels. The group aims to develop a common reporting system for carbon dioxide emissions per aircraft by the end of 2005, and pressed for the early inclusion of aviation in the European Union's carbon emission trading scheme.

Investor Response

Climate change is also a concern for large institutional investors who have a long term time horizon and potentially large exposure to the negative impacts of global warming because of the large geographic footprint of their multi-national holdings. SRI (Socially responsible investing) Funds allow investors to invest in funds that meet high ESG (environmental, social, governance) standards as such funds invest in companies that are aligned with these goals. Proxy firms can be used to draft guidelines for investment managers that take these concerns into account.

Legal Action

In some countries, those affected by climate change may be able to sue major producers. Attempts at litigation have been initiated by entire peoples such as Palau and the Inuit, as well as non-governmental organizations such as the Sierra Club. Although proving that particular weather events are due specifically to global warming may never be possible, methodologies have been developed to show the increased risk of such events caused by global warming.

For a legal action for negligence (or similar) to succeed, "Plaintiffs ... must show that, more probably than not, their individual injuries were caused by the risk factor in question, as opposed to any other cause. This has sometimes been translated to a requirement of a relative risk of at least two." Another route (though with little legal bite) is the World Heritage Convention, if it can be shown that climate change is affecting World Heritage Sites like Mount Everest.

Besides countries suing one another, there are also cases where people in a country have taken legal steps against their own government. Legal action for instance has been taken to try to force the U.S. Environmental Protection Agency to regulate greenhouse gas emissions under the Clean Air Act,and against the Export-Import Bank and OPIC for failing to assess environmental impacts (including global warming impacts) under NEPA.

In the Netherlands and Belgium, organisations as Urgenda and the vzw Klimaatzaak in Belgium have also sued their governments as they believe their governments aren't meeting the emission reductions they agreed to. Urgenda has all ready won their case against the Dutch government.

According to a 2004 study commissioned by Friends of the Earth, ExxonMobil and its predecessors caused 4.7 to 5.3 percent of the world's man-made carbon dioxide emissions between 1882 and 2002. The group suggested that such studies could form the basis for eventual legal action. In 2015, Exxon, received a subpoena. According to the Washington Post and confirmed by the company, the attorney general of New York, Eric Schneiderman, opened an investigation into the possibility that the company had mislead the public and investors about the risks of climate change.

ENVIRONMENTAL MIGRANT

Drought refugees from Oklahoma camping by the roadside, California, 1936

Environmental migrants are people who are forced to leave their home region due to sudden or long-term changes to their local environment which compromise their well being or secure livelihood, such changes are held to include increased droughts, desertification, sea level rise, and disruption of seasonal weather patterns such as monsoons. Environmental migrants may flee to or migrate to another country or they may migrate internally within their own country. However, the term "environmental migrant" is used somewhat interchangeably with a range of similar terms, such as **environmental refugee** or **climate refugee**, although the distinction between these terms is contested. Despite problems in formulating a uniform and clear-cut definition of 'environmental migration', such a concept has increased as an issue of concern in the 2000s as policy-makers, environmental and social scientists attempt to conceptualise the potential societal effects of climate change and general environmental degradation.

Definition and Concept

The term "environmental refugee" was first proposed by Lester Brown in 1976, since then there has been a proliferation in the use of the term at which "environmental migrant" and a cluster of similar categories, including "forced environmental migrant", "environmentally motivated migrant", "climate refugee", "climate change refugee", "environmentally displaced person (EDP)», "disaster refugee", "environmental displacee", "eco-refugee", "ecologically displaced person" and "environmental-refugee-to-be (ERTB)" have been utilized. The differences between these terms are less important than what they have in common: they all suggest that there is a determinable relationship between environmental drivers and human migration which is analytically useful, policy-relevant and possibly grounds for the expansion of refugee law.

Under the United Nations Convention Relating to the Status of Refugees of 1951, a refugee is more narrowly defined (in Article 1A) as a person who "owing to a well-founded fear of being persecuted for reasons of race, religion, nationality, membership of a particular social group, or political opinion, is outside the country of his nationality, and is unable to or, owing to such fear, is unwilling to avail himself of the protection of that country". While the concept of a refugee was expanded by the Convention's 1967 Protocol and by regional conventions in Africa and Latin America to include persons who had fled war or other violence in their home country, in its present state the convention does not provide long-term legal protection to refugees due to environmental change.

The International Organization for Migration proposes the following definition for environmental migrants:

"Environmental migrants are persons or groups of persons who, for compelling reasons of sudden or progressive changes in the environment that adversely affect their lives or living conditions, are obliged to leave their habitual homes, or choose to do so, either temporarily or permanently, and who move either within their country or abroad."

The term *climate refugees* or *climate migrants* refers to the subset of environmental migrants forced to move "due to sudden or gradual alterations in the natural environment related to at least one of three impacts of climate change: sea-level rise, extreme weather events, and drought and water scarcity" (Global Governance Project 2012). The term *climate exiles* has been used to refer to those climate migrants who may be in danger of becoming stateless.

However, there is yet to be a universally accepted definition of "environmental migration" or "climate refugee". Thus, the International Organization for Migration formulated a working definition which encompasses the complexity of the topic.

This working definition recognizes that

- Environmental migrants are not only those displaced by the environmental event but also those whose migration is triggered by deteriorating environmental conditions
- Environmentally induced movement can take place within as well as across international borders;
- It can be both long and short term; and
- Population movements triggered by environmental forces can be forced as well as a matter of choice

Types

The International Organisation for Migration proposes three types of environmental migrants:

- *Environmental emergency migrants*: people who flee temporarily due to an environmental disaster or sudden environmental event. (Examples: someone forced to leave due to hurricane, tsunami, earthquake, etc.)
- *Environmental forced migrants*: people who have to leave due to deteriorating environmental conditions. (Example: someone forced to leave due to a slow deterioration of their environment such as deforestation, coastal deterioration, etc.)
- *Environmental motivated migrants* also known as *environmentally induced economic migrants*: people who choose to leave to avoid possible future problems. (Example: someone who leaves due to declining crop productivity caused by desertification)

Enumeration

There have been a number of attempts over the decades to enumerate environmental migrants and refugees. Jodi Jacobson (1988) is cited as the first researcher to enumerate the issue, stating that there were already up to 10 million 'Environmental Refugees'. Drawing on 'worst-case scenarios' about sea-level rise, she argued that all forms of 'Environmental Refugees' would be six times as numerous as political refugees. (1988: 38). By 1989, Mustafa Tolba, Executive Director of UNEP, was claiming that 'as many as 50 million people could become environmental refugees' if the world did not act to support sustainable development (Tolba 1989: 25). In 1990, the Intergovernmental Panel on Climate Change (IPCC 1990: 20) declared that the greatest single consequence of climate change could be migration, 'with millions of people displaced by shoreline erosion, coastal flooding and severe drought' (Warner & Laczko: 2008: 235). In the mid-1990s, British environmentalist, Norman Myers, became the most prominent proponent of this 'maximalist' school (Suhrke 1993). Noting, that "environmental refugees will soon become the largest group of involuntary refugees". Additionally, he stated that there were 25 million environmental refugees in the mid-1990s, further claiming that this figure could double by 2010, with an upper limit of 200 million by 2050 (Myers 1997). Myers argued that the causes of environmental displacement would include desertification, lack of water, salination of irrigated lands and the depletion of bio-diversity. He also hypothesised that displacement would amount to 30m in China, 30m in India, 15m in Bangladesh, 14m in Egypt, 10m in other delta areas and coastal zones, 1m in island states, and with otherwise agriculturally displaced people totalling 50m (Myers & Kent 1995) by 2050. More recently, Myers has suggested that the figure by 2050 might be as high as 250 million (Christian Aid 2007: 6).

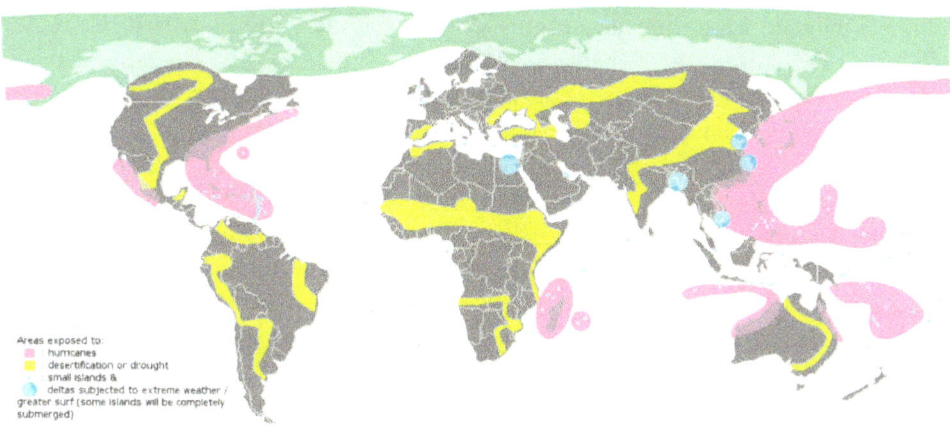

A map showing where natural disasters caused/aggravated by global warming may occur, and thus where environmental refugees would be created

These claims have gained significant currency, with the most common projection being that the world will have 150-200 million climate change refugees by 2050. Variations of this claim have been made in influential reports on climate change by the IPCC (Brown 2008: 11) and the Stern Review on the Economics of Climate Change (Stern et al. 2006: 3), as well as by NGOs such as Friends of the Earth, Greenpeace Germany (Jakobeit and Methmann 2007) and Christian Aid; and inter-governmental organisations such as the Council of Europe, UNESCO, IOM (Brown 2008) and UNHCR.

Norman Myers is the most cited researcher in this field, who found that 25 million environmental migrants existed in 1995 in his work (Myers & Kent 1995), which drew upon over 1000 sources.However, Vikram Kolmannskog has stated that Myers' work can be 'criticized for being inconsistent, impossible to check and failing to take proper account of opportunities to adapt' (2008: 9).Furthermore, Myers himself has acknowledged that his figures are based upon 'heroic extrapolation' (Brown 2008: 12). More generally, Black has argued that there is 'surprisingly little scientific evidence' that indicates that the world is 'filling-up with environmental refugees' (1998: 23). Indeed, Francois Gemenne has stated that: ‹When it comes to predictions, figures are usually based on the number of people living in regions at risk, and not on the number of people actually expected to migrate. Estimates do not account for adaptation strategies [or] different levels of vulnerability' (Gemenne 2009: 159).

Asia and the Pacific

According to the Internal Displacement Monitoring Centre, more than 42 million people were displaced in Asia and the Pacific during 2010 and 2011, more than twice the population of Sri Lanka. This figure includes those displaced by storms, floods, and heat and cold waves. Still others were displaced by drought and sea-level rise. Most of those compelled to leave their homes eventually returned when conditions improved, but an undetermined number became migrants, usually within their country, but also across national borders.

Climate-induced migration is a highly complex issue which needs to be understood as part of global migration dynamics. Migration typically has multiple causes, and environmental factors are intertwined with other social and economic factors, which themselves can be influenced by environmental changes. Environmental migration should not be treated solely as a discrete category, set apart from other migration flows. A 2012 Asian Development Bank study argues that climate-induced migration should be addressed as part of a country's development agenda, given the major implications of migration on economic and social development. The report recommends interventions both to address the situation of those who have migrated, as well as those who remain in areas subject to environmental risk. It says: "To reduce migration compelled by worsening environmental conditions, and to strengthen resilience of at-risk communities, governments should adopt polices and commit financing to social protection, livelihoods development, basic urban infrastructure development, and disaster risk management."

Additionally, it is maintained that the poor populate areas that are most at risk for environmental destruction and climate change, including coastlines, flood-lines and steep slopes. As a result, climate change threatens areas already suffering from extreme poverty. "The issue of equity is crucial. Climate affects us all, but does not affect us all equally," UN Secretary-General Ban Ki-moon told delegates at a climate conference in Indonesia. Africa is also one of the world regions where environmental displacement is critical largely due to droughts and other climate related eventualities.

In Minqin county, Gansu Province, "10,000 people have left the area and have become *shengtai yimin*, 'ecological migrants'".

In 2013 a claim of a Kiribati man of being a "climate change refugee" under the Convention relating to the Status of Refugees (1951) was determined by the New Zealand High Court to be untenable.The Refugee Convention did not apply as there is no persecution or serious harm related to any of the five stipulated convention grounds. The Court rejected the argument that the international community itself (or countries which can be said to have been historically high emitters of carbon dioxide or other greenhouse gases) were the "persecutor" for the purposes of the Refugee Convention. This analysis of the need for the person to identify persecution of the type descried in the Refugee Convention does not exclude the possibility that a people for countries experiencing severe impacts of climate change can come with the Refugee Convention. However it is not the climate change event itself, rather the social and political response to climate change, which is likely to create the pathway for a successful claim. The New Zealand Immigration and Protection Tribunal and the High Court, "there is a complex inter-relationship between natural disasters, environmental degradation and human vulnerability. Sometimes a tenable pathway to international protection under the Refugee Convention can result. Environmental issues sometimes lead to armed conflict. There may be ensuing violence towards or direct repression of an entire section of a population. Humanitarian relief can become politicised, particularly in situations where some group inside a disadvantaged country is the target of direct discrimination." The New Zealand Court of Appeal also rejected the claim in a 2014 decision. On further appeal the New Zealand Supreme Court confirmed the earlier adverse rulings against the application for refugee status, with the Supreme Court also rejecting the proposition "that environmental degradation resulting from climate change or other natural disasters could never create a pathway into the Refugee Convention or protected person jurisdiction."

In 2014 attention was drawn to an appeal to the New Zealand Immigration and Protection Tribunal against the deportation of a Tuvaluan family on the basis that they were "climate change refugees", who would suffer hardship resulting from the environmental degradation of Tuvalu. However the subsequent grant of residence permits to the family was made on grounds unrelated to the refugee claim. The family was successful in their appeal because, under the relevant immigration legislation, there were "exceptional circumstances of a humanitarian nature" that justified the grant of resident permits as the family was integrated into New Zealand society with a sizeable extended family which had effectively relocated to New Zealand.

North America

Isle de Jean Charles, Louisiana, home of a tribe of Biloxi-Chitimacha-Choctaw Indians, is being depopulated with federal grant money, due to saltwater intrusion and sea level rise. The states of New York and New Jersey implemented buyout programs for some homeowners displaced by Hurricane Sandy, which was worsened by global warming's effect on sea surface temperatures.

Political and Legal Perspectives

The International Organization for Migration (IOM) expects the scale of global migration to rise as a result of accelerated climate change. It therefore recommends policymakers around the world to take a proactive stance on the matter. The IOM is composed of 146 member states and 13 observer states and "works closely with governments in promoting migration management that ensures humane and orderly migration that is beneficial to migrants and societies." Additionally, When interviewing Oliver- Smith, an anthropologist and member of the UN group, National Geographic Magazine noted that "there are at least 20 million environmental refugees worldwide, the [UN] group says – more than those displaced by war and political repression combined." Therefore, it is imperative that we begin to recognize this recent division of refugee.

The Environmental Justice Foundation (EJF) has argued that the people who will be forced to move due to climate change currently have no adequate recognition in international law. The EJF contends that a new multilateral legal instrument is required to specifically address the needs of 'climate refugees' in order to confer protection to those fleeing environmental degradation and climate change. They have also asserted that additional funding is needed to enable developing countries to adapt to climate change. Sujatha Byravan and Sudhir Chella Rajan have argued for the use of the term ‹climate exiles' and for international agreements to provide them political and legal rights, including citizenship in other countries, bearing in mind those countries' responsibilities and capabilities.

In some cases, climate change may lead to conflict arising between countries that as a result of flooding or other conditions produce a large number of refugees, and bordering countries that build fences to keep out these refugees. The Bangladesh–India border is largely separated via a fence, and case studies suggest the possibility of violent conflict arising due to people fleeing from areas suffering from destruction of arable land. Current migration has already resulted in low-scale conflicts.

Popular Culture

German artist Hermann Josef Hack's *World Climate Refugee Camp* in Hannover displaying 600 small climate refugee tents.

The notion of ‹environmental migrant', and particularly ‹climate refugee', has gained traction in popular culture. A documentary entitled *Climate Refugees* has been released. "Climate Refugees" is an Official Selection for the 2010 Sundance Film Festival.More recently, Short Documentary Academy Award Nominee, Sun Come Up (2011), tells the story of Carteret islanders who are forced to leave their ancestral land in response to climate change and migrate to war-torn Bougainville

Since 2007, German artist Hermann Josef Hack has shown his World Climate Refugee Camp in the centers of various European cities. The model camp, made of roughly 1000 miniature tents, is a public art intervention that depicts the social impacts of climate change.

ABRUPT CLIMATE CHANGE

An **abrupt climate change** occurs when the climate system is forced to transition to a new climate state at a rate that is determined by the climate system energy-balance, and which is more rapid than the rate of change of the external forcing. Past events include the end of the Carboniferous Rainforest Collapse, Younger Dryas, Dansgaard-Oeschger events, Heinrich events and possibly also the Paleocene-Eocene thermal maximum. The term is also used within the context of global warming to describe sudden climate change that is detectable over the time-scale of a human lifetime. One proposed reason for the observed abrupt climate change is that feedback loops within the climate system both enhance small perturbations and cause a variety of stable states.

Timescales of events described as ‹abrupt' may vary dramatically. Changes recorded in the climate of Greenland at the end of the Younger Dryas, as measured by ice-cores, imply a sudden warming of +10 °C (+18 °F) within a timescale of a few years. Other abrupt changes are the +4 °C (+7.2 °F) on Greenland 11,270 years ago or the abrupt +6 °C (11 °F) warming 22,000 years ago on Antarctica. By contrast, the Paleocene-Eocene thermal maximum may have initiated anywhere between a few decades and several thousand years. Finally, Earth Systems models project that under ongoing greenhouse gas emissions as early as 2047, the Earth's near surface temperature could depart from the range of variability in the last 150 years, affecting over 3 billion people and most places of great species diversity on Earth.

According to the Committee on Abrupt Climate Change of the National Research Council:

There are essentially two definitions of abrupt climate change:

- In terms of physics, it is a *transition of the climate system into a different mode on a time scale that is faster than the responsible forcing.*
- In terms of impacts, *"an abrupt change is one that takes place so rapidly and unexpectedly that human or natural systems have difficulty adapting to it".*

These definitions are complementary: the former gives some insight into how abrupt climate change comes about ; the latter explains why there is so much research devoted to it.

Current Situation

The IPCC states that global warming "could lead to some effects that are abrupt or irreversible".

In an article in *Science*, Richard Alley et al. said "it is conceivable that human forcing of climate change is increasing the probability of large, abrupt events. Were such an event to recur, the economic and ecological impacts could be large and potentially serious."

A 2013 report from the U.S. National Research Council called for attention to the abrupt impacts of climate change, stating that even steady, gradual change in the physical climate system can have abrupt impacts elsewhere—in human infrastructure and ecosystems for example—if critical thresholds are crossed. The report emphasizes the need for an early warning system that could help society better anticipate sudden changes and emerging impacts.

Regional Changes

Lenton et al. investigated tipping elements in the climate system. These were regional effects of global warming, some of which had abrupt onset and may therefore be regarded as abrupt climate change. They found that "Our synthesis of present knowledge suggests that a variety of tipping elements could reach their critical point within this century under anthropogenic climate change."

Ocean Effects

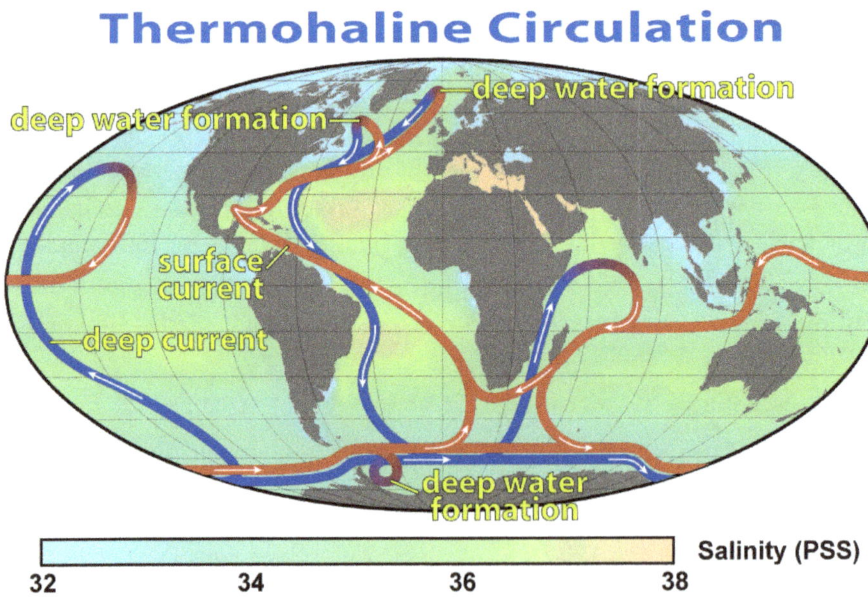

A summary of the path of the thermohaline circulation. Blue paths represent deep-water

Shutdown of thermohaline circulation

Cold blob (North Atlantic)

Global oceans have established patterns of currents. Several potential disruptions to this system of currents have been identified as a result of global warming:

- Increasing frequency of El Niño events
- Potential disruption to the thermohaline circulation, such as that which may have occurred during the Younger Dryas event.
- Changes to the North Atlantic oscillation

Effects on Weather

Hansen et al. 2015 found, that the shutdown or substantial slowdown of the Atlantic Meridional Overturning Circulation (AMOC), besides possibly contributing to extreme end-Eemian events, will cause a more general increase of severe weather. Additional surface cooling from ice melt increases surface and lower tropospheric temperature gradients, and causes in model simulations a large increase of mid-latitude eddy energy throughout the midlatitude troposphere. This in turn leads to an increase of baroclinicity produced by stronger temperature gradients, which provides energy for more severe weather events.

Many of the most memorable and devastating storms in eastern North America and western Europe, popularly known as superstorms, have been winter cyclonic storms, though sometimes occurring in late fall or early spring, that generate near-hurricane-force winds and often large amounts of snowfall. Continued warming of low latitude oceans in coming decades will provide more water vapor to strengthen such storms. If this tropical warming is combined with a cooler North Atlantic Ocean from AMOC slowdown and an increase in midlatitude eddy energy, we can anticipate more severe baroclinic storms.

Our results at least imply that strong cooling in the North Atlantic from AMOC shutdown does create higher wind speed. The increment in seasonal mean wind speed of the northeasterlies relative to preindustrial conditions is as much as 10–20 %. Such a percentage increase of wind speed in a storm translates into an increase of storm power dissipation by a factor ~1.4–2, because wind power dissipation is proportional to the cube of wind speed. However, our simulated changes refer to seasonal mean winds averaged over large grid-boxes, not individual storms.

Climate Feedback Effects

One source of abrupt climate change effects is a feedback process, in which a warming event causes a change which leads to further warming. This can also apply to cooling. Example of such feedback processes are:

- Ice-albedo feedback, where the advance or retreat of ice cover alters the ‹whiteness' of the earth, and its ability to absorb the sun's energy
- The dying and burning of forests, as a result of global warming

Past Events

The Younger Dryas period of abrupt climate change is named after the Alpine flower, Dryas.

Several periods of abrupt climate change have been identified in the paleoclimatic record. Notable examples include:

- About 25 climate shifts, called Dansgaard-Oeschger cycles, which have been identified in the ice core record during the glacial period over the past 100,000 years.
- The Younger Dryas event, notably its sudden end. It is the most recent of the Dansgaard-Oeschger cycles and began 12,900 years ago and moved back into a warm-and-wet climate regime about 11,600 years ago. It has been suggested that: "The extreme rapidity of these changes in a variable that directly represents regional climate implies that the events at the end of the last glaciation may have been responses to some kind of threshold or trigger in the North Atlantic climate system." A model for this event based on disruption to the thermohaline circulation has been supported by other studies.
- The Paleocene-Eocene Thermal Maximum, timed at 55 million years ago, which may have been caused by the clathrate gun effect, although potential alternative mechanisms have been identified. This was associated with rapid ocean acidification
- The Permian-Triassic Extinction Event, also known as the great dying, in which up to 95% of all species became extinct, has been hypothesized to be related to

a rapid change in global climate. Life on land took 30 million years to recover.

- The Carboniferous Rainforest Collapse occurred 300 million years ago, at which time tropical rainforests were devastated by climate change. The cooler, drier climate had a severe effect on the biodiversity of amphibians, the primary form of vertebrate life on land.

There are also abrupt climate changes associated with the catastrophic draining of glacial lakes. One example of this is the 8.2 kiloyear event, which associated with the draining of Glacial Lake Agassiz.Another example is the Antarctic Cold Reversal, c. 14,500 years before present (BP), which is believed to have been caused by a meltwater pulse from the Antarctic ice sheet. These rapid meltwater release events have been hypothesized as a cause for Dansgaard-Oeschger cycles,

Abrupt Climate Shifts Since 1976

Had the 1997 El Niño lasted twice as long, the rain forests of the Amazon basin and Southeast Asia could have quickly added much additional carbon dioxide to the air from burning and rotting, with heat waves and extreme weather quickly felt around the world (The "Burn Locally, Crash Globally" scenario.)

Most abrupt climate shifts, however, are likely due to sudden circulation shifts, analogous to a flood cutting a new river channel. The best-known examples are the several dozen shutdowns of the North Atlantic Ocean's Meridional Overturning Circulation during the last ice age, affecting climate worldwide. But there have been a series of less dramatic abrupt climate shifts since 1976, along with some near misses.

- The circulation shift in the western Pacific in the winter of 1976-1977 proved to have much wider impacts.
- Since 1950, El Niños had been weak and short, but La Niñas were often big and long, This pattern reversed after 1977.
- Land temperatures had remained relatively trendless from 1950 to 1976, despite the CO_2 rising from 310 to 332 ppm as fossil fuel emissions tripled. Then in 1977 there was a marked shift in observed global mean surface temperature to a rising fever of about 2 °C/century.
- The expansion of the tropics from overheating is usually thought to be gradual, but the percentage of the land surface in the two most extreme classifications of drought suddenly doubled in 1982 and stayed there until 1997 when it jumped to triple (after six years, it stepped down to double). While their inceptions correlate with the particularly large El Niños of 1982 and 1997, the global drought steps far outlast the 13-month durations of those El Niños.
- There were near-misses for "Burn Locally, Crash Globally" in Amazonia in 1998, 2005, and 2007, each with higher flammability than its predecessor.
- There have also been two occasions when the Atlantic's Meridional Overturning Circulation lost a crucial safety factor. The Greenland Sea flushing at 75 °N shut down in 1978, recovering over the next decade. Then the second-largest flushing site, the Labrador Sea, shut down in 1997 for ten years. While shutdowns

overlapping in time have not been seen during the fifty years of observation, previous total shutdowns had severe worldwide climate consequences.

This makes abrupt climate shifts more like a heart attack than like a chronic disease whose course can be extrapolated. Like heart attacks, some abrupt climate shifts are minor, some are catastrophic—and one cannot predict which or when. The recent track record, however, is that there have been several sudden shifts and several near-misses in each decade since 1976.

Consequential Effects

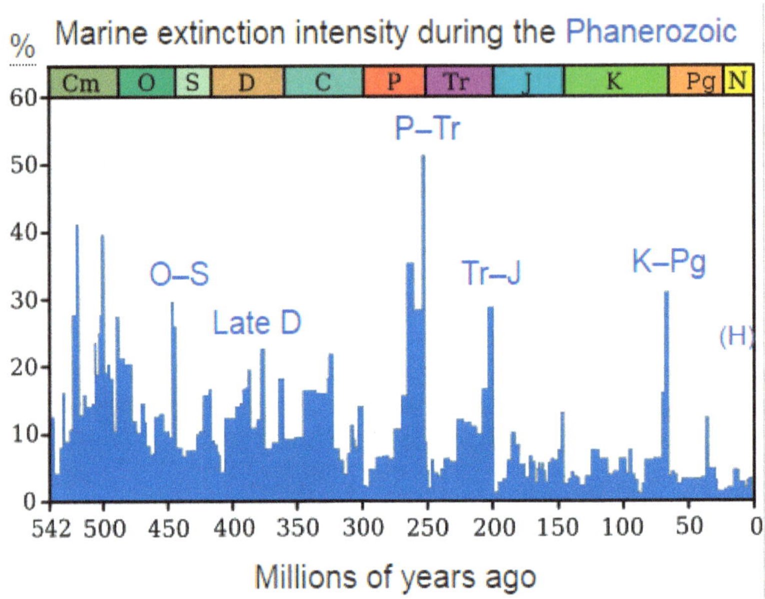

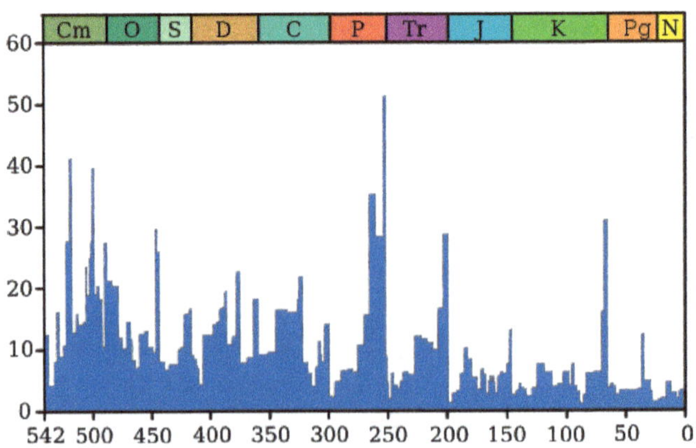

The Permian–Triassic extinction event, labelled "P-Tr" here, is the most significant extinction event in this plot for marine genera.

Abrupt climate change has likely been the cause of wide ranging and severe effects:

- Mass extinctions in the past, most notably the Permian-Triassic Extinction event (often referred to as the great dying) and the Carboniferous Rainforest Collapse, have been suggested as a consequence of abrupt climate change.
- Loss of biodiversity. Without interference from abrupt climate change and other extinction events the biodiversity of this planet would continue to grow.
- Rapid Ocean acidification, which can harm marine life (such as corals).

REFERENCES

1. Marland, G., T.A. Boden, and R. J. Andres. 2007. Global, Regional, and National CO2 Emissions. In Trends: A Compendium of Data on Global Change. Carbon Dioxide Information Analysis Center, Oak Ridge National Laboratory, United States Department of Energy, Oak Ridge, Tenn., U.S.A.
2. Fisher, B.S.; et al., "Ch. 3: Issues related to mitigation in the long-term context", Contribution of Working Group III to the Fourth Assessment Report of the Intergovernmental Panel on Climate Change, 2007, 3.5 Interaction between mitigation and adaptation, in the light of climate change impacts and decision-making under long-term uncertainty, in IPCC AR4 WG3 2007
3. c IPCC, "Summary for policymakers", Climate Change 2007: Working Group III: Mitigation of Climate Change, Table SPM.3, C. Mitigation in the short and medium term (until 2030), in IPCC AR4 WG3 2007
4. Oppenheimer, M., et al., Section 19.7.1: Relationship between Adaptation Efforts, Mitigation Efforts, and Residual Impacts, in: Chapter 19: Emergent risks and key vulnerabilities (archived July 8 2014), pp.46–49, in IPCC AR5 WG2 A 2014
5. "Social, Economic, and Ethical Concepts and Methods, Executive Summary" (PDF), Climate Change 2014: Mitigation of Climate Change, in IPCC AR5 WG3 2014, p. 211
6. Sec 5.5 Technology flows and development", Climate Change 2007: Synthesis Report, in IPCC AR4 SYR 2007, p. 68
7. Levine, M.; et al., "Ch 6: Residential and commercial buildings", Climate Change 2007: Working Group III: Mitigation of Climate Change, Sec 6.4.2 Thermal envelope, in IPCC AR4 WG3 2007
8. Oppenheimer, M., et al., FAQ 19.1, in: Chapter 19: Emergent risks and key vulnerabilities (archived July 8 2014), p.52, in IPCC AR5 WG2 A 2014
9. UNFCCC. Conference of the Parties (COP) (15 March 2011), Report of the Conference of the Parties on its sixteenth session, held in Cancun from 29 November to 10 December 2010. Addendum. Part two: Action taken by the Conference of the Parties at its sixteenth session (PDF), Geneva, Switzerland: United Nations, p.3, paragraph 4. Document available in UN languages and text format.
10. Sutter, John D.; Berlinger, Joshua (12 December 2015). "Final draft of climate

deal formally accepted in Paris". CNN. Cable News Network, Turner Broadcasting System, Inc. Retrieved 12 December 2015.

11. Victor, D., et al., Executive summary, in: Chapter 1: Introductory Chapter, p.4 (archived 3 July 2014), in IPCC AR5 WG3 2014
12. The Radical Emissions Reduction Conference, 2013. Held at the Royal Society, Carlton House Terrace, London. 10–11 Dec.
13. The Guardian, 2013. 'Every little helps' is a dangerous mantra for climate change. By Adam Corner, 13 Dec.
14. New Statesman, 2013. How science is telling us all to revolt. By Naomi Klein, 29 Oct.
15. Anderson, K. & Bows, A., 2011. Beyond 'dangerous' climate change: emission scenarios for a new world. Philos. Trans. Royal Society A.
16. Anderson K. (2011, July). Climate Change: Going beyond dangerous – Brutal numbers, tenuous hope, or cognitive dissonance? Video of presentation made to UK Dept. for Internal Development.
17. 4 Degrees and Beyond Conference, official website: Programme, abstracts, presentations, and audio recordings.
18. Meehl, G.A.; et al., "Ch. 10: Global Climate Projections", Climate Change 2007: Working Group I: The Physical Science Basis, FAQ 10.3: If Emissions of Greenhouse Gases are Reduced, How Quickly do Their Concentrations in the Atmosphere Decrease?, in IPCC AR4 WG1 2007, pp. 824–825
19. Rogner, H.-H.; et al. (2007). "1.2 Ultimate objective of the UNFCCC". In B. Metz et al. Introduction. Climate Change 2007: Mitigation. Contribution of Working Group III to the Fourth Assessment Report of the Intergovernmental Panel on Climate Change. Print version: Cambridge University Press, Cambridge, United Kingdom and New York, NY, USA. This version: IPCC website. Retrieved 2011-06-07.
20. Forster, P.; et al., "Ch. 2: Changes in Atmospheric Constituents and in Radiative Forcing", Contribution of Working Group I to the Fourth Assessment Report of the Intergovernmental Panel on Climate Change, 2007, FAQ 2.1 How do Human Activities Contribute to Climate Change and How do They Compare with Natural Influences?, in IPCC AR4 WG1 2007, p. 135
21. IPCC, "Summary for Policymakers", Contribution of Working Group I to the Fourth Assessment Report of the Intergovernmental Panel on Climate Change, 2007, Human and Natural Drivers of Climate Change, in IPCC AR4 WG1 2007
22. U.S. Climate Change Science Program and the Subcommittee on Global Change Research (January 2009). Granger Morgan; H. Dowlatabadi; M. Henrion; D. Keith; R. Lempert; S. McBride; M. Small; T. Wilbanks, eds. "Best practice approaches for characterizing, communicating, and incorporating scientific uncertainty in decisionmaking". National Oceanic and Atmospheric Administration, Washington D.C., USA. pp. 10–11. Retrieved 2010-06-07.
23. Sterman, J.D.; L.B. Sweeney (2007). "Understanding public complacency about

climate change: adults' mental models of climate change violate conservation of matter" (PDF). Climatic Change. 80 (3–4): 221–222. doi:10.1007/s10584-006-9107-5. Retrieved 2011-05-10.

24. 2. Stabilization and Climate Change of the Next Few Decades and Next Several Centuries, p.21, in: Summary, in US NRC 2011
25. Anderson, Kevin; Bows, Alice (13 January 2011). "Beyond ‹dangerous' climate change: emission scenarios for a new world". Philosophical transactions. Series A, Mathematical, physical, and engineering sciences. 369 (1934): 20–44. doi:10.1098/rsta.2010.0290. PMID 21115511.
26. Anderson, Kevin; Bows, Alice (2012). "A new paradigm for climate change". Nature Climate Change. 2 (9): 639–640. doi:10.1038/nclimate1646.
27. Anderson K. (2012). Real clothes for the Emperor: Facing the challenges of climate change. The Cabot annual lecture, Univ. of Bristol. Video, Transcript
28. The Radical Emission Reduction Conference: 10–11 December 2013, sponsored by the Tyndall Centre. Video proceedings on-line.
29. Fisher, B.S.; et al., "Ch 3: Issues related to mitigation in the long-term context", Contribution of Working Group III to the Fourth Assessment Report of the Intergovernmental Panel on Climate Change, 2007, Sec 3.1 Emissions scenarios, in IPCC AR4 WG3 2007
30. Rogner, H.-H.; et al., "Ch 1: Introduction", Contribution of Working Group III to the Fourth Assessment Report of the Intergovernmental Panel on Climate Change, 2007, Sec 1.3.2.4 Total GHG emissions, in IPCC AR4 WG3 2007, p. 111
31. Fisher, B.S.; et al., "Ch 3: Issues related to mitigation in the long-term context", Contribution of Working Group III to the Fourth Assessment Report of the Intergovernmental Panel on Climate Change, 2007, Sec 3.3 Mitigation scenarios, in IPCC AR4 WG3 2007
32. Fisher, B.S.; et al., "Ch 3: Issues related to mitigation in the long-term context", Contribution of Working Group III to the Fourth Assessment Report of the Intergovernmental Panel on Climate Change, 2007, Table 3.5, in: Sec 3.3.5 Long-term stabilization scenarios, in IPCC AR4 WG3 2007
33. BP: Statistical Review of World Energy, Workbook (xlsx), London, 2012
34. World energy outlook 2012 (IEA)
35. World Consumption of Primary Encrgy by Energy Type and Selected Country Groups 31 December 2008 Microsoft Excel file format
36. Sec 5.5 Technology flows and development", Climate Change 2007: Synthesis Report, in IPCC AR4 SYR 2007
37. Issues in Science & Technology Online; "Promoting Low-Carbon Electricity Production"
38. Sec 4.3 Mitigation options", Climate Change 2007: Synthesis Report, in IPCC AR4 SYR 2007
39. Kahn Ribeiro, S.; et al., "Ch 5: Transport and its infrastructure", Contribution of Working Group III to the Fourth Assessment Report of the Intergovernmental

Panel on Climate Change, 2007, Non-motorized transport (NMT), in: Sec 5.3.1.5 Road transport: mode shifts, in IPCC AR4 WG3 2007

40. Impacts assessment of plug-in hybrid vehicles on electric utilities and regional u.s. power grids" (PDF). Pacific Northwest National Laboratory. 2010.
41. Table 4.2, in: Sec 4.3 Mitigation options", Climate Change 2007: Synthesis Report, in IPCC AR4 SYR 2007
42. Pacala, Stephen; Socolow, Robert H. (2004). "Stabilization Wedges: Solving the Climate Problem for the Next 50 Years with Current Technologies" (PDF). Science. AAAS. 305 (5686): 968–972. Bibcode:2004Sci...305..968P. doi:10.1126/science.1100103. PMID 15310891. See also: "Stabwedge". CMI (Carbon Mitigation Initiative) at Princeton University. Resources for Pacala & Socolow(2004)
43. Romm, Joe (19 June 2008). "Cleaning up on carbon". Nature Reports Climate Change. Retrieved 2 January 2013.
44. Sathaye, J.; et al., "Ch 12: Sustainable Development and mitigation", Contribution of Working Group III to the Fourth Assessment Report of the Intergovernmental Panel on Climate Change, 2007, Sec 12.2.1.1 Development paths as well as climate policies determine GHG emissions, in IPCC AR4 WG3 2007, pp. 701–703
45. Morita, T.; et al., "Ch 2. Greenhouse Gas Emission Mitigation Scenarios and Implications", Climate Change 2001: Working Group III: Mitigation, Sec 2.5.2.2 Storylines of Post-SRES Mitigation Scenarios, in IPCC TAR WG3 2001, pp. 149–150
46. Edwin Cartlidge (18 November 2011). "Saving for a rainy day". Science (Vol 334). pp. 922–924.
47. IEA Renewable Energy Working Party (2002). Renewable Energy... into the mainstream, p. 9.
48. HM Treasury (2006). Stern Review on the Economics of Climate Change.
49. International Energy Agency. IEA urges governments to adopt effective policies based on key design principles to accelerate the exploitation of the large potential for renewable energy 29 September 2008.
50. REN21 (2006). Changing climates: The Role of Renewable Energy in a Carbon-constrained World(PDF) p. 2.
51. New UN report points to power of renewable energy to mitigate carbon emissions UN News Centre, 8 December 2007.
52. Joel Makower, Ron Pernick and Clint Wilder (2008). Clean Energy Trends 2008, Clean Edge, p. 2.
53. United Nations Environment Programme and New Energy Finance Ltd. (2007). Global Trends in Sustainable Energy Investment 2007: Analysis of Trends and Issues in the Financing of Renewable Energy and Energy Efficiency in OECD and Developing Countries (PDF) p. 3.
54. REN21 (2010). Renewables 2010 Global Status Report p. 15.
55. Conclusion". Worldwide electricity production from renewable energy sources.

Paris: Observ'ER. 2012. Retrieved 28 March 2013.

56. http://www.ren21.net/Portals/0/documents/Resources/GSR/2014/GSR2014_full%20report_low%20res.pdf
57. Paul Gipe (4 April 2013). "100 Percent Renewable Vision Building". Renewable Energy World.
58. IPCC (2011). "Special Report on Renewable Energy Sources and Climate Change Mitigation" (PDF). Cambridge University Press, Cambridge, United Kingdom and New York, NY, USA. p. 17. Archived from the original (PDF) on 2014-01-11.
59. International Renewable Energy Agency (2012). "Renewable Power Generation Costs in 2012: An Overview" (PDF).
60. Donald W. Aitken. Transitioning to a Renewable Energy Future, International Solar Energy Society, January 2010, p. 3.
61. REN21 (2012). Renewables Global Status Report 2012 Archived December 15, 2012, at the Wayback Machine. p. 17.
62. REN21 (2011). Renewables 2011: Global Status Report" (PDF). pp. 11–13. Archived from the original (PDF) on 2011-09-05.
63. Top of the list, Renewable Energy World, 2 January 2006.
64. Keith Johnson, Wind Shear: GE Wins, Vestas Loses in Wind-Power Market Race, Wall Street Journal, March 25th 2009, accessed on January 7th 2010.
65. Mark A. Delucchi; Mark Z. Jacobson (2011). "Providing all global energy with wind, water, and solar power, Part II: Reliability, system and transmission costs, and policies" (PDF). Energy Policy. Elsevier Ltd. pp. 1170–1190.
66. Ben Sills (August 29, 2011). "Solar May Produce Most of World's Power by 2060, IEA Says". Bloomberg.
67. Smil, Vaclav (2012). "A Skeptic Looks at Alternative Energy". Spectrum magazine. IEEE. Retrieved 28 March 2013.
68. Japan's nuclear disaster boosts renewables". UPI.com. March 21, 2011.
69. John Vidall (15 March 2011). "Japan nuclear crisis prompts surging investor confidence in renewables". The Guardian. London.
70. Lindsay Morris (25 January 2012). "Obama: Sticking to "Promise of Clean Energy"". Renewable Energy World.
71. REN21 (2010). Renewables 2010 Global Status Report Archived April 16, 2012, at the Wayback Machine. p. 9 & 34.
72. REN21 (2010). Renewables 2010 Global Status Report p. 53.
73. The Nuclear Renaissance – World Nuclear Association".
74. Nuclear Renaissance Threatened as Japan's Reactor Struggles Bloomberg, published March 2011, accessed 2011-03-14
75. Analysis: Nuclear renaissance could fizzle after Japan quake Reuters, published 2011-03-14, accessed 2011-03-14

76. Japan nuclear woes cast shadow over U.S. energy policy Reuters, published 2011-03-13, accessed 2011-03-14
77. NEWS ANALYSIS: Japan crisis puts global nuclear expansion in doubt". Platts. 21 March 2011.
78. WNA (20 June 2013). "Nuclear power down in 2012". World Nuclear News.
79. Sims, R.E.H.; et al., "Ch. 4: Energy supply", Climate Change 2007: Working Group III: Mitigation of Climate Change, Executive summary, in IPCC AR4 WG3 2007
80. IAEA 2008, p. 42
81. IEA (16 June 2010), Technology Roadmap: Nuclear Energy. 2010 Edition (PDF), Organization for Economic Co-operation and Development (OECD) / International Energy Agency (IEA) and OECD / Nuclear Energy Agency (NEA), pp.5–6. Also available in Chinese and Italian
82. Kharecha, Pushker A.; Hansen, James E. (7 May 2013). "Prevented Mortality and Greenhouse Gas Emissions from Historical and Projected Nuclear Power". Environ. Sci. Technol. 47 (9): 4889–4895. doi:10.1021/es3051197.
83. Brian Martin. Opposing nuclear power: past and present, Social Alternatives, Vol. 26, No. 2, Second Quarter 2007, pp. 43–47.
84. M.V. Ramana (July 2011). Nuclear power and the public". Bulletin of the Atomic Scientists. p. 44.
85. Mark Cooper (July 2011). The implications of Fukushima: The US perspective". Bulletin of the Atomic Scientists. p. 9.
86. Choi, Hangbok; Baxter, Alan (1 May 2010). "A comparative study on recycling spent fuels in gas-cooled fast reactors". Annals of Nuclear Energy. 37 (5): 723–729. doi:10.1016/j.anucene.2010.01.014.
87. Publication: Key World Energy Statistics 2016" (PDF). p. 25.
88. World Power Reactors". Archived from the original on March 3, 2008. Retrieved 2016-05-06.
89. Warner, E. S.; Heath, G. A. (2012). "Life Cycle Greenhouse Gas Emissions of Nuclear Electricity Generation". Journal of Industrial Ecology. 16: S73. doi:10.1111/j.1530-9290.2012.00472.x.
90. IPCC Working Group III – Mitigation of Climate Change, Annex II I: Technology – specific cost and performance parameters" (PDF). IPCC. 2014. p. 10. Retrieved 1 August 2014.
91. Free exchange: Sun, wind and drain". The Economist. 26 July 2014. Sun, wind and drain Wind and solar power are even more expensive than is commonly thought
92. Comparing the Costs of Intermittent and Dispatchable Electricity-Generating Technologies", by Paul Joskow, Massachusetts Institute of Technology, September 2011
93. Archived copy" (PDF). Archived from the original (PDF) on October 17, 2008.

Retrieved November 21, 2012.

94. Qvist, Staffan A.; Brook, Barry W. (13 May 2015). "Potential for Worldwide Displacement of Fossil-Fuel Electricity by Nuclear Energy in Three Decades Based on Extrapolation of Regional Deployment Data". PLoS ONE. 10 (5): e0124074. doi:10.1371/journal.pone.0124074. PMC 4429979. PMID 25970621.

95. Brook, Barry W. (1 March 2012). "Could nuclear fission energy, etc., solve the greenhouse problem? The affirmative case". Energy Policy. 42: 4–8. doi:10.1016/j.enpol.2011.11.041.

96. Loftus, Peter J.; Cohen, Armond M.; Long, Jane C. S.; Jenkins, Jesse D. (1 January 2015). "A critical review of global decarbonization scenarios: what do they tell us about feasibility?". WIREs Clim Change. 6 (1): 93–112. doi:10.1002/wcc.324.

97. A critical review of global decarbonization scenarios: what do they tell us about feasibility? Open access PDF

98. A critical review of global decarbonization scenarios: what do they tell us about feasibility? Open access PDF. Figure 6

99. Trevor Findlay. The Future of Nuclear Energy to 2030 and its Implications for Safety, Security and Nonproliferation February 4, 2010.

100. Trevor Findlay (2010). The Future of Nuclear Energy to 2030 and its Implications for Safety, Security and Nonproliferation: Overview, The Centre for International Governance Innovation (CIGI), Waterloo, Ontario, Canada, pp. 10–11.

101. M.V. Ramana. Nuclear Power: Economic, Safety, Health, and Environmental Issues of Near-Term Technologies, Annual Review of Environment and Resources, 2009, 34, pp. 144–145.

102. International Energy Agency, World Energy Outlook, 2009, p. 160.

103. James Kanter. In Finland, Nuclear Renaissance Runs Into Trouble New York Times, May 28, 2009.

104. James Kanter. Is the Nuclear Renaissance Fizzling? Green, 29 May 2009.

105. Rob Broomby. Nuclear dawn delayed in Finland BBC News, 8 July 2009.

106. China Builds Nuclear Reactor for 40% Less Than Cost in France, Areva Says". Bloomberg.

107. Author: Marion Brünglinghaus, ENS, European Nuclear Society. "Nuclear power plants, world-wide". Euronuclear.org. Retrieved 2016-05-06.

108. World Nuclear Association (December 10, 2010). Nuclear Power in China

109. Nuclear Power in the USA". World Nuclear Association. June 2008. Retrieved 2008-07-25.

110. Matthew L. Wald (December 7, 2010). Nuclear 'Renaissance' Is Short on Largess The New York Times.

111. Michael Dittmar. Taking stock of nuclear renaissance that never was Sydney Morning Herald, August 18, 2010.

112. Ipsos (9 March 2012), After Fukushima: Global Opinion on Energy Policy

(PDF), p.7. Survey website: After Fukushima: Global Opinion on Energy Policy: Ipsos Public Affairs

113. UK popular support for nuclear power rises -poll". Reuters. 2 July 2016.
114. Annika Breidthardt (May 30, 2011). "German government wants nuclear exit by 2022 at latest". Reuters.
115. Beyond ITER". The ITER Project. Information Services, Princeton Plasma Physics Laboratory. Archived from the original on 7 November 2006. Retrieved 5 February 2011. – Projected fusion power timeline
116. Stewart, C. L.; Stacey, W. M. (1 July 2014). "The SABrR Concept for A Fission-Fusion Hybrid ^{238}U-to-^{239}Pu Fissile Production Reactor". NT. 187 (1). doi:10.13182/NT13-102.
117. Climate Control: a proposal for controlling global greenhouse gas emissions" (PDF). Sustento Institute. Retrieved 2007-12-10.
118. Hackney, Thomas (July 2009). "#7: Moratorium on New Projects for Fossil Fuel Production & Declining Cap on Existing Production" (PDF). BCSEA's Climate Action Portfolio. BC Sustainable Energy Association. Retrieved 2008-04-24.
119. http://www.usclimatenetwork.org/resource-database/report-coal-to-gas-the-influence-of-methane-leakage
120. Greater focus needed on methane leakage from natural gas infrastructure". Pnas.org. Retrieved 2016-05-06.
121. Brandt, A. R.; Heath, G. A.; Kort, E. A.; O'Sullivan, F.; Pétron, G.; Jordaan, S. M.; Tans, P.; Wilcox, J.; Gopstein, A. M.; Arent, D.; others (2014). "Methane leaks from North American natural gas systems"(PDF). Science. 343 (6172): 733–735. Bibcode:2014Sci...343..733B. doi:10.1126/science.1247045. Retrieved 2015-04-29.
122. Staffell Iain; et al. (2012). "A review of domestic heat pumps". Energy and Environmental Science. 5 (11): 9291–9306. doi:10.1039/c2ee22653g.
123. Carvalho et al, Ground source heat pump carbon emissions and primary energy reduction potential for heating in buildings in Europe—results of a case study in Portugal. Renewable and Sustainable Energy Reviews 2015; 45,, 755–768 doi:10.1016/j.rser.2015.02.034.
124. Sternberg André, Bardow André (2015). "Power-to-What? – Environmental assessment of energy storage systems. In". Energy and Environmental Science. 8 (2): 389–400. doi:10.1039/c4ee03051f.
125. Dodge, Edward (December 6, 2014). "Power-to-Gas Enables Massive Energy Storage". TheEnergyCollective.com. Retrieved 25 May 2015.
126. Ground, Lukas; Schulze, Paula; Holstein, Johan (June 20, 2013). Systems Analysis Power to Gas(PDF). Groningen: DNV, KEMA Nederland B.V. Retrieved 25 May 2015.
127. Shell Pearl GTL – Andy Brown," Royal Dutch Shell Company video, March 18, 2011.
128. Scott, Mark (October 7, 2014). "Energy for a Rainy Day, or a Windless One".

New York Times. Retrieved 26 May 2015.

129. Randall, Tom (January 30, 2015). "Seven Reasons Cheap Oil Can't Stop Renewables Now". BloombergBusiness. Bloomberg L.P. Retrieved 26 May 2015.

130. Home – 5th Conference on Carbon Dioxide as Feedstock for Fuels, Chemistry and Polymers". Co2-chemistry.eu. 2008-12-01. Retrieved 2016-05-06.

131. Philips Tornado Asian Compact Fluorescent". Philips. Retrieved 2007-12-24.

132. Diesendorf, Mark (2007). Greenhouse Solutions with Sustainable Energy, UNSW Press, p. 86.

133. Loading Order White Paper" (PDF). Retrieved 2010-07-16.

134. Diesendorf, Mark (2007). Greenhouse Solutions with Sustainable Energy, UNSW Press, p. 87.

135. Sophie Hebden (2006-06-22). "Invest in clean technology says IEA report". Scidev.net. Retrieved 2010-07-16.

136. Climate Change". California Air Resources Board. Retrieved 2015-04-29.

137. Roadmap for moving to a low-carbon economy in 2050 – European Commission". Retrieved 2015-04-29.

138. Wei, Max; Nelson, James H; Greenblatt, Jeffery B; Mileva, Ana; Johnston, Josiah; Ting, Michael; Yang, Christopher; Jones, Chris; McMahon, James E; Kammen, Daniel M (2013-03-01). "Deep carbon reductions in California require electrification and integration across economic sectors". Environmental Research Letters. 8 (1): 014038. Bibcode:2013ERL.....8a4038W. doi:10.1088/1748-9326/8/1/014038. ISSN 1748-9326. Retrieved 2015-03-21.

139. Williams, James H. (2013). "The technology path to deep greenhouse gas emissions cuts by 2050: The pivotal role of electricity" (PDF). Science, 335, 6064, 53–59, 2012. Retrieved 2015-03-21.

140. Natural Gas and the Environment". Naturalgas.org. Retrieved 2011-02-06.

141. Annual Energy Outlook 2015 – Energy Information Administration". Eia.gov. 2015-04-14. Retrieved 2016-05-06.

142. E3 Decarbonizing Pipeline 01-27-2015.pdf (PDF), retrieved 2015-04-14

143. Edenhofer, Ottmar; Pichs-Madruga, Ramón; et al. (2014). "Summary for Policymakers". In IPCC. Climate change 2014: mitigation of climate change. Contribution of Working Group III to the Fifth Assessment Report of thc Intergovernmental Panel on Climate Change (PDF). Cambridge, UK and New York, NY, USA: Cambridge University Press. ISBN 978-1-107-65481-5. Retrieved 2016-06-21.

144. Fleurbaey, Marc; Kartha, Sivan; et al. (2014). "Chapter 4: Sustainable Development and Equity". In IPCC. Climate change 2014: mitigation of climate change. Contribution of Working Group III to the Fifth Assessment Report of the Intergovernmental Panel on Climate Change (PDF). Cambridge, UK and New York, NY, USA: Cambridge University Press. ISBN 978-1-107-65481-5. Retrieved 2016-06-21.

145. Scarborough, Peter; Appleby, Paul N.; Mizdrak, Anja; Briggs, Adam D.M.; Travis, Ruth C.; Bradbury, Kathryn E.; Key, Timothy J. (July 2014). "Dietary greenhouse gas emissions of meat-eaters, fish-eaters, vegetarians and vegans in the UK" (PDF). Climatic Change. 125 (2): 179–192. doi:10.1007/s10584-014-1169-1. PMC 4372775. PMID 25834298. Retrieved 2016-06-21.
146. Harvey, Fiona (21 March 2016). "Eat less meat to avoid dangerous global warming, scientists say". The Guardian. Retrieved 2016-06-20.
147. Milman, Oliver (20 June 2016). "China's plan to cut meat consumption by 50% cheered by climate campaigners". The Guardian. Retrieved 2016-06-20.
148. Carrington, Damian (7 November 2016). "Tax meat and dairy to cut emissions and save lives, study urges". The Guardian. London, United Kingdom. ISSN 0261-3077. Retrieved 2016-11-07.
149. Springmann, Marco; Mason-D'Croz, Daniel; Robinson, Sherman; Wiebe, Keith; Godfray, H Charles J; Rayner, Mike; Scarborough, Peter (7 November 2016). "Mitigation potential and global health impacts from emissions pricing of food commodities". Nature Climate Change. doi:10.1038/nclimate3155. ISSN 1758-678X.
150. Archived copy". Archived from the original on August 11, 2013. Retrieved July 21, 2013.
151. Archived copy". Archived from the original on May 14, 2013. Retrieved July 21, 2013.
152. Archived copy". Archived from the original on August 11, 2013. Retrieved July 21, 2013.
153. Stern, N. (2006). Stern Review on the Economics of Climate Change: Part III: The Economics of Stabilisation. HM Treasury, London: http://hm-treasury.gov.uk/sternreview_index.htm
154. ^ Jump up to:[a] [b] India should follow China to find a way out of the woods on saving forest people". The Guardian. 22 July 2016. Retrieved 2 November 2016.
155. ^ Jump up to:[a] [b] China's forest tenure reforms". rightsandresources.org. Retrieved 7 August 2016.
156. Ding, Helen; Veit, Peter; Gray, Erin; Reytar, Katie; Altamirano, Juan-Carlos; Blackman, Allen; Hodgdon, Benjamin (October 2016). "Climate benefits, tenure costs: The economic case for securing indigenous land rights in the Amazon". World Resources Institute (WRI). Washington DC, USA. Retrieved 2016-11-02.
157. Ding, Helen; Veit, Peter G; Blackman, Allen; Gray, Erin; Reytar, Katie; Altamirano, Juan-Carlos; Hodgdon, Benjamin (2016). Climate benefits, tenure costs: The economic case for securing indigenous land rights in the Amazon (PDF). Washington DC, USA: World Resources Institute (WRI). ISBN 978-1-56973-894-8. Retrieved 2016-11-02.
158. New Jungles Prompt a Debate on Rain Forests". New York Times. 29 January 2009. Retrieved 18 July 2016.
159. Young, E. (2008). IPCC Wrong On Logging Threat to Climate. New Scientist,

August 5, 2008. Retrieved on August 18, 2008, from http://environment.newscientist.com/article/dn14466-ipcc-wrong-on-logging-threat-toclimate.html

160. In Latin America, Forests May Rise to Challenge of Carbon Dioxide". New York Times. 16 May 2016. Retrieved 18 July 2016.
161. How cows could repair the world". nationalgeographic.com. March 6, 2013. Retrieved May 5, 2013.
162. Restoring soil carbon can reverse global warming, desertification and biodiversity". mongabay.com. February 21, 2008. Retrieved May 5, 2013.
163. How eating grass-fed beef could help fight climate change". time.com. January 25, 2010. Retrieved May 11, 2013.
164. Global Status of CCS Report:2011". Global CCS Institute. Retrieved 2011-12-14.
165. OECD Environmental Outlook to 2050, Climate Change Chapter, pre-release version" (PDF). OECD. 2011. Retrieved 2012-04-23.
166. IEA Technology Roadmap Carbon Capture and Storage 2009" (PDF). OECD/IEA. 2009. Retrieved 2012-04-23.
167. Geoengineering the climate: science, governance and uncertainty". The Royal Society. 2009. Retrieved 2012-04-23.
168. Hare, B.; Meinshausen, M. (2006). "How Much Warming are We Committed to and How Much can be Avoided?". Climatic Change. 75: 111–149. doi:10.1007/s10584-005-9027-9.
169. Azar, C., Lindgren, K., Larson, E.D. and Möllersten, K.: (2006)"Carbon capture and storage from fossil fuels and biomass – Costs and potential role in stabilising the atmosphere", Climatic Change, 74, 47–79.
170. OECD Environmental Outlook to 2050, Climate Change Chapter, pre-release version" (PDF). OECD. 2011. Retrieved 2012-01-16.
171. Barker, T.; et al. (2007). 11.2.2 Ocean fertilization and other geo-engineering options. In (book chapter): Mitigation from a cross-sectoral perspective. In: Climate Change 2007: Mitigation. Contribution of Working Group III to the Fourth Assessment Report of the Intergovernmental Panel on Climate Change (B. Metz et al.(eds.)). Print version: Cambridge University Press, Cambridge, U.K., and New York, N.Y., U.S.A.. This version: IPCC website. ISBN 978-0-521-88011-4. Retrieved 2010-04-05.
172. IPCC (2007). C. Mitigation in the short and medium term (until 2030). In (book section): Summary for Policymakers. In: Climate Change 2007: Mitigation. Contribution of Working Group III to the Fourth Assessment Report of the Intergovernmental Panel on Climate Change (B. Metz et al. (eds.)). Print version: Cambridge University Press, Cambridge, U.K., and New York, N.Y., U.S.A.. This version: IPCC website. ISBN 978-0-521-88011-4. Retrieved 2010-05-15.
173. Policy Implications of Greenhouse Warming: Mitigation, Adaptation, and the Science Base (1992), Committee on Science, Engineering, and Public Policy (COSEPUP)
174. GAO (2011). Technical status, future directions, and potential responses. July

2011. GAO-11-71

175. The Royal Society, (2009) Geoengineering the climate: science, governance and uncertainty". Retrieved 2009-09-12.

176. Statoil, Shell in plan to raise oil output by injecting CO_2 – report, AFX News via Forbes, March 8, 2006, checked 2009-01-15

177. MIT Energy Research Council : Research Spotlight". Web.mit.edu. Retrieved 2010-08-26.

178. GreenFuel Technologies Closing Down : Greentech Media". Greentechmedia.com. 2009-05-13. Retrieved 2010-08-26.

179. Launder B.; J.M.T. Thompson (2008). "Global and Arctic climate engineering: numerical model studies". Phil. Trans. R. Soc. A. 366 (1882): 4039–4056. Bibcode:2008RSPTA.366.4039C. doi:10.1098/rsta.2008.0132. PMID 18757275.

180. Shindell, Drew (2012-01-13). "Simultaneously Mitigating Near-Term Climate Change and Improving Human Health and Food Security | Science". Sciencemag.org. Retrieved 2016-05-06.

181. Grubb, M. (July–September 2003). "The Economics of the Kyoto Protocol" (PDF). World Economics. 4 (3): 146–147. Retrieved 2010-03-25.

182. Methane vs. Carbon Dioxide: A Greenhouse Gas Showdown". One Green Planet. Retrieved 2015-11-15.

183. Burp vaccine cuts greenhouse gas emissions Rachel Nowak for NewScientist September 2004

184. Albrittion, D.L.; et al. (2001). "Halocarbons and related compounds". In J.T. Houghton; et al. Technical summary. Climate Change 2001: The physical science basis. Contribution of Working Group I to the Third Assessment Report of the Intergovernmental Panel on Climate Change. Print version: Cambridge University Press, Cambridge, United Kingdom and New York, NY, USA. This version: GRID-Arendal website. p. 43. Retrieved 2011-06-07.

185. Glossary: A.P.M. Baede. of the main report: S. Solomon et al., eds. (2007). "Definition of "Global Warming Potential"". Annex I: Glossary. Climate Change 2007: The Physical Science Basis. Contribution of Working Group I to the Fourth Assessment Report of the Intergovernmental Panel on Climate Change. Print version: Cambridge University Press, Cambridge, United Kingdom and New York, NY, USA. This version: IPCC website. Retrieved 2011-06-07.

186. Velders, G.J.M.; et al. (20 March 2007). "The importance of the Montreal Protocol in protecting climate". PNAS. 104 (12): 4814–4819. Bibcode:2007PNAS..104.4814V. doi:10.1073/pnas.0610328104. PMC 1817831. PMID 17360370. Retrieved 2011-06-07.

187. The Future of the Canals" (PDF). London Canal Museum. Retrieved 8 September 2013.

188. Lowe, Marcia D. (April 1994). "Back on Track: The Global Rail Revival". Retrieved 2007-02-15.

189. Schwartzman, Peter. "TRUCKS VS. TRAINS—WHO WINS?". Retrieved

2007-02-15.

190. Fulton, William; Pendall, Rolf; Nguyen, Mai; Harrison, Alicia (2001). "Who Sprawls Most? How Growth Patterns Differ Across the U.S" (PDF). Survey Series. Washington D.C.: The Brookings Institution Center on Urban and Metropolitan Policy.

191. Energy Saving Trust: Home and the environment". Energy Saving Trust. Retrieved 2010-08-26.

192. Osborne, Hilary (2005-08-02). "Energy efficiency ‹saves £350m a year'". Guardian Unlimited. London.

193. Fetcher, Ned (December 2009). "Effects of climate and institution size on greenhouse gas emissions from colleges and universities in the United States". Sustainability: the Journal of Record. 2 (6): 362–367. doi:10.1089/SUS.2009.9820. Retrieved 2 March 2011.

194. Greenhouse Gas Emissions of Colleges and Universities". Retrieved 2 March 2011.

195. Rosenfeld, Arthur H.; Romm, Joseph J.; Akbari, Hashem; Lloyd, Alan C. (February–March 1997). "Technology Review". Painting the Town White – and Green. Massachusetts Institute of Technology.

196. Committee on Science, Engineering; Public Policy (1992). Policy Implications of Greenhouse Warming: Mitigation, Adaptation, and the Science Base. Washington, D.C.: National Academy Press. ISBN 0-309-04386-7.

197. Agriculture: Sources of Greenhouse Gas Emissions". EPA. 2015.

198. Susan S. Lang (13 July 2005). "Organic farming produces same corn and soybean yields as conventional farms, but consumes less energy and no pesticides, study finds". Retrieved 8 July 2008.

199. Pimentel, David; Hepperly, Paul; Hanson, James; Douds, David; Seidel, Rita (2005). "Environmental, Energetic, and Economic Comparisons of Organic and Conventional Farming Systems". BioScience. 55 (7): 573–82. doi:10.1641/0006-3568(2005)055[0573:EEAECO]2.0.CO;2.

200. Lal, Rattan; Griffin, Michael; Apt, Jay; Lave, Lester; Morgan, M. Granger (2004). "Ecology: Managing Soil Carbon". Science. 304 (5669): 393. doi:10.1126/science.1093079. PMID 15087532.

201. A. N. (Thanos) Papanicolaou, Kenneth M. Wacha, Benjamin K. Abban, Christopher G. Wilson, Jerry L. Hatfield, Charles O. Stanier, Timothy R. Filley (2015). "Conservation Farming Shown to Protect Carbon in Soil". Journal of Geophysical Research: Biogeosciences. 120 (11): 2375–2401. doi:10.1002/2015JG003078.

202. Cover Crops, a Farming Revolution With Deep Roots in the Past". The New York Times. 2016.

203. Facing a changing world: women, population and climate, United Nations Population Fund

204. Population and Global Warming Factsheet from Sierra Club

205. Population and Global Warming National Wild Life Federation
206. Population and the Environment Fact Sheet Population Connection
207. Population Connection Statement of Policy
208. To the point of farce: a martian view of the hardinian taboo—the silence that surrounds population controlMaurice King, Charles Elliott BMJ
209. Who is Heating Up the Planet? A Closer Look at Population and Global Warming from Sierra Club
210. Jowit, Juliette; Wintour, Patrick (26 June 2008). "Cost of tackling global climate change has doubled, warns Stern". The Guardian. London.
211. CIA World Factbook". US CIA. 18 Oct 2011. Retrieved 21 Oct 2011.
212. Yohe, G.W.; et al. (2007). Executive summary. In (book chapter): Perspectives on climate change and sustainability. In: Climate Change 2007: Impacts, Adaptation and Vulnerability. Contribution of Working Group II to the Fourth Assessment Report of the Intergovernmental Panel on Climate Change (M.L. Parry et al., (eds.)). Print version: Cambridge University Press, Cambridge, U.K., and New York, N.Y., U.S.A.. Web version: IPCC website. ISBN 978-0-521-88010-7. Retrieved 2010-05-15.
213. Toth, F.L.; et al. (2001). 10.4.7 Emerging Conclusions with Respect to Policy-relevant Scientific Questions. In (book chapter): Decision-making Frameworks. In: Climate Change 2001: Mitigation. Contribution of Working Group III to the Third Assessment Report of the Intergovernmental Panel on Climate Change (B. Metz et al. Eds.). Print version: Cambridge University Press, Cambridge, U.K., and New York, N.Y., U.S.A.. This version: GRID-Arendal website]. ISBN 978-0-521-01502-8. Retrieved 2010-01-10.
214. Rogner, H.-H.; et al. (2007). Executive Summary. In (book chapter): Introduction. In: Climate Change 2007: Mitigation. Contribution of Working Group III to the Fourth Assessment Report of the Intergovernmental Panel on Climate Change (B. Metz et al. (eds)). Print version: Cambridge University Press, Cambridge, United Kingdom and New York, NY, USA. Web version: IPCC website. ISBN 978-0-521-88011-4. Retrieved 2010-05-05.
215. Banuri, T.; et al. (1996). Equity and Social Considerations. In: Climate Change 1995: Economic and Social Dimensions of Climate Change. Contribution of Working Group III to the Second Assessment Report of the Intergovernmental Panel on Climate Change (J.P. Bruce et al. Eds.) (PDF). This version: Printed by Cambridge University Press, Cambridge, U.K., and New York, N.Y., U.S.A.. PDF version: IPCC website. doi:10.2277/0521568544. ISBN 978-0-521-56854-8.
216. Goldemberg, J.; et al. (1996). Introduction: scope of the assessment. In: Climate Change 1995: Economic and Social Dimensions of Climate Change. Contribution of Working Group III to the Second Assessment Report of the Intergovernmental Panel on Climate Change (J.P. Bruce et al. Eds.) (PDF). This version: Printed by Cambridge University Press, Cambridge, U.K., and New York, N.Y., U.S.A.. Web version: IPCC website. doi:10.2277/0521568544. ISBN 978-0-521-56854-8.
217. Article Adaptation. If you can't stand the heat", The Economist, special report

on “Climate change”, 28 November 2015, page 10-12.

218. World Bank (2010). World Development Report 2010: Development and Climate Change. The International Bank for Reconstruction and Development / The World Bank, 1818 H Street NW, Washington DC 20433. doi:10.1596/978-0-8213-7987-5. ISBN 978-0-8213-7987-5. Retrieved 2010-04-06.

219. PBL Netherlands Environment Agency (15 June 2012), “Figure 6.14, in: Chapter 6: The energy and climate challenge”, in van Vuuren, D.; M. Kok, Roads from Rio+20 (PDF), ISBN 978-90-78645-98-6, p.177, Report no: 500062001. Report website.

220. Jaeger, C.C.; J. Jaeger (2011), “Three views of two degrees” (PDF), Regional Environmental Change, Springer-Verlag, 11 (1), pp. 15–26, doi:10.1007/s10113-010-0190-9, ISSN 1436-3798

221. Rijsberman, F.J.; R.J. Stewart, eds. (1990), Targets and Indicators of Climate Change, Stockholm, Sweden: Stockholm Environment Institute, ISBN 91-88116-21-2. Summary available from the Climate Emergency Institute.

222. UNFCCC (15 March 2011), FCCC/CP/2010/7/Add.1: Report of the Conference of the Parties on its sixteenth session, held in Cancun from 29 November to 10 December 2010. Addendum. Part two: Action taken by the Conference of the Parties at its sixteenth session (PDF), Geneva, Switzerland: UN Office. Available as a PDF in English, Spanish, French, Arabic, and Russian.

223. UNFCCC (3 May 2012), Essential Background, UNFCCC

224. Oliver Geden (2013), Modifying the 2°C Target. Climate Policy Objectives in the Contested Terrain of Scientific Policy Advice, Political Preferences, and Rising Emissions, SWP Research Paper 5/13

225. Oliver Geden (2010), What Comes After the Two-Degree Target?, SWP Comments 19

226. EU climate change target “unfeasible””. EurActiv.com. 2006-02-01. Retrieved 2007-02-21.

227. Adam, David (14 April 2009). “World will not meet 2C warming target, climate change experts agree”. London: Guardian News and Media Limited. Retrieved 2009-04-14.

228. United States Department of Energy World Trends

229. Oppenheimer, M., et al., Section 19.7.2: Limits to Mitigation, in: Chapter 19: Emergent risks and key vulnerabilities (archived July 8 2014), pp.49–50, in IPCC AR5 WG2 A 2014

230. Oliver Geden/Silke Beck: Renegotiating the global climate stabilization target. In: Nature Climate Change, 4, 2014, pp. 747–748

231. Anderson, K. & Bows, A., 2011. Beyond ‹dangerous’ climate change: emission scenarios for a new world. Philos. Trans. Royal Society A.

232. “Implication for Carbon Emissions Target,” in: Hansen & others 2013, p. 15

233. “4. Discussion and conclusions,” in: Luderer & others 2013, p. 6

234. Report on the structured expert dialogue on the 2013–2015 review" (PDF). UNFCCC, Subsidiary Body for Scientific and Technological Advice & Subsidiary Body for Implementation. 2015-04-04. Retrieved 2016-06-21.

235. 1.5°C temperature limit – key facts". Climate Analytics. Retrieved 2016-06-21.

236. Gupta, S.; et al. (2007). "13.2.1.2 Taxes and charges". In B. Metz; et al. Policies, instruments, and co-operative arrangements. Climate Change 2007: Mitigation. Contribution of Working Group III to the Fourth Assessment Report of the Intergovernmental Panel on Climate Change. Print version: Cambridge University Press, Cambridge, U.K., and New York, N.Y., U.S.A.. This version: IPCC website. Retrieved 2010-03-18.

237. Jacobson, M.Z. (2009). "Review of solutions to global warming, air pollution, and energy security" (PDF). Energy and Environmental Science. 2 (2): 148–73. doi:10.1039/b809990c.

238. Jacobson, M.Z.; Delucchi, M.A. (2009). "A Plan to Power 100 Percent of the Planet with Renewables" (originally published as "A Path to Sustainable Energy by 2030")". Scientific American. 301 (5): 58–65. doi:10.1038/scientificamerican1109-58. PMID 19873905.

239. Farah, Paolo Davide (2015). "Sustainable Energy Investments and National Security: Arbitration and Negotiation Issues". Journal of World Energy Law and Business. 8 (6). Retrieved 26 November 2015.

240. How high-pressure politics threatens action on climate The Observer June 2005

241. StoryOfStuff.com (2009) The Story of Cap and Trade"

242. Success of Northeast Cap-and-Trade System Shows Market-Based Climate Policy Is Well Within Reach".

243. Emission Trading Scheme (EU ETS) from ec.europa.eu

244. The $20,000,000,000,000 question Robins, Nick for Opendemocracy

245. State and Trends of the Carbon Market International Emissions Trading Association 2005

246. Statement of G8 Climate Change Roundtable Archived May 8, 2013, at the Wayback Machine. Convened by the World Economic Forum June 2005

247. Evans. J (forthcoming 2012) Environmental Governance, Routledge, Oxon

248. Biesbroek. G.R, Termeer. C.J.A.M, Kabat. P, Klostermann.J.E.M (unpublished) Institutional governance barriers for the development and implementation of climate adaptation strategies, Working paper for the International Human Dimensions Programme (IHDP) conference "Earth System Governance: People, Places, and the Planet", December 2–4, Amsterdam, the Netherlands

249. Mee. L. D, Dublin. H. T, Eberhard. A. A (2008) Evaluating the Global Environment Facility: A goodwill gesture or a serious attempt to deliver global benefits?, Global Environmental Change 18, 800–810

250. http://www10.iadb.org/intal/intalcdi/pe/2009/04268.pdf

251. Preston. B. L, Westaway. R. M, Yuen. E. Y (2004) Climate adaptation planning

in practice: an evaluation of adaptation plans from three developed nations, European Management Journal, 22(3) 304–314

252. UNFCCC (2011) Report on the twentieth meeting of the Least Developed Countries Expert Group, Subsidiary Body for Implementation, United Nations Framework Convention on Climate Change
253. UNFCCC (2011) Annual report of the Joint Implementation Supervisory Committee to the Conference of the Parties serving as the meeting of the Parties to the Kyoto Protocol, United Nations Framework Convention on Climate Change
254. Industrial Technologies Program: BestPractices". Eere.energy.gov. Retrieved 2010-08-26.
255. Clinton Hails Global Warming Pact". All Politics. CNN. 1997-12-11. Retrieved 2006-11-05.
256. How the White House Worked to Scuttle California's Climate Law", San Francisco Chronicle, September 25, 2007
257. Reuters, January 30, 2007, free archived version at http://www.commondreams.org/headlines07/0130-10.htm, last visited Jan. 30, ‹07
258. Written testimony of Dr. Grifo before the Committee on Oversight and Government Reform of the U.S. House of Representatives on January 30, 2007, archived at "Archived copy" (PDF). Archived from the original (PDF) on 2009-08-05. Retrieved 2009-12-15.
259. written testimony of Rick Piltz before the Committee on Oversight and Government Reform of the U.S. House of Representatives on January 30, 2007, archived at http://oversight.house.gov/Documents/20070130113813-92288.pdf last visited Jan. 30, 07
260. Barringer, Felicity (2012-10-13). "In California, a Grand Experiment to Rein in Climate Change". The New York Times.
261. Prototype Carbon Fund from the World Bank Carbon Finance Unit
262. Jessica Brown, Neil Bird and Liane Schalatek (2010) Climate finance additionality: emerging definitions and their implications Overseas Development Institute
263. Free trade can help combat global warming, finds UN report UN News Centre, 26 June 2009
264. Andrew Biggin (16 August 2007). "Scientific bodies must take own action on emissions". Nature. 448 (7155): 749. Bibcode:2007Natur.448..749B. doi:10.1038/448749a. PMID 17700677.
265. Anderson, K; Bows, A (2008). "Reframing the climate change challenge in light of post-2000 emission trends". Philosophical Transactions of the Royal Society A: Mathematical, Physical and Engineering Sciences. 366 (1882): 3863–3882. Bibcode:2008RSPTA.366.3863A. doi:10.1098/rsta.2008.0138.
266. Anderson, K (June 17, 2008). "Reframing climate change: from long-term targets to emission pathways". (esp. slide 24 onward)
267. Gössling S, Ceron JP, Dubois G, Hall CM, Gössling IS, Upham P, Earthscan London (2009). Hypermobile travellers. and Implications for Carbon Dioxide

Emissions Reduction. In: Climate Change and Aviation: Issues, Challenges and Solutions, London. The chapter: Chapter 6

268. Green Electric? GE unveils eco-strategy". MSNBC.
269. Aviation groups set targets to limit their environmental impact". FT.com.
270. 5 Mutual Funds for Socially Responsible Investors". Kiplinger.
271. Investing to Curb Climate Change" (PDF). USSIF. p. 2.
272. Video: Paradise lost? - Need to Know". PBS. Palau suing the industrialized countries over global warming
273. Inuit suing the US in regards to global warming Archived August 25, 2010, at the Wayback Machine.
274. Environmental Integrity Project, Sierra Club Announce Plans to Sue EPA Unless It Revises Nitrogen Oxide Emissions Standard, Curbs Nitrous Oxide Pollution Linked to Global Warming – NewsOn6.com – Tulsa, OK – News, Weather, Video and Sports – KOTV.com -".
275. Edward Lorenz (1982): "Climate is what you expect, weather is what you get"
276. Stott, et al. (2004), "Human contribution to the European heatwave of 2003", Nature, Vol. 432, 2 December 2004
277. Press release (29 January 2004). Archived press release: Exxonmobil's contribution to global warming revealed. Friends of the Earth Trust. Retrieved May 25, 2015.
278. New York is investigating Exxon Mobil for allegedly misleading the public about climate change". The Washington Post. 2015-11-05. Retrieved 2015-12-29.

CHAPTER 7

Global Warming and Climate Change

Chapter Contents

Global warming is a phenomenon characterized by a slow warming of the surface temperature of the earth. The basic understanding is usually that the atmosphere immediately surrounding the planet is getting incrementally hotter each year, which in turn leads the surface of both land and sea to warm, as well. The term itself does not imply a cause or speak to a cause; rather, it is a statement of fact that most in the scientific community acknowledge as true and provable. What causes more disagreement is discussions of *why* the temperature seems to be rising, what potential negatives are likely to result, and what human communities can or should do to slow or reverse the trend. Most theories posit that human emissions of carbon dioxide (CO_2) gas, whether through industrial processing or vehicle emissions, are chiefly to blame, while others suggest that the warming is simply part of the earth's natural rhythm and will right itself with time. A lot of research around the world is devoted to the topic, and it's often a hotly debated concept in political arenas and research institutes globally.

INTRODUCTION

Global warming and climate change are terms for the observed century-scale rise in the average temperature of the Earth's climate system and its related effects. Multiple lines of scientific evidence show that the climate system is warming. Although the increase of near-surface atmospheric temperature is the measure of global warming often reported in the popular press, most of the additional energy stored in the climate system since 1970 has gone into the oceans. The rest has melted ice and warmed the continents and atmosphere. Many of the observed changes since the 1950s are unprecedented over tens to thousands of years.

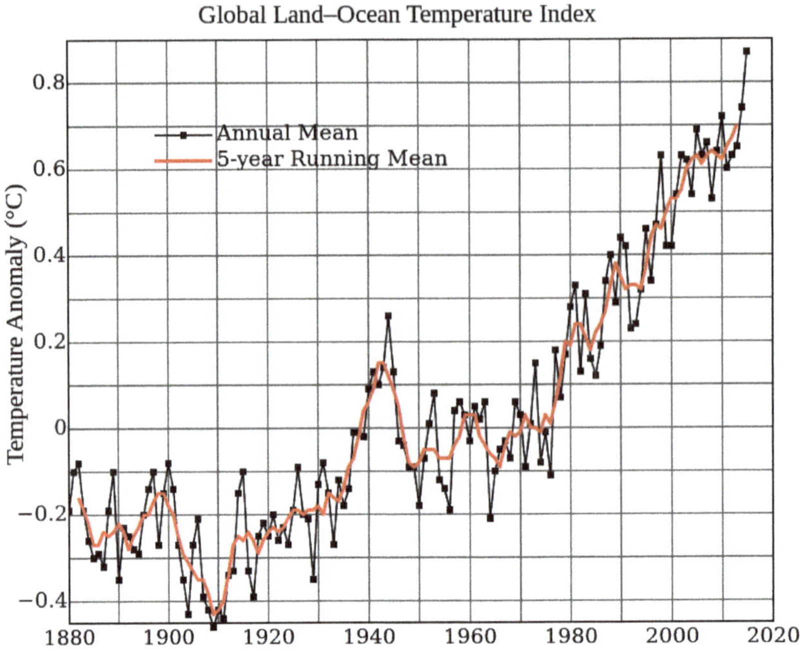

Global mean surface temperature change from 1880 to 2015, relative to the 1951–1980 mean. The black line is the annual mean and the red line is the 5-year running mean. Source: NASA GISS.

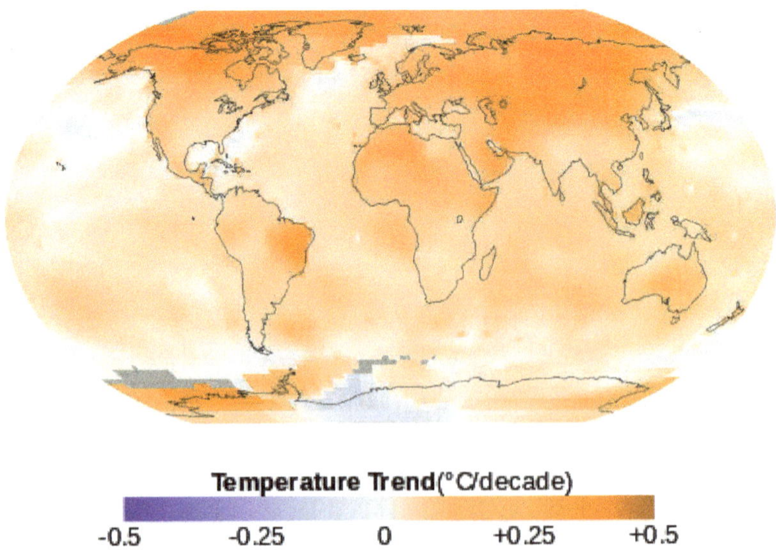

World map showing surface temperature trends (°C per decade) between 1950 and 2014. Source: NASA GISS.

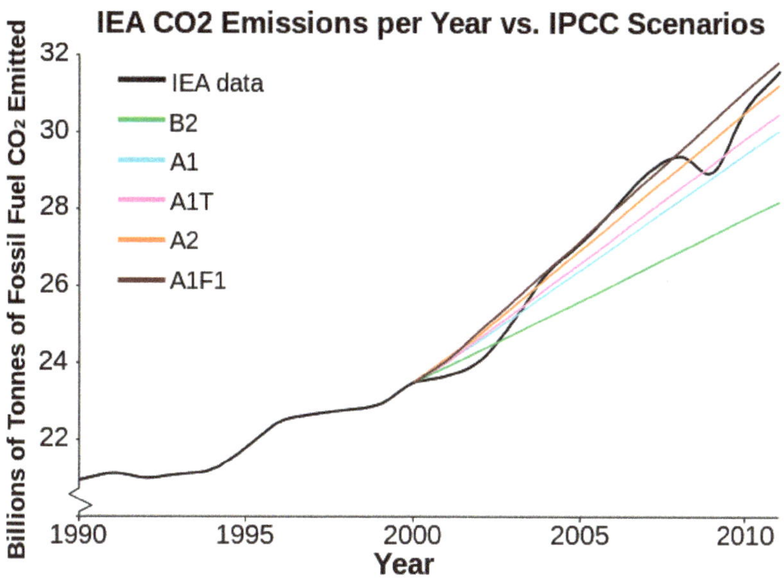

Fossil fuel related carbon dioxide (CO_2) emissions compared to five of the IPCC's "SRES" emissions scenarios, published in 2000. The dips are related to global recessions. Image source: Skeptical Science.

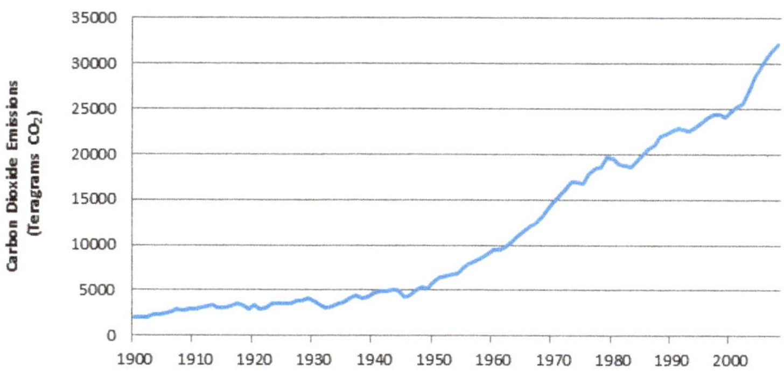

Fossil fuel related carbon dioxide emissions over the 20th century. Image source: EPA.

Scientific understanding of global warming is increasing. The Intergovernmental Panel on Climate Change (IPCC) reported in 2014 that scientists were more than 95% certain that global warming is mostly being caused by human (anthropogenic) activities, mainly increasing concentrations of greenhouse gases such as methane and carbon dioxide (CO_2). Human activities have led to carbon dioxide concentrations above levels not seen in hundreds of thousands of years. Methane and other, often much more potent, greenhouse gases are also rising along with CO_2. Currently, about half of the carbon dioxide released from the burning of fossil fuels remains in the atmosphere. The rest is absorbed by vegetation and the oceans.Climate model projections summarized in the report indicated that during the 21st century the global surface temperature is likely to rise a further 0.3 to 1.7 °C (0.5 to 3.1 °F) for their lowest emissions scenario and 2.6 to 4.8 °C (4.7 to 8.6 °F) for the highest emissions scenario. These findings have been recognized by the national science academies of the major industrialized nations[b] and are not disputed by any scientific body of national or international standing.

Future climate change and associated impacts will differ from region to region around the globe. Anticipated effects include warming global temperature, rising sea levels, changing precipitation, and expansion of deserts in the subtropics. Warming is expected to be greater over land than over the oceans and greatest in the Arctic, with the continuing retreat of glaciers, permafrost and sea ice. Other likely changes include more frequent extreme weather events including heat waves, droughts, heavy rainfall with floods and heavy snowfall; ocean acidification; and species extinctions due to shifting temperature regimes. Effects significant to humans include the threat to food security from decreasing crop yields and the abandonment of populated areas due to rising sea levels. Because the climate system has a large "inertia" and greenhouse gases will stay in the atmosphere for a long time, many of these effects will not only exist for decades or centuries, but will persist for tens of thousands of years.

Possible societal responses to global warming include mitigation by emissions reduction, adaptation to its effects, building systems resilient to its effects, and possible future climate engineering. Most countries are parties to the United Nations Framework Convention on Climate Change (UNFCCC),whose ultimate objective is to prevent dangerous anthropogenic climate change. Parties to the UNFCCC have agreed that deep cuts in emissions are required and that global warming should be limited to well below 2.0 °C (3.6 °F) relative to pre-industrial levels,[c] with efforts made to limit warming to 1.5 °C (2.7 °F).

Public reactions to global warming and concern about its effects are also increasing. A global 2015 Pew Research Center report showed a median of 54% consider it "a very serious problem". There are significant regional differences, with Americans and Chinese (whose economies are responsible for the greatest annual CO_2 emissions) among the least concerned.

OBSERVED TEMPERATURE CHANGES

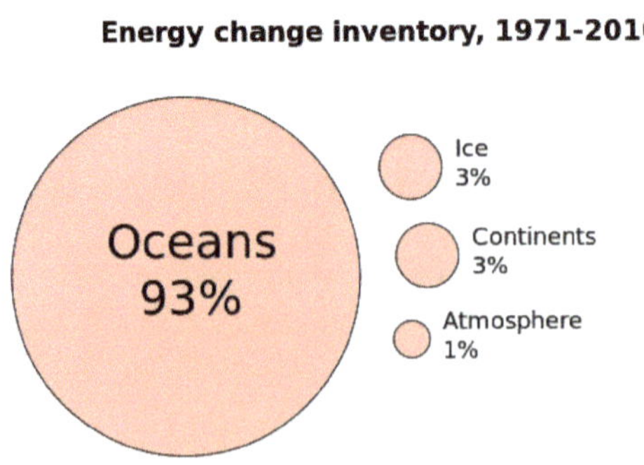

Earth has been in *radiative imbalance* since at least the 1970s, where less energy leaves the atmosphere than enters it. Most of this extra energy has been absorbed by the oceans. It is very likely that human activities substantially contributed to this increase in ocean heat content.

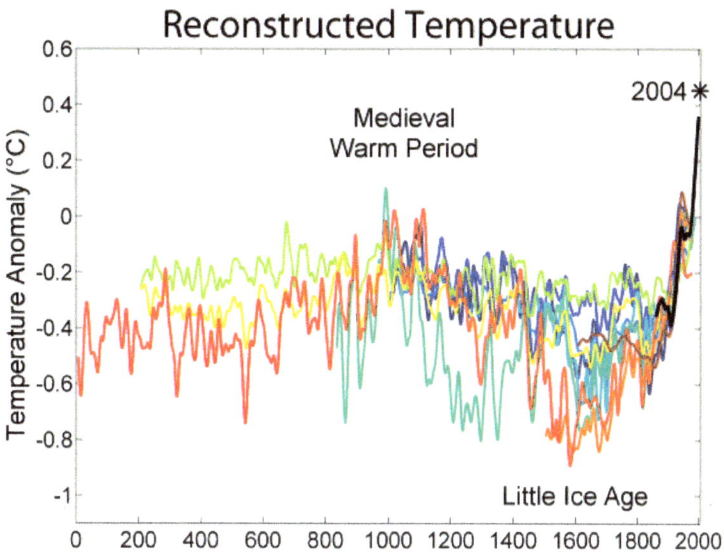

Two millennia of mean surface temperatures according to different reconstructions from climate proxies, each smoothed on a decadal scale, with the instrumental temperature record overlaid in black.

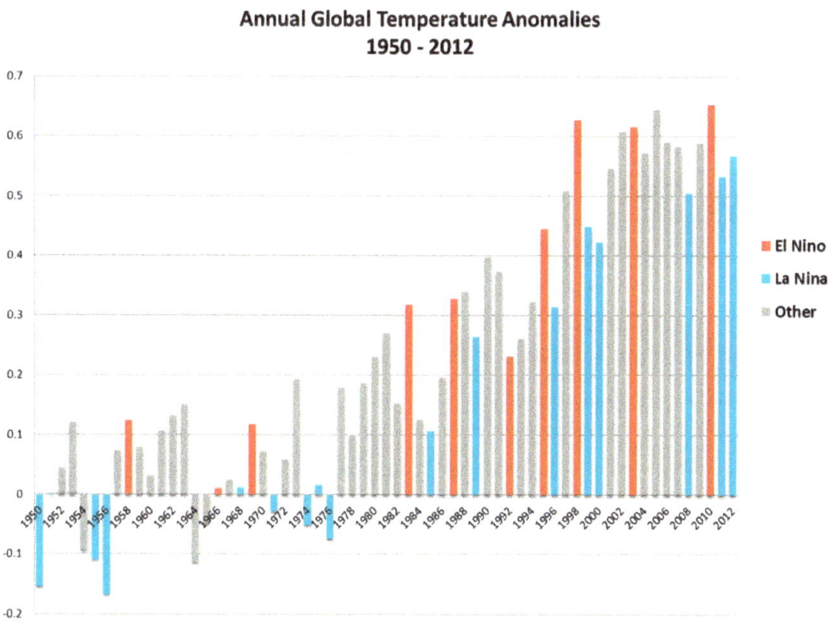

NOAA graph of Global Annual Temperature Anomalies 1950–2012, showing the El Niño Southern Oscillation.

The global average (land and ocean) surface temperature shows a warming of 0.85 [0.65 to 1.06] °C in the period 1880 to 2012, based on multiple independently produced datasets. Earth's average surface temperature rose by 0.74±0.18 °C over the period 1906–2005. The rate of warming almost doubled for the last half of that period (0.13±0.03 °C per decade, versus 0.07±0.02 °C per decade).

The average temperature of the lower troposphere has increased between 0.13 and 0.22 °C (0.23 and 0.40 °F) per decade since 1979, according to satellite temperature measurements. Climate proxies show the temperature to have been relatively stable over the one or two thousand years before 1850, with regionally varying fluctuations such as the Medieval Warm Period and the Little Ice Age.

The warming that is evident in the instrumental temperature record is consistent with a wide range of observations, as documented by many independent scientific groups.Examples include sea level rise, widespread melting of snow and land ice, increased heat content of the oceans, increased humidity, and the earlier timing of spring events, e.g., the flowering of plants. The probability that these changes could have occurred by chance is virtually zero.

Trends

Temperature changes vary over the globe. Since 1979, land temperatures have increased about twice as fast as ocean temperatures (0.25 °C per decade against 0.13 °C per decade). Ocean temperatures increase more slowly than land temperatures because of the larger effective heat capacity of the oceans and because the ocean loses more heat by evaporation. Since the beginning of industrialisation the temperature difference between the hemispheres has increased due to melting of sea ice and snow in the North.Average arctic temperatures have been increasing at almost twice the rate of the rest of the world in the past 100 years; however arctic temperatures are also highly variable.Although more greenhouse gases are emitted in the Northern than Southern Hemisphere this does not contribute to the difference in warming because the major greenhouse gases persist long enough to mix between hemispheres.

The thermal inertia of the oceans and slow responses of other indirect effects mean that climate can take centuries or longer to adjust to changes in forcing. One climate commitment study concluded that if greenhouse gases were stabilized at year 2000 levels, surface temperatures would still increase by about one-half degree Celsius, and another found that if they were stabilized at 2005 levels surface warming could exceed a whole degree Celsius. Some of this surface warming will be driven by past natural forcings which are still seeking equilibrium in the climate system. One study using a highly simplified climate model indicates these past natural forcings may account for as much as 64% of the committed 2050 surface warming and their influence will fade with time compared to the human contribution.

Global temperature is subject to short-term fluctuations that overlay long-term trends and can temporarily mask them. The relative stability in surface temperature from 2002 to 2009, which has been dubbed the global warming hiatus by the media and some scientists, is consistent with such an episode.2015 updates to account for differing methods of measuring ocean surface temperature measurements show a positive trend over the recent decade.

Warmest Years

Fifteen of the top 16 warmest years have occurred since 2000. While record-breaking years can attract considerable public interest, individual years are less significant than the overall trend. Some climatologists have criticized the attention that the popular press gives to "warmest year" statistics; for example, Gavin Schmidt stated "the long-term trends or the expected sequence of records are far more important than whether any single year is a record or not." Ocean oscillations such as the El Niño Southern Oscillation (ENSO) can affect global average temperatures, causing temperatures of a given year to be abnormally warm or cold for reasons not directly related to the overall trend of climate change. For example, 1998 and 2015 temperatures were significantly warmed by strong El Niño conditions.

Initial causes of temperature changes (external forcings)

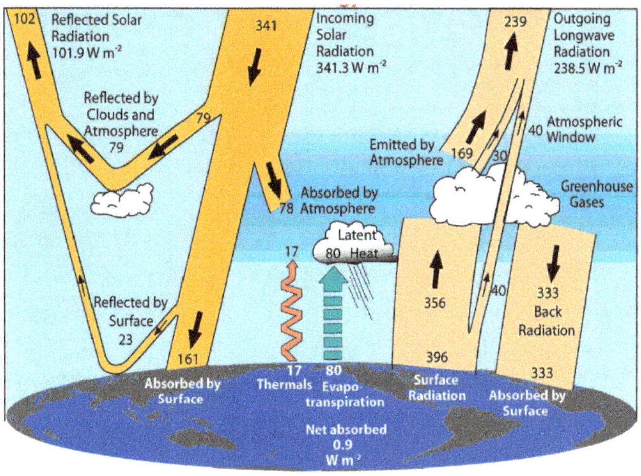

Greenhouse effect schematic showing energy flows between space, the atmosphere, and Earth's surface. Energy exchanges are expressed in watts per square metre (W/m^2).

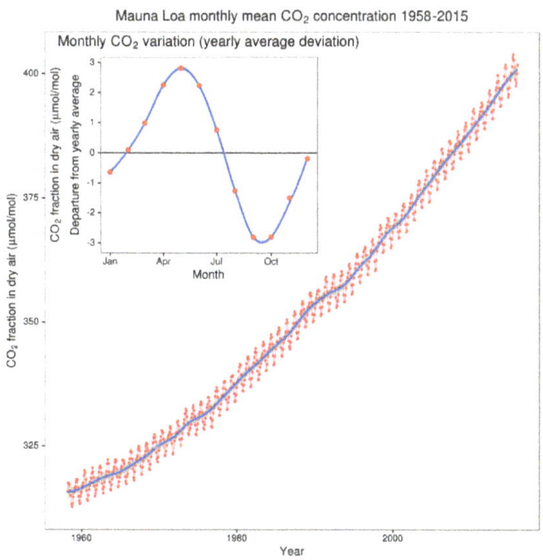

This graph, known as the Keeling Curve, documents the increase of atmospheric carbon dioxide concentrations from 1958–2015. Monthly CO_2 measurements display seasonal oscillations in an upward trend; each year's maximum occurs during the Northern Hemisphere's late spring, and declines during its growing season as plants remove some atmospheric CO_2.

The climate system can spontaneously generate changes in global temperature for years to decades at a time but long-term changes in global temperature require *external forcings*. These forcings are "external" to the climate system but not necessarily external to Earth. Examples of external forcings include changes in atmospheric composition (e.g., increased concentrations of greenhouse gases), solar luminosity, volcanic eruptions, and variations in Earth's orbit around the Sun.

OBSERVED AND EXPECTED EFFECTS ON SOCIAL SYSTEMS

The effects of climate change on human systems, mostly due to warming or shifts in precipitation patterns, or both, have been detected worldwide. Production of wheat and maize globally has been impacted by climate change. While crop production has increased in some mid-latitude regions such as the UK and Northeast China, economic losses due to extreme weather events have increased globally. There has been a shift from cold- to heat-related mortality in some regions as a result of warming. Livelihoods of indigenous peoples of the Arctic have been altered by climate change, and there is emerging evidence of climate change impacts on livelihoods of indigenous peoples in other regions. Regional impacts of climate change are now observable at more locations than before, on all continents and across ocean regions. The future social impacts of climate change will be uneven. Many risks are expected to increase with higher magnitudes of global warming. All regions are at risk of experiencing negative impacts.Low-latitude, less developed areas face the greatest risk. A study from 2015 concluded that economic growth (gross domestic product) of poorer countries is much more impaired with projected future climate warming, than previously thought.

A meta-analysis of 56 studies concluded in 2014 that each degree of temperature rise will increase violence by up to 20%, which includes fist fights, violent crimes, civil unrest or wars.

Examples of impacts include:

- *Food*: Crop production will probably be negatively affected in low latitude countries, while effects at northern latitudes may be positive or negative. Global warming of around 4.6 °C relative to pre-industrial levels could pose a large risk to global and regional food security.
- *Health*: Generally impacts will be more negative than positive. Impacts include: the effects of extreme weather, leading to injury and loss of life; and indirect effects, such as undernutrition brought on by crop failures.

Habitat inundation

In small islands and mega deltas, inundation as a result of sea level rise is expected to threaten vital infrastructure and human settlements. This could lead to issues of homelessness in countries with low-lying areas such as Bangladesh, as well as statelessness for populations in countries such as the Maldives and Tuvalu.

Economy

Estimates based on the IPCC A1B emission scenario from additional CO_2 and CH_4 greenhouse gases released from permafrost, estimate associated impact damages by US$43 trillion.

Infrastructure

Continued permafrost degradation will likely result in unstable infrastructure in Arctic regions, or Alaska before 2100. Thus, impacting roads, pipelines and buildings, as well as water distribution, and cause slope failures.

Possible responses to global warming

Mitigation

The graph on the right shows three "pathways" to meet the UNFCCC's 2 °C target, labelled "global technology", "decentralized solutions", and "consumption change". Each pathway shows how various measures (e.g., improved energy efficiency, increased use of renewable energy) could contribute to emissions reductions. Image credit: PBL Netherlands Environmental Assessment Agency.

Mitigation of climate change are actions to reduce greenhouse gas emissions, or enhance the capacity of carbon sinks to absorb GHGs from the atmosphere. There is a large potential for future reductions in emissions by a combination of activities, including: energy conservation and increased energy efficiency; the use of low-carbon energy technologies, such as renewable energy, nuclear energy, and carbon capture and storage; and enhancing carbon sinks through, for example, reforestation and preventing deforestation. A 2015 report by Citibank concluded that transitioning to a low carbon economy would yield positive return on investments.

Near- and long-term trends in the global energy system are inconsistent with limiting global warming at below 1.5 or 2 °C, relative to pre-industrial levels. Pledges made as part of the Cancún agreements are broadly consistent with having a likely chance (66 to 100% probability) of limiting global warming (in the 21st century) at below 3 °C, relative to pre-industrial levels.

In limiting warming at below 2 °C, more stringent emission reductions in the near-term would allow for less rapid reductions after 2030. Many integrated models are unable to meet the 2 °C target if pessimistic assumptions are made about the availability of mitigation technologies.

Adaptation

Other policy responses include adaptation to climate change. Adaptation to climate change may be planned, either in reaction to or anticipation of climate change, or spontaneous, i.e., without government intervention. Planned adaptation is already occurring on a limited basis. The barriers, limits, and costs of future adaptation are not fully understood.

A concept related to adaptation is *adaptive capacity*, which is the ability of a system (human, natural or managed) to adjust to climate change (including climate variability and extremes) to moderate potential damages, to take advantage of opportunities, or to cope with consequences. Unmitigated climate change (i.e., future climate change without efforts to limit greenhouse gas emissions) would, in the long term, be likely to exceed the capacity of natural, managed and human systems to adapt.

Environmental organizations and public figures have emphasized changes in the climate and the risks they entail, while promoting adaptation to changes in infrastructural needs and emissions reductions.

Climate Engineering

Climate engineering (sometimes called *geoengineering* or *climate intervention*) is the deliberate modification of the climate. It has been investigated as a possible response to global warming, e.g. by NASA and the Royal Society. Techniques under research fall generally into the categories solar radiation management and carbon dioxide removal, although various other schemes have been suggested. A study from 2014 investigated the most common climate engineering methods and concluded they are either ineffective or have potentially severe side effects and cannot be stopped without causing rapid climate change.

DISCOURSE ABOUT GLOBAL WARMING

Political Discussion

Most countries in the world are parties to the United Nations Framework Convention on Climate Change (UNFCCC). The ultimate objective of the Convention is to prevent dangerous human interference of the climate system. As stated in the Convention, this requires that GHG concentrations are stabilized in the atmosphere at a level where ecosystems can adapt naturally to climate change, food production is not threatened, and economic development can proceed in a sustainable fashion. The Framework Convention was agreed in 1992, but since then, global emissions have risen.

During negotiations, the G77 (a lobbying group in the United Nations representing 133 developing nations)[4] pushed for a mandate requiring developed countries to "[take] the lead" in reducing their emissions. This was justified on the basis that: the developed world's emissions had contributed most to the cumulation of GHGs in the atmosphere; per-capita emissions (i.e., emissions per head of population) were still relatively low in developing countries; and the emissions of developing countries would grow to meet their development needs.

This mandate was sustained in the Kyoto Protocol to the Framework Convention, which entered into legal effect in 2005. In ratifying the Kyoto Protocol, most developed countries accepted legally binding commitments to limit their emissions. These first-round commitments expired in 2012. United States President George W. Bush rejected the treaty on the basis that "it exempts 80% of the world, including major population centres such as China and India, from compliance, and would cause serious harm to the US economy."

At the 15th UNFCCC Conference of the Parties, held in 2009 at Copenhagen, several UNFCCC Parties produced the Copenhagen Accord. Parties associated with the Accord (140 countries, as of

November 2010)[9] aim to limit the future increase in global mean temperature to below 2 °C. The 16th Conference of the Parties (COP16) was held at Cancún in 2010. It produced an agreement, not a binding treaty, that the Parties should take urgent action to reduce greenhouse gas emissions to meet a goal of limiting global warming to 2 °C above pre-industrial temperatures. It also recognized the need to consider strengthening the goal to a global average rise of 1.5 °C.

Scientific Discussion

There is continuing discussion through published peer-reviewed scientific papers, which are assessed by scientists working in the relevant fields taking part in the Intergovernmental Panel on Climate Change. The scientific consensus as of 2013 stated in the IPCC Fifth Assessment Report is that it "is extremely likely that human influence has been the dominant cause of the observed warming since the mid-20th century". A 2008 report by the U.S. National Academy of Sciences stated that most scientists by then agreed that observed warming in recent decades was primarily caused by human activities increasing the amount of greenhouse gases in the atmosphere. In 2005 the Royal Society stated that while the overwhelming majority of scientists were in agreement on the main points, some individuals and organizations opposed to the consensus on urgent action needed to reduce greenhouse gas emissions have tried to undermine the science and work of the IPCC. National science academies have called on world leaders for policies to cut global emissions.

In the scientific literature, there is a strong consensus that global surface temperatures have increased in recent decades and that the trend is caused mainly by human-induced emissions of greenhouse gases. No scientific body of national or international standing disagrees with this view.

Discussion by the Public and in Popular Media

Global warming was the cover story in this 2007 issue of *Ms. magazine.*

The global warming controversy refers to a variety of disputes, substantially more pronounced in the popular media than in the scientific literature, regarding the nature, causes, and consequences of global warming. The disputed issues include the causes of increased global average air temperature, especially since the mid-20th century, whether this warming trend is unprecedented or within normal climatic variations, whether humankind has contributed significantly to it, and whether the increase is completely or partially an artefact of poor measurements. Additional disputes concern estimates of climate sensitivity, predictions of additional warming, and what the consequences of global warming will be.

By 1990, American conservative think tanks had begun challenging the legitimacy of global warming as a social problem. They challenged the scientific evidence, argued that global warming will have benefits, and asserted that proposed solutions would do more harm than good. Some people dispute aspects of climate change science. Organizations such as the libertarian Competitive Enterprise Institute, conservative commentators, and some companies such as ExxonMobil have challenged IPCC climate change scenarios, funded scientists who disagree with the scientific consensus, and provided their own projections of the economic cost of stricter controls.On the other hand, some fossil fuel companies have scaled back their efforts in recent years, or even called for policies to reduce global warming. Global oil companies have begun to acknowledge climate change exists and is caused by human activities and the burning of fossil fuels.

Surveys of Public Opinion

The world public, or at least people in economically advanced regions, became broadly aware of the global warming problem in the late 1980s. Polling groups began to track opinions on the subject, at first mainly in the United States. The longest consistent polling, by Gallup in the US, found relatively small deviations of 10% or so from 1998 to 2015 in opinion on the seriousness of global warming, but with increasing polarization between those concerned and those unconcerned.

The first major worldwide poll, conducted by Gallup in 2008-2009 in 127 countries, found that some 62% of people worldwide said they knew about global warming. In the advanced countries of North America, Europe and Japan, 90% or more knew about it (97% in the U.S., 99% in Japan); in less developed countries, especially in Africa, fewer than a quarter knew about it, although many had noticed local weather changes. Among those who knew about global warming, there was a wide variation between nations in belief that the warming was a result of human activities.

By 2010, with 111 countries surveyed, Gallup determined that there was a substantial decrease since 2007–08 in the number of Americans and Europeans who viewed global warming as a serious threat. In the US, just a little over half the population (53%) now viewed it as a serious concern for either themselves or their families; this was 10 points below the 2008 poll (63%). Latin America had the biggest rise in concern: 73% said global warming is a serious threat to their families. This global poll also found that people are more likely to attribute global warming to human activities than to natural causes, except in the US where nearly half (47%) of the population attributed global warming to natural causes.

A March–May 2013 survey by Pew Research Center for the People & the Press polled 39 countries about global threats. According to 54% of those questioned, global warming featured top of the perceived global threats. In a January 2013 survey, Pew found that 69% of Americans say there is solid evidence that the Earth's average temperature has got warmer over the past few decades, up six points since November 2011 and 12 points since 2009.

A 2010 survey of 14 industrialized countries found that skepticism about the danger of global warming was highest in Australia, Norway, New Zealand and the United States, in that order, correlating positively with per capita emissions of carbon dioxide.

ETYMOLOGY

In the 1950s, research suggested increasing temperatures, and a 1952 newspaper reported "climate change". This phrase next appeared in a November 1957 report in *The Hammond Times* which described Roger Revelle's research into the effects of increasing human-caused CO_2 emissions on the greenhouse effect, "a large scale global warming, with radical climate changes may result". Both phrases were only used occasionally until 1975, when Wallace Smith Broecker published a scientific paper on the topic; "Climatic Change: Are We on the Brink of a Pronounced Global Warming?" The phrase began to come into common use, and in 1976 Mikhail Budyko's statement that "a global warming up has started" was widely reported. Other studies, such as a 1971 MIT report, referred to the human impact as "inadvertent climate modification", but an influential 1979 National Academy of Sciences study headed by Jule Charney followed Broecker in using *global warming* for rising surface temperatures, while describing the wider effects of increased CO_2 as *climate change.*

In 1986 and November 1987, NASA climate scientist James Hansen gave testimony to Congress on global warming. There were increasing heatwaves and drought problems in the summer of 1988, and when Hansen testified in the Senate on 23 June he sparked worldwide interest. He said: "global warming has reached a level such that we can ascribe with a high degree of confidence a cause and effect relationship between the greenhouse effect and the observed warming." Public attention increased over the summer, and *global warming* became the dominant popular term, commonly used both by the press and in public discourse.

In a 2008 NASA article on usage, Erik M. Conway defined *Global warming* as "the increase in Earth's average surface temperature due to rising levels of greenhouse gases", while *Climate change* was "a long-term change in the Earth's climate, or of a region on Earth." As effects such as changing patterns of rainfall and rising sea levels would probably have more impact than temperatures alone, he considered *global climate change* a more scientifically accurate term, and like the Intergovernmental Panel on Climate Change, the NASA website would emphasize this wider context.

LONG-TERM EFFECTS OF GLOBAL WARMING

There are expected to be various **long-term effects of global warming**. Most discussion and research, including that by the Intergovernmental Panel on Climate Change (IPCC) reports, concentrates on the effects of global warming up to 2100, with only an outline of the effects beyond this.

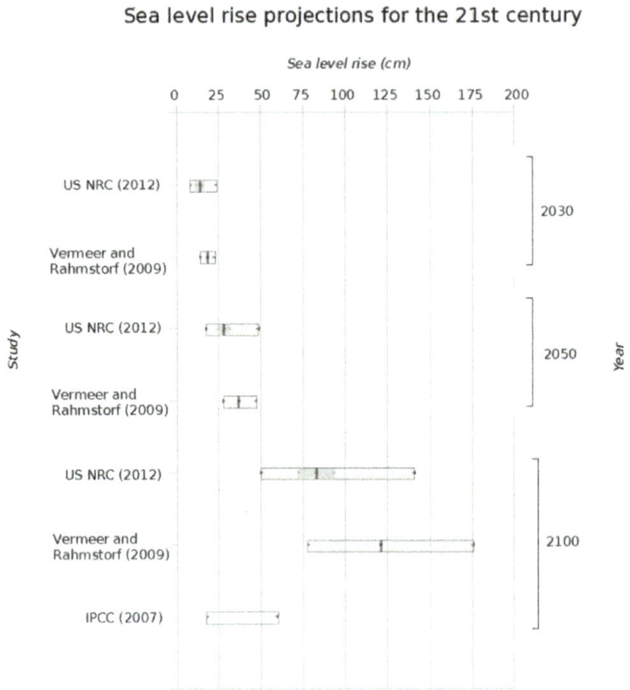

Sea level rise projections for the 21st century.

Ice Loss and sea Level Rise

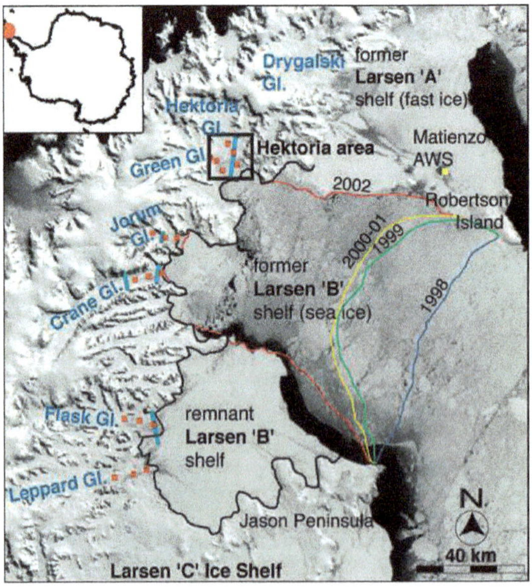

The collapse of Larsen B Shelf, showing the diminishing extent of the shelf from 1998 to 2002

Meltwater from melting ice sheets and glacier retreat contributes to a rise in the future sea level.

Antarctica

Concern has been expressed about the stability of the West Antarctic ice sheet (WAIS). In 2002 Vaughan & Spouge argued that the "WAIS most likely will not collapse in the next few centuries." In an inaugural article for members of the National Academy of Sciences elected in 2005, Timothy Lenton and others suggest that a collapse of the West Antarctic ice sheet could occur within a millennium. Specifically they state "Although the timescale is highly uncertain, a qualitative WAIS change could occur within this millennium, with collapse within 300 years being a worst-case scenario. Rapid sea-level rise (greater than 1 m per century) is more likely to come from the WAIS than from the Greenland ice sheet."

Greenland

Greenland's ice sheet contains enough fresh water as ice to raise sea level worldwide by 7 metres (23 ft). Greenland may become warm enough by 2100 to begin an almost complete melt over more than 1,000 years. James E. Hansen suggests that inadequate attention is being given to this issue.

One study suggests it would take 3,000 years to completely melt the Greenland ice sheet. This figure was derived from the assumed levels of greenhouse gases over the duration of the experiment.

As the Greenland ice sheet loses mass from calving of icebergs as well as by melting of ice, any such processes tend to accelerate the loss of the ice sheet.

Millennial-timescale events

Some long-term effects happen over thousands, not hundreds, of years.

Disruption to Thermohaline Circulation

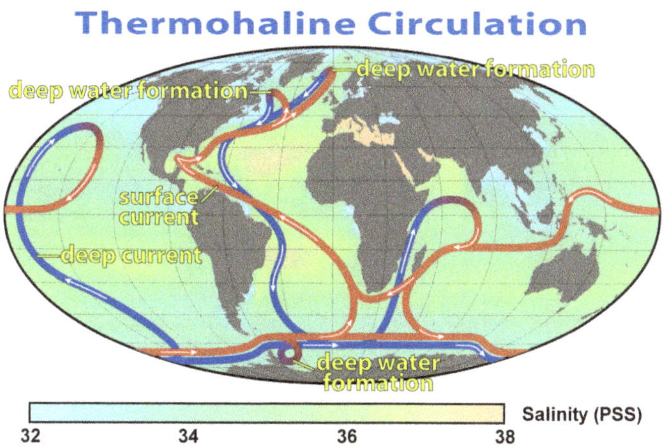

A schematic of the path of the thermohaline circulation. Blue paths represent deep-water currents, while red paths represent surface currents

Early work with simplified models suggested that global warming could cause a shutdown of the thermohaline circulation. This effect is not replicated in more sophisticated coupled ocean-atmosphere global climate models, which do not shutdown but do display varying degrees of slowdown.

This may take a considerable time to occur, as Knutti and Stocker found, again from a simplified model, when :"...thermohaline shutdowns can occur thousands of years after the warming has stopped."

Ocean Anoxia

One study suggests that the amount of oxygen dissolved in the oceans may decline, with adverse consequences for ocean life. This effect was determined using a model run of 100,000 years. The researchers predicted:

...severe, long-term ocean oxygen depletion, as well as a great expansion of ocean oxygen-minimum zones for scenarios with high emissions or high climate sensitivity. We find that climate feedbacks within the Earth system amplify the strength and duration of global warming, ocean heating and oxygen depletion. Decreased oxygen solubility from surface-layer warming accounts for most of the enhanced oxygen depletion in the upper 500 m of the ocean. Possible weakening of ocean overturning and convection lead to further oxygen depletion, also in the deep ocean.

Other research by the Leibniz Institute for Marine Sciences predicts that minimum oxygen zones, particularly in the pacific are expanding, with an estimated total reduction of ocean oxygen levels from 1% to 7% by the end of the century.

Clathrate Decomposition

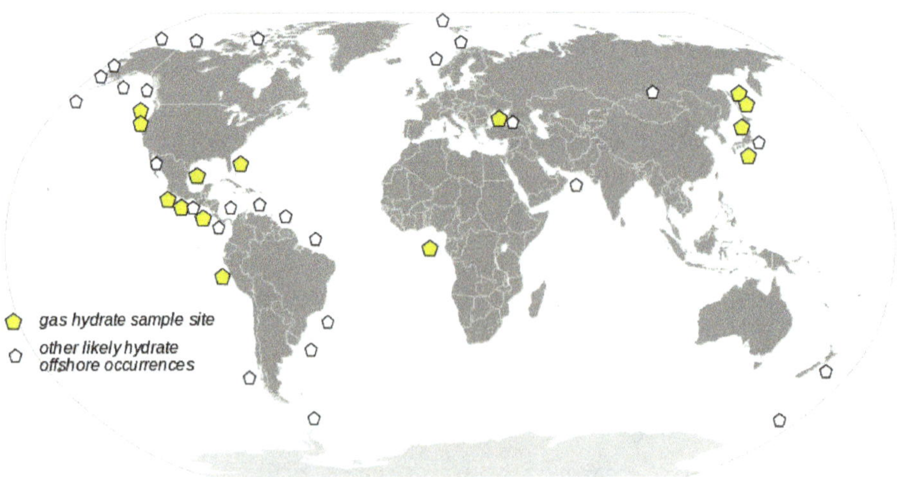

Worldwide distribution of confirmed or inferred offshore gas hydrate-bearing sediments, 1996.
Source: USGS

Methane clathrate, also called methane hydrate, is a form of water ice that contains a large amount of methane within its crystal structure. Extremely large deposits of methane clathrate have been found under sediments on the ocean floors of Earth (estimated at 3000–11,000 Gton C).

MacDonald suggests clathrate volumes are "about 11,000 Gt of carbon for ocean sediments, and about 400 Gt for sediments under permafrost regions". Buffett and Archer predict "eventual releases of 2000–4000 Gton C in response to a ~2000 Gton C anthropogenic carbon release", for which they suggest no timescale.

Archer considered the timescale and determined that "...on longer time scales of 1–10 ky, there may be a positive feedback with ocean temperature, amplifying the long-term climate impact of anthropogenic CO2 release."

Long-term Return to Equilibrium

After the PETM peak there was an extended period of cooling or "climate whiplash".

While the warm surface waters of the oceans have limited ability to absorb anthropogenic carbon dioxide, the coldest surface waters near the poles (2–3% of ocean surfaces) can transfer significant amounts of carbon dioxide to deep-ocean reserves. Over a period of many centuries, this process and the process of calcium carbonate absorption of carbon dioxide on land and in the oceans will remove 60–80% of the excess carbon dioxide.

Igneous rock when exposed to a near surface environment absorbs carbon dioxide through a very slow weathering rate, but weathering increases in a warmer, higher rainfall climate, speeding the process. This geological weathering will absorb the remaining 20–40% of anthropogenic carbon dioxide over the period of tens of thousands to hundreds of thousands of years.

REFERENCES

1. 16 January 2015: NASA GISS: NASA GISS: NASA, NOAA Find 2014 Warmest Year in Modern Record, in: Research News. NASA Goddard Institute for Space Studies, New York, US. Retrieved 20 February 2015
2. Gillis, Justin (28 November 2015). "Short Answers to Hard Questions About Climate Change". The New York Times. Retrieved 29 November 2015.
3. Simpson, John. Oxford English Dictionary. (Global warming is a) gradual increase in the overall temperature of the earth's atmosphere generally attributed to the greenhouse effect caused by increased levels of carbon dioxide, CFC's, and other pollutants.
4. Hartmann, D. L.; Klein Tank, A. M. G.; Rusticucci, M. (2013). "2: Observations: Atmosphere and Surface" (PDF). IPCC WGI AR5 (Report). p. 198. Evidence for a warming world comes from multiple independent climate indicators, from high up in the atmosphere to the depths of the oceans. They include changes in surface, atmospheric and oceanic temperatures; glaciers; snow cover; sea ice; sea level and atmospheric water vapour. Scientists from all over the world have independently verified this evidence many times.
5. Myth vs Facts....". EPA (US). 2013.The U.S. Global Change Research Program, the

National Academy of Sciences, and the Intergovernmental Panel on Climate Change (IPCC) have each independently concluded that warming of the climate system in recent decades is ‹unequivocal'. This conclusion is not drawn from any one source of data but is based on multiple lines of evidence, including three worldwide temperature datasets showing nearly identical warming trends as well as numerous other independent indicators of global warming (e.g., rising sea levels, shrinking Arctic sea ice).

6. Borenstein, Seth (29 November 2015). "Earth is a wilder, warmer place since last climate deal made". Retrieved 29 November 2015.

7. Rhein, M.; Rintoul, S.R. (2013). "3: Observations: Ocean" (PDF). IPCC WGI AR5 (Report). p. 257. Ocean warming dominates the global energy change inventory. Warming of the ocean accounts for about 93% of the increase in the Earth's energy inventory between 1971 and 2010 (high confidence), with warming of the upper (0 to 700 m) ocean accounting for about 64% of the total. Melting ice (including Arctic sea ice, ice sheets and glaciers) and warming of the continents and atmosphere account for the remainder of the change in energy.

8. IPCC, Climate Change 2013: The Physical Science Basis - Summary for Policymakers, Observed Changes in the Climate System, p. 2, in IPCC AR5 WG1 2013. "Warming of the climate system is unequivocal, and since the 1950s, many of the observed changes are unprecedented over decades to millennia."

9. ^ Jump up to:[a] [b] CLIMATE CHANGE 2014: Synthesis Report. Summary for Policymakers" (PDF). IPCC. Retrieved 1 November 2015. The following terms have been used to indicate the assessed likelihood of an outcome or a result: virtually certain 99–100% probability, very likely 90–100%, likely 66–100%, about as likely as not 33–66%, unlikely 0–33%, very unlikely 0–10%, exceptionally unlikely 0–1%. Additional terms (extremely likely: 95–100%, more likely than not >50–100%, more unlikely than likely 0–<50% and extremely unlikely 0–5%) may also be used when appropriate.

10. CLIMATE CHANGE 2014: Synthesis Report. Summary for Policymakers" (PDF). IPCC. Retrieved 7 March 2015. The evidence for human influence on the climate system has grown since the Fourth Assessment Report (AR4). It is extremely likely that more than half of the observed increase in global average surface temperature from 1951 to 2010 was caused by the anthropogenic increase in greenhouse gas concentrations and other anthropogenic forcings together

11. EPA". EPA. EPA.

12. Buis, Alan; Ramsayer, Kate; Rasmussen, Carol (12 November 2015). "A Breathing Planet, Off Balance". NASA. Retrieved 13 November 2015.

13. Stocker et al., Technical Summary, in IPCC AR5 WG1 2013.

14. Joint Science Academies' Statement" (PDF). Retrieved 6 January 2014.

15. Kirby, Alex (17 May 2001). "Science academies back Kyoto". BBC News. Retrieved 27 July 2011.

16. DiMento, Joseph F. C.; Doughman, Pamela M. (2007). Climate Change: What It Means for Us, Our Children, and Our Grandchildren. The MIT Press. p. 68. ISBN 978-0-262-54193-0.

17. Parry, M.L.; et al., "Technical summary", Box TS.6. The main projected impacts for regions, in IPCC AR4 WG2 2007, pp. 59–63

18. Solomon et al., Technical Summary, Section TS.5.3: Regional-Scale Projections, in

IPCC AR4 WG1 2007.

19. Lu, Jian; Vechhi, Gabriel A.; Reichler, Thomas (2007). "Expansion of the Hadley cell under global warming" (PDF). Geophysical Research Letters. **34** *(6): L06805. Bibcode*:2007GeoRL..3406805L. doi:10.1029/2006GL028443.
20. On snowfall:
21. Christopher Joyce (15 February 2010). "Get This: Warming Planet Can Mean More Snow". NPR.
22. "Global warming means more snowstorms: scientists". 1 March 2011.
23. "Does record snowfall disprove global warming?". 9 July 2010. Retrieved 14 December 2014.
24. Battisti, David; Naylor, Rosamund L. (2009). "Historical warnings of future food insecurity with unprecedented seasonal heat". Science. **323** *(5911): 240–4. doi:10.1126/science.1164363. PMID 19131626. Retrieved 13 April 2012.*
25. Peter, U.; et al. "Clark et al. 2016 Consequences of twenty-first-century policy for multi-millennial climate and sea-level change". Nature Climate Change. **6**: 360–369. doi:10.1038/NCLIMATE2923.
26. United Nations Framework Convention on Climate Change (UNFCCC) (2011). "Status of Ratification of the Convention". UNFCCC Secretariat: Bonn, Germany: UNFCCC.. Most countries in the world are Parties to the United Nations Framework Convention on Climate Change (UNFCCC), which has adopted the 2 °C limit. As of 25 November 2011, there are 195 parties (194 states and 1 regional economic integration organization (the European Union)) to the UNFCCC.
27. America's Climate Choices. Washington, D.C.: The National Academies Press. 2011. p. 15. ISBN 978-0-309-14585-5. The average temperature of the Earth's surface increased by about 1.4 °F (0.8 °C) over the past 100 years, with about 1.0 °F (0.6 °C) of this warming occurring over just the past three decades.
28. Sutter, John D.; Berlinger, Joshua (12 December 2015). "Final draft of climate deal formally accepted in Paris". CNN. Cable News Network, Turner Broadcasting System, Inc. Retrieved 12 December 2015.
29. Vaughan, A. (12 December 2015). "Paris climate deal: key points at a glance". The Guardian. London and Manchester, UK. Retrieved 12 December 2015.. Archived 1 August 2016.
30. Stokes, Bruce; Wike, Richard; Carle, Jill (5 November 2015). "Global Concern about Climate Change, Broad Support for Limiting Emissions: U.S., China Less Worried; Partisan Divides in Key Countries". Pew Research Center. Retrieved 18 June 2016.
31. Brown, Dwayne; Cabbage, Michael; McCarthy, Leslie; Norton, Karen (20 January 2016). "NASA, NOAA Analyses Reveal Record-Shattering Global Warm Temperatures in 2015". NASA. Retrieved 21 January 2016.
32. Rhein, M., et al. (June 7, 2013): Box 3.1, in: Chapter 3: Observations: Ocean (final draft accepted by IPCC Working Group I), pp.11-12 (pp.14-15 of PDF chapter), in: IPCC AR5 WG1 2013
33. IPCC (November 11, 2013): D.3 Detection and Attribution of Climate Change, in: Summary for Policymakers (finalized version), p.15, in: IPCC AR5 WG1 2013

34. Climate Change 2013: The Physical Science Basis, IPCC Fifth Assessment Report (WGI AR5)" (PDF). WGI AR5. IPCC AR5. 2013. p. 5.
35. Climate Change 2007: Working Group I: The Physical Science Basis". IPCC AR4. 2007.
36. Jansen et al., Ch. 6, Palaeoclimate, Section 6.6.1.1: What Do Reconstructions Based on Palaeoclimatic Proxies Show?, pp. 466–478, in IPCC AR4 WG1 2007.
37. Kennedy, J.J.; et al. (2010). "How do we know the world has warmed? in: 2. Global Climate, in: State of the Climate in 2009". Bull. Amer. Meteor. Soc. **91** *(7): 26.*
38. Kennedy, C. (10 July 2012). "ClimateWatch Magazine >> State of the Climate: 2011 Global Sea Level". NOAA Climate Services Portal.
39. Summary for Policymakers". Direct Observations of Recent Climate Change., in IPCC AR4 WG1 2007
40. Summary for Policymakers". B. Current knowledge about observed impacts of climate change on the natural and human environment., in IPCC AR4 WG2 2007
41. Rosenzweig, C.; et al. "Ch 1: Assessment of Observed Changes and Responses in Natural and Managed Systems". Sec 1.3.5.1 Changes in phenology., in IPCC AR4 WG2 2007, p. 99
42. Trenberth et al., Chap 3, Observations: Atmospheric Surface and Climate Change, Executive Summary, p. 237, in IPCC AR4 WG1 2007.
43. Rowan T. Sutton; Buwen Dong; Jonathan M. Gregory (2007). "Land/sea warming ratio in response to climate change: IPCC AR4 model results and comparison with observations". Geophysical Research Letters. **34** *(2): L02701. Bibcode*:2007GeoRL..3402701S. doi:10.1029/2006GL028164. Retrieved 19 September 2007.
44. Feulner, Georg; Rahmstorf, Stefan; Levermann, Anders; Volkwardt, Silvia (March 2013). "On the Origin of the Surface Air Temperature Difference Between the Hemispheres in Earth's Present-Day Climate". Journal of Climate. **26**: 130325101629005. doi:10.1175/JCLI-D-12-00636.1. Retrieved 25 April 2013.
45. TS.3.1.2 Spatial Distribution of Changes in Temperature, Circulation and Related Variables - AR4 WGI Technical Summary
46. Ehhalt et al., Chapter 4: Atmospheric Chemistry and Greenhouse Gases, Section 4.2.3.1: Carbon monoxide (CO) and hydrogen (H_2), p. 256, in IPCC TAR WG1 2001.
47. Meehl, Gerald A.; Washington, Warren M.; Collins, William D.; Arblaster, Julie M.; Hu, Aixue; Buja, Lawrence E.; Strand, Warren G.; Teng, Haiyan (18 March 2005). "How Much More Global Warming and Sea Level Rise" (PDF). Science. **307** *(5716): 1769–1772. Bibcode*:2005Sci...307.1769M. doi:10.1126/science.1106663. PMID 15774757. Retrieved 11 February 2007.
48. England, Matthew (February 2014). "Recent intensification of wind-driven circulation in the Pacific and the ongoing warming hiatus". Nature Climate Change. **4**: 222–227. Bibcode:2014NatCC...4..222E. doi:10.1038/nclimate2106.
49. Knight, J.; Kenney, J.J.; Folland, C.; Harris, G.; Jones, G.S.; Palmer, M.; Parker, D.; Scaife, A.; Stott, P. (August 2009). "Do Global Temperature Trends Over the Last Decade Falsify Climate Predictions? [in "State of the Climate in 2008"]" (PDF). Bull. Amer. Meteor. Soc. **90** *(8): S75–S79. Retrieved 13 August 2011.*
50. Global temperature slowdown – not an end to climate change. UK Met Office. Retrieved 20 March 2011.

51. Gavin Schmidt (4 June 2015). "NOAA temperature record updates and the ‹hiatus'".

52. NOAA (4 June 2015). "Science publishes new NOAA analysis: Data show no recent slowdown in global warming".

53. Miller, Brandon (20 January 2016). "2015 is warmest year on record, NOAA and NASA say". CNN. Retrieved 27 March 2016.

54. Schmidt, Gavin (22 January 2015). "Thoughts on 2014 and ongoing temperature trends". RealClimate. Retrieved 4 September 2015.

55. Group (28 November 2004). "Forcings (filed under: Glossary)". RealClimate.

56. Pew Center on Global Climate Change / Center for Climate and Energy Solutions (September 2006). "Science Brief 1: The Causes of Global Climate Change" (PDF). Arlington, Virginia, USA: Center for Climate and Energy Solutions., p.2

57. Brown, Patrick T.; Li, Wenhong; Jiang, Jonathan H.; Su, Hui (2015-12-07). "Unforced Surface Air Temperature Variability and Its Contrasting Relationship with the Anomalous TOA Energy Flux at Local and Global Spatial Scales". Journal of Climate. **29** *(3): 925–940. doi:10.1175/JCLI-D-15-0384.1. ISSN 0894-8755.*

58. Hegerl et al., Chapter 9: Understanding and Attributing Climate Change, Section 9.4.1.5: The Influence of Other Anthropogenic and Natural Forcings, in IPCC AR4 WG1 2007, pp. 690–691. "Recent estimates indicate a relatively small combined effect of natural forcings on the global mean temperature evolution of the second half of the 20th century, with a small net cooling from the combined effects of solar and volcanic forcings." p. 690

59. Tyndall, John (1861). "On the Absorption and Radiation of Heat by Gases and Vapours, and on the Physical Connection of Radiation, Absorption, and Conduction" (PDF). Philosophical Magazine. 4. **22**: 169–94, 273–85. Retrieved 8 May 2013.

60. Weart, Spencer (2008). "The Carbon Dioxide Greenhouse Effect". The Discovery of Global Warming. American Institute of Physics. Retrieved 21 April 2009.

61. The Callendar Effect: the life and work of Guy Stewart Callendar (1898–1964) Amer Meteor Soc., Boston. ISBN 978-1-878220-76-9

62. Le Treut; et al. "Chapter 1: Historical Overview of Climate Change Science". FAQ 1.1., p. 97, in IPCC AR4 WG1 2007: "To emit 240 W m–2, a surface would have to have a temperature of around −19 °C. This is much colder than the conditions that actually exist at the Earth's surface (the global mean surface temperature is about 14 °C). Instead, the necessary −19 °C is found at an altitude about 5 km above the surface."

63. Blue, Jessica. "What is the Natural Greenhouse Effect?". *National Geographic. Retrieved 1 Jan 2015.*

64. Kiehl, J.T.; Trenberth, K.E. (1997). "Earth's Annual Global Mean Energy Budget" (PDF). Bulletin of the American Meteorological Society. **78** *(2): 197–208. Bibcode*:1997BAMS...78..197K. doi:10.1175/1520-0477(1997)078<0197:EAGMEB>2.0.CO;2. ISSN 1520-0477. Archived from the original (PDF) on 24 June 2008. Retrieved 21 April 2009.

65. Schmidt, Gavin (6 April 2005). "Water vapour: feedback or forcing?". RealClimate. Retrieved 21 April 2009.

66. Russell, Randy (16 May 2007). "The Greenhouse Effect & Greenhouse Gases". University Corporation for Atmospheric Research Windows to the Universe. Retrieved

27 December 2009.

67. EPA (2007). "Recent Climate Change: Atmosphere Changes". Climate Change Science Program. United States Environmental Protection Agency. Retrieved 21 April 2009.

68. Spahni, Renato; Jérôme Chappellaz; Thomas F. Stocker; Laetitia Loulergue; Gregor Hausammann; Kenji Kawamura; Jacqueline Flückiger; Jakob Schwander; Dominique Raynaud; Valérie Masson-Delmotte; Jean Jouzel (November 2005). "Atmospheric Methane and Nitrous Oxide of the Late Pleistocene from Antarctic Ice Cores". Science. **310** *(5752): 1317–1321. Bibcode*:2005Sci...310.1317S. doi:10.1126/science.1120132. PMID 16311333.

69. Siegenthaler, Urs; et al. (November 2005). "Stable Carbon Cycle–Climate Relationship During the Late Pleistocene" (PDF). Science. **310** *(5752): 1313–1317. Bibcode*:2005Sci...310.1313S. doi:10.1126/science.1120130. PMID 16311332. Retrieved 25 August 2010.

70. Petit, J. R.; et al. (3 June 1999). "Climate and atmospheric history of the past 420,000 years from the Vostok ice core, Antarctica" (PDF). Nature. **399** *(6735): 429–436. Bibcode*:1999Natur.399..429P. doi:10.1038/20859. Retrieved 27 December 2009.

71. Lüthi, D.; Le Floch, M.; Bereiter, B.; Blunier, T.; Barnola, J. M.; Siegenthaler, U.; Raynaud, D.; Jouzel, J.; Fischer, H.; Kawamura, K.; Stocker, T. F. (2008). "High-resolution carbon dioxide concentration record 650,000–800,000 years before present". Nature. **453** *(7193): 379–382. Bibcode*:2008Natur.453..379L. doi:10.1038/nature06949. PMID 18480821.

72. Pearson, PN; Palmer, MR (2000). "Atmospheric carbon dioxide concentrations over the past 60 million years". Nature. **406** *(6797): 695–699.* doi:10.1038/35021000. PMID 10963587.

73. IPCC, Summary for Policymakers, Concentrations of atmospheric greenhouse gases ..., p. 7, in IPCC TAR WG1 2001.

74. IPCC (2007) AR4. Climate Change 2007: Working Group III: Mitigation of Climate Change, section 7.4.5.1. https://www.ipcc.ch/publications_and_data/ar4/wg3/en/ch7s7-4-5.html

75. Le Quéré, C.;; Andres, R.J.; Boden, T.; Conway, T.; Houghton, R.A.; House, J.I.; Marland, G.; Peters, G.P.; van der Werf, G.; Ahlström, A.; Andrew, R.M.; Bopp, L.; Canadell, J.G.; Ciais, P.; Doney, S.C.; Enright, C.; Friedlingstein, P.; Huntingford, C.; Jain, A.K.; Jourdain, C.; Kato, E.; Keeling, R.F.; Klein Goldewijk, K.; Levis, S.; Levy, P.; Lomas, M.; Poulter, B.; Raupach, M.R.; Schwinger, J.; Sitch, S.; Stocker, B.D.; Viovy, N.; Zaehle, S.; Zeng, N. (2 December 2012). "The global carbon budget 1959–2011". Earth System Science Data Discussions. **5** *(2): 1107–1157. Bibcode*:2012ESSDD...5.1107L. doi:10.5194/essdd-5-1107-2012.

76. Carbon dioxide passes symbolic mark". BBC. 10 May 2013. Retrieved 27 May 2013.

77. Pilita Clark (10 May 2013). "CO_2 *at highest level for millions of years". Financial Times. Retrieved 27 May 2013. (*registration required (help)).

78. Climate scientists discuss future of their field". 7 July 2015.

79. Rogner, H.-H., et al., Chap. 1, Introduction, Section 1.3.1.2: Intensities, in IPCC AR4 WG3 2007.

80. NRC (2008). "Understanding and Responding to Climate Change" (PDF). Board on

Atmospheric Sciences and Climate, US National Academy of Sciences. p. 2. Retrieved 9 November 2010.

81. World Bank (2010). World Development Report 2010: Development and Climate Change. The International Bank for Reconstruction and Development / The World Bank, 1818 H Street NW, Washington, D.C. 20433. doi:10.1596/978-0-8213-7987-5. ISBN 978-0-8213-7987-5. Archived from the original on 5 March 2010. Retrieved 6 April 2010.

82. Banuri et al., Chapter 3: Equity and Social Considerations, Section 3.3.3: Patterns of greenhouse gas emissions, and Box 3.1, pp. 92–93 in IPCC SAR WG3 1996.

83. ^ Jump up to:[a] [b] [c] Liverman, D.M. (2008). "Conventions of climate change: constructions of danger and the dispossession of the atmosphere" (PDF). Journal of Historical Geography. **35** *(2): 279–296. doi:10.1016/j.jhg.2008.08.008. Retrieved 10 May 2011.*

84. Fisher et al., Chapter 3: Issues related to mitigation in the long-term context, Section 3.1: Emissions scenarios: Issues related to mitigation in the long term context in IPCC AR4 WG3 2007.

85. Morita, Chapter 2: Greenhouse Gas Emission Mitigation Scenarios and Implications, Section 2.5.1.4: Emissions and Other Results of the SRES Scenarios, in IPCC TAR WG3 2001.

86. Rogner et al., Ch. 1: Introduction, Figure 1.7, in IPCC AR4 WG3 2007.

87. IPCC, Summary for Policymakers, Introduction, paragraph 6, in IPCC TAR WG3 2001.

88. Prentence et al., Chapter 3: The Carbon Cycle and Atmospheric Carbon Dioxide Executive Summary, in IPCC TAR WG1 2001.

89. Newell, P.J., 2000: Climate for change: non-state actors and the global politics of greenhouse. Cambridge University Press, ISBN 0-521-63250-1.

90. Talk of the Nation. "Americans Fail the Climate Quiz". NPR. Retrieved 27 December 2011.

91. Shindell, Drew; Faluvegi, Greg; Lacis, Andrew; Hansen, James; Ruedy, Reto; Aguilar, Elliot (2006). "Role of tropospheric ozone increases in 20th-century climate change". Journal of Geophysical Research. **111** *(D8): D08302. Bibcode:*2006JGRD..11108302S. doi:10.1029/2005JD006348.

92. Solomon, S; D. Qin; M. Manning; Z. Chen; M. Marquis; K.B. Averyt; M. Tignor; H.L. Miller, eds. (2007). "3.4.4.2 Surface Radiation". Climate Change 2007: Working Group I: The Physical Science Basis. ISBN 978-0-521-88009-1.

93. Hansen, J; Sato, M; Ruedy, R; Lacis, A; Oinas, V (2000). "Global warming in the twenty-first century: an alternative scenario". Proc. Natl. Acad. Sci. U.S.A. **97** *(18): 9875–80. Bibcode:*2000PNAS...97.9875H. doi:10.1073/pnas.170278997. PMC 27611. PMID 10944197.

94. Ramanathan, V.; Carmichael, G. (2008). "Global and regional climate changes due to black carbon". Nature Geoscience. **1** *(4): 221–227. Bibcode:*2008NatGe...1..221R. doi:10.1038/ngeo156.

95. V. Ramanathan and G. Carmichael, supra note 1, at 221 (". . . emissions of black carbon are the second strongest contribution to current global warming, after

carbon dioxide emissions.") Numerous scientists also calculate that black carbon may be second only to CO_2 in its contribution to climate change, including Tami C. Bond & Haolin Sun, Can Reducing Black Carbon Emissions Counteract Global Warming, ENVIRON. SCI. TECHN. (2005), at 5921 ("BC is the second or third largest individual warming agent, following carbon dioxide and methane."); and J. Hansen, A Brighter Future, 53 CLIMATE CHANGE 435 (2002), available athttp://pubs.giss.nasa.gov/docs/2002/2002_Hansen_1.pdf (calculating the climate forcing of BC at 1.0±0.5 W/m^2).

96. Twomey, S. (1977). "Influence of pollution on shortwave albedo of clouds". J. Atmos. Sci. **34** *(7): 1149–1152. Bibcode*:1977JAtS...34.1149T. doi:10.1175/1520-0469(1977)034<1149:TIOPOT>2.0.CO;2. ISSN 1520-0469.

97. Albrecht, B. (1989). "Aerosols, cloud microphysics, and fractional cloudiness". Science. **245** *(4923): 1227–1239. Bibcode*:1989Sci...245.1227A. doi:10.1126/science.245.4923.1227. PMID 17747885.

98. IPCC, "Aerosols, their Direct and Indirect Effects", pp. 291–292 in IPCC TAR WG1 2001.

99. Ramanathan, V.; Chung, C.; Kim, D.; Bettge, T.; Buja, L.; Kiehl, J. T.; Washington, W. M.; Fu, Q.; Sikka, D. R.; Wild, M. (2005). "Atmospheric brown clouds: Impacts on South Asian climate and hydrological cycle"(Full free text). Proceedings of the National Academy of Sciences. **102** *(15): 5326–5333. Bibcode*:2005PNAS..102.5326R. doi:10.1073/pnas.0500656102. PMC 552786. PMID 15749818.

100. Ramanathan, V.; et al. (2008). "Report Summary" (PDF). Atmospheric Brown Clouds: Regional Assessment Report with Focus on Asia. United Nations Environment Programme.

101. Ramanathan, V.; et al. (2008). "Part III: Global and Future Implications" (PDF). Atmospheric Brown Clouds: Regional Assessment Report with Focus on Asia. United Nations Environment Programme.

102. ^ Jump up to:[a] [b] IPCC, Summary for Policymakers, Human and Natural Drivers of Climate Change, Figure SPM.2, in IPCC AR4 WG1 2007.

103. US Environmental Protection Agency (2009). "3.2.2 Solar Irradiance". Volume 3: Attribution of Observed Climate Change. Endangerment and Cause or Contribute Findings for Greenhouse Gases under Section 202(a) of the Clean Air Act. EPA's Response to Public Comments. US Environmental Protection Agency. Archived from the original on 16 June 2011. Retrieved June 23, 2011.

104. US NRC 2008, p. 6

105. Hegerl, et al., Chapter 9: Understanding and Attributing Climate Change, Frequently Asked Question 9.2: Can the Warming of the 20th century be Explained by Natural Variability?, in IPCC AR4 WG1 2007.

106. Simmon, R.; D. Herring (November 2009). "Notes for slide number 7, titled "Satellite evidence also suggests greenhouse gas warming," in presentation, "Human contributions to global climate change"". Presentation library on the U.S. National Oceanic and Atmospheric Administration's Climate Services website. Archived from the original on 3 July 2011. Retrieved 23 June 2011.

107. Hegerl et al., Chapter 9: Understanding and Attributing Climate Change, Frequently Asked Question 9.2: Can the Warming of the 20th century be Explained by Natural Variability?, in IPCC AR4 WG1 2007.

108. Randel, William J.; Shine, Keith P.; Austin, John; et al. (2009). "An update of observed stratospheric temperature trends". Journal of Geophysical Research. **114** *(D2): D02107. Bibcode*:2009JGRD..11402107R. doi:10.1029/2008JD010421.

109. USGCRP 2009, p. 20

110. R.S. Bradley; K.R. Briffa; J. Cole; M.K. Hughes; T.J. Osborn (2003). "The climate of the last millennium". In K.D. Alverson; R.S. Bradley; T.F. Pederson. Paleoclimate, global change and the future. Springer. pp. 105–141. ISBN 3-540-42402-4.

111. Kaufman, D. S.; Schneider, D. P.; McKay, N. P.; Ammann, C. M.; Bradley, R. S.; Briffa, K. R.; Miller, G. H.; Otto-Bliesner, B. L.; Overpeck, J. T.; Vinther, B. M.; Abbott, M.; Axford, M.; Bird, Y.; Birks, B.; Bjune, H. J. B.; Briner, A. E.; Cook, J.; Chipman, T.; Francus, M.; Gajewski, P.; Geirsdottir, K.; Hu, A.; Kutchko, F. S.; Lamoureux, B.; Loso, S.; MacDonald, M.; Peros, G.; Porinchu, M.; Schiff, D.; Seppa, C.; Seppa, H.; Arctic Lakes 2k Project Members (2009). "Recent Warming Reverses Long-Term Arctic Cooling". Science. **325** *(5945): 1236–1239. Bibcode*:2009Sci...325.1236K. doi:10.1126/science.1173983. PMID 19729653.

112. Arctic Warming Overtakes 2,000 Years of Natural Cooling". UCAR. 3 September 2009. Retrieved 8 June 2011.

113. Bello, David (4 September 2009). "Global Warming Reverses Long-Term Arctic Cooling". Scientific American. Retrieved 8 June 2011.

114. Mann, M. E.; Zhang, Z.; Hughes, M. K.; Bradley, R. S.; Miller, S. K.; Rutherford, S.; Ni, F. (2008). "Proxy-based reconstructions of hemispheric and global surface temperature variations over the past two millennia". Proceedings of the National Academy of Sciences. **105** *(36): 13252–7. Bibcode*:2008PNAS..10513252M. doi:10.1073/pnas.0805721105. PMC 2527990. PMID 18765811.

115. Berger, A. (2002). "CLIMATE: An Exceptionally Long Interglacial Ahead?". *Science.* **297** *(5585): 1287–8. doi:10.1126/science.1076120. PMID 12193773.*

116. Masson-Delmotte V.M.; et al. (2013). "Information from paleoclimate archives". In Stocker T.F. et al. Climate Change 2013: The Physical Science Basis. Contribution of Working Group I to the Fifth Assessment Report of the Intergovernmental Panel on Climate Change. Cambridge University Press. pp. 383–464. ISBN 978-1-107-66182-0.

117. Jackson, R.; A. Jenkins (17 November 2012). "Vital signs of the planet: global climate change and global warming: uncertainties". Earth Science Communications Team at NASA's Jet Propulsion Laboratory / California Institute of Technology.

118. Riebeek, H. (16 June 2011). "The Carbon Cycle: Feature Articles: Effects of Changing the Carbon Cycle". Earth Observatory, part of the EOS Project Science Office located at NASA Goddard Space Flight Center.

119. US National Research Council (2003). "Ch. 1 Introduction". Understanding Climate Change Feedbacks. Washington, D.C., USA: National Academies Press., p.19

120. Lindsey, R. (14 January 2009). "Earth's Energy Budget (p.4), in: Climate and Earth's Energy Budget: Feature Articles". Earth Observatory, part of the EOS Project Science Office, located at NASA Goddard Space Flight Center.

121. US National Research Council (2006). "Ch. 1 Introduction to Technical Chapters". Surface Temperature Reconstructions for the Last 2,000 Years. Washington, D.C., USA: National Academies Press., pp.26-27

122. AMS Council (20 August 2012). "2012 American Meteorological Society (AMS) Information Statement on Climate Change". Boston, Massachusetts, USA: AMS.

123. Meehl, G.A.; et al. "Ch 10: Global Climate Projections". Sec 10.5.4.6 Synthesis of Projected Global Temperature at Year 2100]., in IPCC AR4 WG1 2007

124. NOAA (January 2007). "Patterns of greenhouse warming" (PDF). GFDL Climate Modeling Research Highlights. Princeton, New Jersey, USA: The National Oceanic and Atmospheric Administration (NOAA) Geophysical Fluid Dynamics Laboratory (GFDL). **1** *(6).*, revision 2 February 2007, 8:50.08 AM.

125. NOAA Geophysical Fluid Dynamics Laboratory (GFDL) (9 October 2012). "NOAA GFDL Climate Research Highlights Image Gallery: Patterns of Greenhouse Warming". NOAA GFDL.

126. IPCC, Glossary A-D: "Climate Model", in IPCC AR4 SYR 2007.

127. Karl, TR; et al., eds. (2009). "Global Climate Change". Global Climate Change Impacts in the United States. Cambridge University Press. ISBN 978-0-521-14407-0.

128. KEVIN SCHAEFER; TINGJUN ZHANG; LORI BRUHWILER; ANDREW P. BARRETT (2011). "Amount and timing of permafrost carbon release in response to climate warming". Tellus Series B. **63** *(2): 165–180. Bibcode*:2011TellB..63..165S. doi:10.1111/j.1600-0889.2011.00527.x.

129. Hansen, James (2000). "Climatic Change: Understanding Global Warming". In Robert Lanza. One World: The Health & Survival of the Human Species in the 21st century. Health Press (New Mexico). pp. 173–190. ISBN 0-929173-33-3. Retrieved 18 August 2007.

130. Stocker et al., Chapter 7: Physical Climate Processes and Feedbacks, Section 7.2.2: Cloud Processes and Feedbacks, in IPCC TAR WG1 2001.

131. Torn, Margaret; Harte, John (2006). "Missing feedbacks, asymmetric uncertainties, and the underestimation of future warming" (PDF). Geophysical Research Letters. **33** *(10): L10703. Bibcode*:2006GeoRL..3310703T. doi:10.1029/2005GL025540. Retrieved 4 March 2007.

132. Harte, John; Saleska, Scott; Shih, Tiffany (2006). "Shifts in plant dominance control carbon-cycle responses to experimental warming and widespread drought". Environmental Research Letters. **1** *(1): 014001. Bibcode*:2006ERL.....1a4001H. doi:10.1088/1748-9326/1/1/014001. Retrieved 2 May 2007.

133. Scheffer, Marten; Brovkin, Victor; Cox, Peter (2006). "Positive feedback between global warming and atmospheric CO_2 *concentration inferred from past climate change" (PDF). Geophysical Research Letters.* **33** *(10): L10702. Bibcode*:2006GeoRL..3310702S. doi:10.1029/2005gl025044. Retrieved 4 May 2007.

134. Randall et al., Chapter 8, Climate Models and Their Evaluation, Sec. FAQ 8.1 in IPCC AR4 WG1 2007.

135. IPCC, Technical Summary, p. 54, in IPCC TAR WG1 2001.

136. Stroeve, J.; et al. (2007). "Arctic sea ice decline: Faster than forecast". Geophysical Research Letters. **34** *(9): L09501. Bibcode*:2007GeoRL..3409501S. doi:10.1029/2007GL029703.

137. Wentz,FJ; et al. (2007). "How Much More Rain Will Global Warming Bring?". *Science.* **317** *(5835): 233–5. Bibcode*:2007Sci...317..233W. doi:10.1126/science.1140746. PMID

17540863.

138. Liepert, Beate G.; Previdi, Michael (2009). "Do Models and Observations Disagree on the Rainfall Response to Global Warming?". *Journal of Climate.* **22** *(11): 3156–3166. Bibcode*:2009JCli...22.3156L. doi:10.1175/2008JCLI2472.1.

139. Rahmstorf, S.; Cazenave, A.; Church, J. A.; Hansen, J. E.; Keeling, R. F.; Parker, D. E.; Somerville, R. C. J. (4 May 2007). "Recent Climate Observations Compared to Projections". Science. **316** *(5825): 709–709. Bibcode*:2007Sci...316..709R. doi:10.1126/science.1136843. PMID 17272686.

140. 4. Global Mean Sea Level Rise Scenarios, in: Main Report, in Parris & others 2012, p. 12

141. Executive Summary, in Parris & others 2012, p. 1

142. Hegerl, G.C.; et al. "Ch 9: Understanding and Attributing Climate Change". Executive Summary., in IPCC AR4 WG1 2007

143. Sahara Desert Greening Due to Climate Change?". *National Geographic. Retrieved 12 June 2010.*

144. Meehl, G.A.; et al. "Ch 10: Global Climate Projections". Box 10.1: Future Abrupt Climate Change, 'Climate Surprises', and Irreversible Changes: Glaciers and ice caps., in IPCC AR4 WG1 2007, p. 776

145. Meehl, G.A.; et al. "Ch 10: Global Climate Projections". Sec 10.3.3.2 Changes in Snow Cover and Frozen Ground., in IPCC AR4 WG1 2007, pp. 770, 772

146. Meehl, G.A.; et al. "Ch 10: Global Climate Projections". Sec 10.3.3.1 Changes in Sea Ice Cover., in IPCC AR4 WG1 2007, p. 770

147. Wang, M.; Overland, J. E. (2009). "A sea ice free summer Arctic within 30 years?". Geophys. Res. Lett. **36** *(7). Bibcode*:2009GeoRL..3607502W. doi:10.1029/2009GL037820. Retrieved 2 May 2011.

148. Met Office. "Arctic sea ice 2012". Exeter, UK: Met Office.

149. IPCC, Glossary A-D: "Detection and attribution", in IPCC AR4 WG1 2007. See also Hegerl et al., Section 9.1.2: What are Climate Change Detection and Attribution?, in IPCC AR4 WG1 2007.

150. Rosenzweig et al., Chapter 1: Assessment of Observed Changes and Responses in Natural and Managed Systems Section 1.2 Methods of detection and attribution of observed changes, in IPCC AR4 WG2 2007 .

151. IPCC, Synthesis Report Summary for Policymakers, Section 3: Projected climate change and its impacts, in IPCC AR4 SYR 2007.

152. NOAA (February 2007). "Will the wet get wetter and the dry drier?" (PDF). GFDL Climate Modeling Research Highlights. Princeton, New Jersey, USA: National Oceanic and Atmospheric Administration (NOAA) Geophysical Fluid Dynamics Laboratory (GFDL). **1** *(5).*, p.1. Revision 15 October 2008, 4:47:16 PM.

153. D. Future Climate Extremes, Impacts, and Disaster Losses, in: Summary for policymakers". MANAGING THE RISKS OF EXTREME EVENTS AND DISASTERS TO ADVANCE CLIMATE CHANGE ADAPTATION., in IPCC SREX 2012, pp. 9–13

154. Justin Gillis (27 April 2015). "New Study Links Weather Extremes to Global Warming". The New York Times. Retrieved 27 April 2015. "The bottom line is that things are not that complicated," Dr. Knutti said. "You make the world a degree or two warmer, and

there will be more hot days. There will be more moisture in the atmosphere, so that must come down somewhere."

155. E. M. Fischer; R. Knutti (27 April 2015). "Anthropogenic contribution to global occurrence of heavy-precipitation and high-temperature extremes" (online). Nature Climate Change. **5**: 560–564. Bibcode:2015NatCC...5..560F. doi:10.1038/nclimate2617.
156. UCI study finds dramatic increase in concurrent droughts, heat waves". UCI. 2015.
157. Indian Monsoons Are Becoming More Extreme". Scientific American. 2014.
158. Christopher S. Watson; Neil J. White; John A. Church; Matt A. King; Reed J. Burgette; Benoit Legresy (11 May 2015). "Unabated global mean sea-level rise over the satellite altimeter era". Nature Climate Change. **5**: 565–568. Bibcode:2015NatCC...5..565W. doi:10.1038/nclimate2635.
159. Churchs, John; Clark, Peter. "Chapter 13: Sea Level Change - Final Draft Underlying Scientific-Technical Assessment" (PDF). IPCC Working Group I. Retrieved 21 January 2015.
160. PROJECTIONS OF FUTURE SEA LEVEL RISE, pp.243-244, in: Ch. 7. Sea Level Rise and the Coastal Environment, in National Research Council 2010
161. BOX SYN-1: SUSTAINED WARMING COULD LEAD TO SEVERE IMPACTS, p.5, in: Synopsis, in National Research Council 2011
162. Anders Levermann; Peter U. Clark; Ben Marzeion; Glenn A. Milne; David Pollard; Valentina Radic; Alexander Robinson (13 June 2013). "The multimillennial sea-level commitment of global warming". PNAS. **110**: 13745–13750. Bibcode:2013PNAS..11013745L. doi:10.1073/pnas.1219414110.
163. Ricarda Winkelmann; Anders Levermann; Andy Ridgwell; Ken Caldeira (11 September 2015). "Combustion of available fossil fuel resources sufficient to eliminate the Antarctic Ice Sheet". doi:10.1126/sciadv.1500589.
164. IPCC, Synthesis Report Summary for Policymakers, Section 1: Observed changes in climate and their effects, in IPCC AR4 SYR 2007.
165. Fischlin, et al., Chapter 4: Ecosystems, their Properties, Goods and Services, Executive Summary, p. 213, in IPCC AR4 WG2 2007. Executive summary not present in on-line text; see pdf.
166. Schneider et al., Chapter 19: Assessing Key Vulnerabilities and the Risk from Climate Change, Section 19.3.4: Ecosystems and biodiversity, in IPCC AR4 WG2 2007.
167. Ocean Acidification, in: Ch. 2. Our Changing Climate, in NCADAC 2013, pp. 69–70
168. Ocean acidification, in: Executive summary, in Good & others 2010, p. 14
169. Deutsch; et al. (2011). "Climate-Forced Variability of Ocean Hypoxia" (PDF). Science. **333**: 336–339. Bibcode:2011Sci...333..336D. doi:10.1126/science.1202422.
170. Summary, pp.14-19, in National Research Council 2011
171. FAQ 12.3, in: Chapter 12: Long-term Climate Change: Projections, Commitments and Irreversibility, in IPCC AR5 WG1 2013, pp.88-89 (pp.90-91 of PDF chapter)
172. BOX 2.1: STABILIZATION AND NON-CO_2 GREENHOUSE GASES (p.65), in: Chapter 2: Emissions, Concentrations, and Related Factors, in National Research Council 2011

173. The world's carbon-dioxide emissions have stabilised". The Economist. 16 March 2016. Retrieved 12 December 2016.

174. Rapid rise in methane emissions in 10 years surprises scientists The Guardian

175. Bill McGuire. "Climate forcing of geological and geomorphological hazards". Philosophical Transactions A. Royal Society. **368**: 2311–2315. Bibcode:2010RSPTA.368.2311M. doi:10.1098/rsta.2010.0077.

176. Jérôme Lopez Saez; Christophe Corona; Markus Stoffel; Frédéric Berger. "Climate change increases frequency of shallow spring landslides in the French Alps". Geology. **41**: 619–622. doi:10.1130/G34098.1.

177. Smith, J.B.; et al. "Ch. 19. Vulnerability to Climate Change and Reasons for Concern: A Synthesis". Sec 19.6. Extreme and Irreversible Effects., in IPCC TAR WG2 2001

178. Smith, J. B.; Schneider, S. H.; Oppenheimer, M.; Yohe, G. W.; Hare, W.; Mastrandrea, M. D.; Patwardhan, A.; Burton, I.; Corfee-Morlot, J.; Magadza, C. H. D.; Füssel, H.-M.; Pittock, A. B.; Rahman, A.; Suarez, A.; van Ypersele, J.-P. (17 March 2009). "Assessing dangerous climate change through an update of the Intergovernmental Panel on Climate Change (IPCC) ‹reasons for concern'". Proceedings of the National Academy of Sciences. **106** *(11): 4133–7. Bibcode*:2009PNAS..106.4133S. doi:10.1073/pnas.0812355106. PMC 2648893. PMID 19251662.

179. Clark, P.U.; et al. (December 2008). "Executive Summary". Abrupt Climate Change. A Report by the U.S. Climate Change Science Program and the Subcommittee on Global Change Research. Reston, Virginia, USA: U.S. Geological Survey., pp. 1–7. Report website

180. Siberian permafrost thaw warning sparked by cave data". BBC. 22 February 2013. Retrieved 24 February 2013.

181. US National Research Council (2010). "Advancing the Science of Climate Change: Report in Brief". Washington, D.C., USA: National Academies Press., p.3. PDF of Report

182. IPCC. "Summary for Policymakers". Sec. 2.6. The Potential for Large-Scale and Possibly Irreversible Impacts Poses Risks that have yet to be Reliably Quantified., in IPCC TAR WG2 2001

183. Cramer, W., et al., Executive summary, in: Chapter 18: Detection and attribution of observed impacts(archived 8 July 2014), pp.3-4, in IPCC AR5 WG2 A 2014

184. FAQ 7 and 8, in: Volume-wide Frequently Asked Questions (FAQs) (archived 8 July 2014), pp.2-3, in IPCC AR5 WG2 A 2014

185. Oppenheimer, M., et al., Section 19.6.3: Updating Reasons for Concern, in: Chapter 19: Emergent risks and key vulnerabilities (archived 8 July 2014), pp.39-46, in IPCC AR5 WG2 A 2014

186. Field, C., et al., B-3: Regional Risks and Potential for Adaptation, in: Technical Summary (archived 8 July 2014), pp.27-30, in IPCC AR5 WG2 A 2014

187. Oppenheimer, M., et al., Section 19.6.3: Updating Reasons for Concern, in: Chapter 19: Emergent risks and key vulnerabilities (archived 8 July 2014), pp.42-43, in IPCC AR5 WG2 A 2014

188. Dana Nuccitelli (26 January 2015). "Climate change could impact the poor much more than previously thought". The Guardian.

189. Chris Mooney (22 October 2014). "There's a surprisingly strong link between climate change and violence". The Washington Post.

190. Porter, J.R., et al., Executive summary, in: Chapter 7: Food security and food production systems(archived 8 July 2014), p.3, in IPCC AR5 WG2 A 2014

191. Reference temperature period converted from late-20th century to pre-industrial times (approximated in the source as 1850-1900).

192. Assessment Box SPM-1 (p.14) and B-2. Sectoral Risks and Potential for Adaptation: Food security and food production systems (p.18), in: Summary for Policymakers (archived 8 July 2014), in IPCC AR5 WG2 A 2014

193. Smith, K.R., et al., FAQ 11.2, in: Chapter 11: Human health: impacts, adaptation, and co-benefits(archived 8 July 2014), p.37, in IPCC AR5 WG2 A 2014

194. Costello, Anthony; Abbas, Mustafa; Allen, Adriana; Ball, Sarah; Bell, Sarah; Bellamy, Richard; Friel, Sharon; Groce, Nora; Johnson, Anne; Kett, Maria; Lee, Maria; Levy, Caren; Maslin, Mark; McCoy, David; McGuire, Bill; Montgomery, Hugh; Napier, David; Pagel, Christina; Patel, Jinesh; de Oliveira, Jose Antonio Puppim; Redclift, Nanneke; Rees, Hannah; Rogger, Daniel; Scott, Joanne; Stephenson, Judith; Twigg, John; Wolff, Jonathan; Patterson, Craig (May 2009). "Managing the health effects of climate change". The Lancet. **373** *(9676): 1693–1733.* doi:10.1016/S0140-6736(09)60935-1.

195. ^ Jump up to:[a] [b] Watts, Nick; Adger, W Neil; Agnolucci, Paolo; Blackstock, Jason; Byass, Peter; Cai, Wenjia; Chaytor, Sarah; Colbourn, Tim; Collins, Mat; Cooper, Adam; Cox, Peter M; Depledge, Joanna; Drummond, Paul; Ekins, Paul; Galaz, Victor; Grace, Delia; Graham, Hilary; Grubb, Michael; Haines, Andy; Hamilton, Ian; Hunter, Alasdair; Jiang, Xujia; Li, Moxuan; Kelman, Ilan; Liang, Lu; Lott, Melissa; Lowe, Robert; Luo, Yong; Mace, Georgina; Maslin, Mark; Nilsson, Maria; Oreszczyn, Tadj; Pye, Steve; Quinn, Tara; Svensdotter, My; Venevsky, Sergey; Warner, Koko; Xu, Bing; Yang, Jun; Yin, Yongyuan; Yu, Chaoqing; Zhang, Qiang; Gong, Peng; Montgomery, Hugh; Costello, Anthony (November 2015). "Health and climate change: policy responses to protect public health". The Lancet. **386** *(10006): 1861–1914.* doi:10.1016/S0140-6736(15)60854-6. PMID 26111439. Retrieved 4 January 2016.

196. Smith, K.R., et al., Section 11.4: Direct Impacts of Climate and Weather on Health, in: Chapter 11: Human health: impacts, adaptation, and co-benefits (archived 8 July 2014), pp.10-13, in IPCC AR5 WG2 A 2014

197. Smith, K.R., et al., Section 11.6.1. Nutrition, in: Chapter 11: Human health: impacts, adaptation, and co-benefits (archived 8 July 2014), pp.10-13, in IPCC AR5 WG2 A 2014

198. IPCC AR4 SYR 2007. 3.3.3 Especially affected systems, sectors and regions. Synthesis report.

199. Mimura, N.; et al. (2007). "Executive summary". In Parry, M.L. et al. Chapter 16: Small Islands. Climate change 2007: impacts, adaptation and vulnerability: contribution of Working Group II to the fourth assessment report of the Intergovernmental Panel on Climate Change (IPCC). Cambridge University Press (CUP): Cambridge, UK: Print version: CUP. This version: IPCC website. ISBN 0521880106. Retrieved 15 September 2011.

200. Climate change and the risk of statelessness" (PDF). May 2011. Retrieved 13 April 2012.

201. Chris Hope; Kevin Schaefer (2015). "Economic impacts of carbon dioxide and methane released from thawing permafrost". Nature. **6**: 56–59. Bibcode:2016NatCC...6...56H. doi:10.1038/nclimate2807.

202. North Slope permafrost thawing sooner than expected". University of Alaska Fairbanks.

2015.

203. PBL Netherlands Environment Agency (June 15, 2012). "Figure 6.14, in: Chapter 6: The energy and climate challenge". In van Vuuren, D.; M. Kok. Roads from Rio+20 (PDF). ISBN 978-90-78645-98-6., p.177, Report no: 500062001. Report website.

204. Mitigation, in USGCRP 2015

205. IPCC, Synthesis Report Summary for Policymakers, Section 4: Adaptation and mitigation options, in IPCC AR4 SYR 2007.

206. Edenhofer, O., et al., Table TS.3, in: Technical summary (archived 30 December 2014), in: IPCC AR5 WG3 2014, p. 68

207. Citi report: slowing global warming would save tens of trillions of dollars". The Guardian. 2015.

208. Clarke, L., et al., Executive summary, in: Chapter 6: Assessing Transformation Pathways (archived 30 December 2014), in: IPCC AR5 WG3 2014, p. 418

209. SPM4.1: Long-term mitigation pathways, in: Summary for Policymakers (archived 27 December 2014), in: IPCC AR5 WG3 2014, pp. 10–13

210. Edenhofer, O., et al., TS.3.1.2: Short- and long-term requirements of mitigation pathways, in: Technical summary (archived 30 December 2014), in: IPCC AR5 WG3 2014, pp. 55–56

211. Edenhofer, O., et al., TS.3.1.3: Costs, investments and burden sharing, in: Technical summary (archived 30 December 2014), in: IPCC AR5 WG3 2014, p. 58

212. Smit et al., Chapter 18: Adaptation to Climate Change in the Context of Sustainable Development and Equity, Section 18.2.3: Adaptation Types and Forms, in IPCC TAR WG2 2001.

213. Appendix I. Glossary". Adaptive capacity., in IPCC AR4 WG2 2007

214. Synthesis report". Sec 6.3 Responses to climate change: Robust findings]., in IPCC AR4 SYR 2007

215. New Report Provides Authoritative Assessment of National, Regional Impacts of Global Climate Change" (Press release). U.S. Global Change Research Program. 16 June 2009. Retrieved 14 January 2016.

216. Workshop on managing solar radiation" (PDF). NASA. April 2007. Retrieved 23 May 2009.

217. Stop emitting CO2 or geoengineering could be our only hope" (Press release). The Royal Society. 28 August 2009. Retrieved 14 June 2011.

218. P. Keller, David; Feng, Ellias Y.; Oschlies, Andreas (January 2014). "Potential climate engineering effectiveness and side effects during a high carbon dioxide-emission scenario". Nature. **5**: 3304. Bibcode:2014NatCo...5E3304K. doi:10.1038/ncomms4304. Retrieved 31 March 2014. We find that even when applied continuously and at scales as large as currently deemed possible, all methods are, individually, either relatively ineffective with limited (<8%) warming reductions, or they have potentially severe side effects and cannot be stopped without causing rapid climate change.

219. Quoted in IPCC SAR SYR 1996, "Synthesis of Scientific-Technical Information Relevant to Interpreting Article 2 of the UN Framework Convention on Climate Change", paragraph 4.1, p. 8 (pdf p. 18.)

220. Granger Morgan, M. (Lead Author), H. Dowlatabadi, M. Henrion, D. Keith, R. Lempert, S. McBride, M. Small and T. Wilbanks (Contributing Authors) (2009). "Non-Technical Summary: BOX NT.1 Summary of Climate Change Basics". Synthesis and Assessment Product 5.2: Best practice approaches for characterizing, communicating, and incorporating scientific uncertainty in decisionmaking. A Report by the U.S. Climate Change Science Program and the Subcommittee on Global Change Research (PDF). Washington, D.C., USA.: National Oceanic and Atmospheric Administration. p. 11. Retrieved June 1, 2011.

221. UNFCCC (n.d.). "Essential Background". UNFCCC website. Retrieved 18 May 2010.

222. UNFCCC (n.d.). "Full text of the Convention, Article 2". UNFCCC website. Retrieved 18 May 2010.

223. Rogner et al., Chapter 1: Introduction, Executive summary, in IPCC AR4 WG3 2007.

224. Raupach, R.; Marland, G.; Ciais, P.; Le Quere, C.; Canadell, G.; Klepper, G.; Field, B. (Jun 2007). "Global and regional drivers of accelerating CO2 emissions" (Free full text). Proceedings of the National Academy of Sciences. **104** *(24): 10288–10293. Bibcode*:2007PNAS..10410288R. doi:10.1073/pnas.0700609104. ISSN 0027-8424. PMC 1876160. PMID 17519334.

225. Dessai, S. (2001). "The climate regime from The Hague to Marrakech: Saving or sinking the Kyoto Protocol?" (PDF). Tyndall Centre Working Paper 12. Tyndall Centre website. Retrieved 5 May 2010.

226. Grubb, M. (July–September 2003). "The Economics of the Kyoto Protocol" (PDF). World Economics. **4** *(3): 144–145. Retrieved 25 March 2010.*

227. UNFCCC (n.d.). "Kyoto Protocol". UNFCCC website. Retrieved 21 May 2011.

228. Müller, Benito (February 2010). Copenhagen 2009: Failure or final wake-up call for our leaders? EV 49(PDF). Oxford Institute for Energy Studies. p. i. ISBN 978-1-907555-04-6. Retrieved 18 May 2010.

229. Rudd, Kevin (25 May 2015). "Paris Can't Be Another Copenhagen". The New York Times. Retrieved 26 May 2015.

230. United Nations Environment Programme (November 2010). "Technical summary". The Emissions Gap Report: Are the Copenhagen Accord pledges sufficient to limit global warming to 2 °C or 1.5 °C? A preliminary assessment (advance copy) (PDF). UNEP website. Retrieved 11 May 2011. This publication is also available in e-book format

231. UNFCCC (30 March 2010). "Decision 2/CP. 15 Copenhagen Accord. In: Report of the Conference of the Parties on its fifteenth session, held in Copenhagen from 7 to 19 December 2009. Addendum. Part Two: Action taken by the Conference of the Parties at its fifteenth session" (PDF). United Nations Office at Geneva, Switzerland. p. 5. Retrieved 17 May 2010.

232. Outcome of the work of the Ad Hoc Working Group on long-term Cooperative Action under the Convention" (PDF). PRESIDENCIA DE LA REPÚBLICA, MÉXICO. 11 December 2010. p. 2. Retrieved 12 January 2011.

233. IPCC, "Summary for Policymakers", Detection and Attribution of Climate Change, "It is extremely likely that human influence has been the dominant cause of the observed warming since the mid-20th century" (page 15) and "In this Summary for Policymakers, the following terms have been used to indicate the assessed likelihood of an outcome or a result: (...) extremely likely: 95–100%" (page 2)., in IPCC AR5 WG1 2013.

234. John Cook; et al. (13 April 2016). "Consensus on consensus: a synthesis of consensus estimates on human-caused global warming". Environmental Research Letters. **11** *(4): 048002. doi:10.1088/1748-9326/11/4/048002. Retrieved 21 July 2016.*

235. Julie Brigham-Grette; et al. (September 2006). "Petroleum Geologists' Award to Novelist Crichton Is Inappropriate" (PDF). Eos. **87** *(36): 364. Bibcode*:2006EOSTr..87..364B. doi:10.1029/2006EO360008. Retrieved 23 January 2007. The AAPG stands alone among scientific societies in its denial of human-induced effects on global warming.

236. Boykoff, M.; Boykoff, J. (July 2004). "Balance as bias: global warming and the US prestige press". Global Environmental Change Part A. **14** *(2): 125–136. doi:10.1016/j.gloenvcha.2003.10.001.*

237. Oreskes, Naomi; Conway, Erik. Merchants of Doubt: How a Handful of Scientists Obscured the Truth on Issues from Tobacco Smoke to Global Warming (first ed.). Bloomsbury Press. ISBN 978-1-59691-610-4.

238. Aaron M. McCright and Riley E. Dunlap, "Challenging Global Warming as a Social Problem: An Analysis of the Conservative Movement's Counter-Claims", Social Problems, November 2000, Vol. 47 Issue 4, pp 499–522 in JSTOR

239. Weart, S. (July 2009). "The Public and Climate Change (cont. – since 1980). Section: After 1988". American Institute of Physics website. Retrieved 5 May 2010.

240. SEPP (n.d.). "Frequently Asked Questions About Climate Change". Science & Environmental Policy Project (SEPP) website. Archived from the original on 11 May 2008. Retrieved 5 May 2010.

241. Begley, Sharon (13 August 2007). "The Truth About Denial". Newsweek. Retrieved 13 August 2007.

242. Adams, David (20 September 2006). "Royal Society tells Exxon: stop funding climate change denial". The Guardian. London. Retrieved 9 August 2007.

243. Exxon cuts ties to global warming skeptics". MSNBC. 12 January 2007. Retrieved 2 May 2007.

244. Sandell, Clayton (3 January 2007). "Report: Big Money Confusing Public on Global Warming". ABC. Retrieved 27 April 2007.

245. Greenpeace: Exxon still funding climate skeptics". USA Today. Reuters. 18 May 2007. Retrieved 21 January 2010.

246. Global Warming Resolutions at U.S. Oil Companies Bring Policy Commitments from Leaders, and Record High Votes at Laggards" (Press release). Ceres. 13 May 2004. Retrieved 4 March 2010.

247. Oil Company Positions on the Reality and Risk of Climate Change". Environmental Studies. University of Oshkosh - Wisconsin. Retrieved 27 March 2016.

248. Weart, S. (February 2015). "The Public and Climate Change (cont. – since 1980). Section: after 1988". American Institute of Physics website. Retrieved 18 August 2015.

249. Environment". Gallup. 2015. Retrieved 18 August 2015.

250. Pelham, Brett (2009). "Awareness, Opinions about Global Warming Vary Worldwide". Gallup. Retrieved 18 August 2015.

251. Pugliese, Anita (20 April 2011). "Fewer Americans, Europeans View Global Warming as a Threat". Gallup. Retrieved 22 April 2011.

252. Ray, Julie; Anita Pugliese (22 April 2011). "Worldwide, Blame for Climate Change Falls on Humans". Gallup.Com. Retrieved 3 May 2011. People nearly everywhere, including majorities in developed Asia and Latin America, are more likely to attribute global warming to human activities rather than natural causes. The U.S. is the exception, with nearly half (47%) – and the largest percentage in the world – attributing global warming to natural causes.

253. Climate Change and Financial Instability Seen as Top Global Threats". Pew Research Center for the People & the Press.

254. Climate Change: Key Data Points from Pew Research | Pew Research Center

255. Tranter, Bruce; Booth, Kate (July 2015). "Scepticism in a Changing Climate: A Cross-national Study". Global Environmental Change. **33**: 54–164. doi:10.1016/j.gloenvcha.2015.05.003.

256. Weart, Spencer R. (February 2014). "The Discovery of Global Warming; The Public and Climate Change: Suspicions of a Human-Caused Greenhouse (1956-1969)". American Institute of Physics. Retrieved 12 May 2015., and footnote 27

257. Erik Conway. "What's in a Name? Global Warming vs. Climate Change", NASA, 5 December 2008

258. Weart, Spencer R. (February 2014). "The Discovery of Global Warming; The Public and Climate Change: The Summer of 1988". American Institute of Physics. Retrieved 12 May 2015.

259. U.S. Senate, Committee on Energy and Natural Resources, "Greenhouse Effect and Global Climate Change, part 2" 100th Cong., 1st sess., 23 June 1988, p. 44.

CHAPTER 8

Adaptation to Environment

Chapter Contents

Species adaptation is the way in which organisms change or evolve to meet the needs of their environment. These changes typically involve behavior, body parts, or body coverings. Adapting to their habitat is generally geared toward protecting the species from predators, finding ways to attract or capture prey, and attracting a mate.

Some species adaptation is behavioral. The hognose snake does not slither away when startled like many nocturnal snakes do. Since it usually hunts prey in the daytime, it is often disturbed by other animals or humans. When startled, the hognose snake often rears up like a cobra and strikes at the intruder with its mouth closed.

If this behavior does not scare off the threat, this snake often rolls onto its back and plays dead, usually with its tongue hanging out of its mouth. This behavior is similar to the kind exhibited by the only North American marsupial, the opossum. This animal "plays possum" whenever a predator comes near.

INTRODUCTION

In biology, an adaptation, also called an adaptive trait, is a trait with a current functional role in the life of an organism that is maintained and evolved by means of natural selection. Adaptation refers to both the current state of being adapted and to the dynamic evolutionary process that leads to the adaptation. Adaptations enhance the fitness and survival of individuals. Organisms face a succession of environmental challenges as they grow and develop and are equipped with an adaptive plasticity as the phenotype of traits develop in response to the imposed conditions. The developmental norm of reaction for any given trait is essential to the correction of adaptation as it affords a kind of biological insurance or resilience to varying environments.

GENERAL PRINCIPLES

Adaptation is, first of all, a *process*, rather than a part of a body. An internal parasite (such as a liver fluke) can illustrate the distinction: such a parasite may have a very simple bodily structure, but nevertheless the organism is highly adapted to its specific environment. From this we see that adaptation is not just a matter of visible traits: in such parasites critical adaptations take place in the life cycle, which is often quite complex. However, as a practical term, adaptation is often used for the *product*: those features of a species which result from the process. Many aspects of an animal or plant can be correctly called adaptations, though there are always some features whose function is in doubt. By using the term *adaptation* for the evolutionary *process*, and *adaptive trait* for the bodily part or function (the product), the two senses of the word may be distinguished.

Adaptation is one of the two main processes that explain the diverse species we see in biology, such as the different species of Darwin's finches. The other is speciation (species-splitting or cladogenesis), caused by geographical isolation or some other mechanism. A favorite example used today to study the interplay of adaptation and speciation is the evolution of cichlid fish in African lakes, where the question of reproductive isolation is much more complex.

Adaptation is not always a simple matter, where the ideal phenotype evolves for a given external environment. An organism must be viable at all stages of its development and at all stages of its evolution. This places *constraints* on the evolution of development, behavior and structure of organisms. The main constraint, over which there has been much debate, is the requirement that each genetic and phenotypic change during evolution should be relatively small, because developmental systems are so complex and interlinked. However, it is not clear what "relatively small" should mean, for example polyploidy in plants is a reasonably common large genetic change. The origin of eukaryotic symbiosis is a more dramatic example.

All adaptations help organisms survive in their ecological niches. These adaptive traits may be structural, behavioral or physiological. Structural adaptations are physical features of an organism (shape, body covering, armament; and also the internal organization). Behavioral adaptations are composed of inherited behavior chains and/or the ability to learn: behaviors may be inherited in detail (instincts), or a capacity for learning may be inherited (see neuropsychology). Examples: searching for food, mating, vocalizations. Physiological adaptations permit the organism to perform special functions (for instance, making venom, secreting slime, phototropism); but also more general functions such as growth and development, temperature regulation, ionic balance and other aspects of homeostasis. Adaptation, then, affects all aspects of the life of an organism.

Definitions

The following definitions are mainly due to Theodosius Dobzhansky.

- *Adaptation* is the evolutionary process whereby an organism becomes better able to live in its habitat or habitats.
- *Adaptedness* is the state of being adapted: the degree to which an organism is able to live and reproduce in a given set of habitats.
- An *adaptive trait* is an aspect of the developmental pattern of the organism which enables or enhances the probability of that organism surviving and reproducing.

Adaptedness and Fitness

From the above definitions, it is clear that there is a relationship between adaptedness and *fitness* (a key population genetics concept). Differences in fitness between genotypes predict the rate of evolution by natural selection. Natural selection changes the relative frequencies of alternative phenotypes, insofar as they are heritable. Although the two are connected, the one does not imply the other: a phenotype with high adaptedness may not have high fitness. Dobzhansky mentioned the example of the Californian redwood, which is highly adapted, but a relict species in danger of extinction. Elliott Sober commented that adaptation was a retrospective concept since it implied something about the history of a trait, whereas fitness predicts a trait's future.

- Relative fitness. The average contribution to the next generation by a genotype or a class of genotypes, relative to the contributions of other genotypes in the population. This is also known as *Darwinian fitness*, *selection coefficient*, and other terms.
- Absolute fitness. The absolute contribution to the next generation by a genotype or a class of genotypes. Also known as the Malthusian parameter when applied to the population as a whole.
- Adaptedness. The extent to which a phenotype fits its local ecological niche. This can sometimes be tested through a reciprocal transplant.

BRIEF HISTORY

Adaptation is a fact of life that has been accepted by many of the great thinkers who have tackled the world of living organisms. It is their explanations of how adaptation arises that separates these thinkers. A few of the most significant ideas:

- Empedocles did not believe that adaptation required a final cause (~ purpose), but "came about naturally, since such things survived." Aristotle, however, did believe in final causes.

Jean-Baptiste Lamarck (1744–1829).

- In natural theology, adaptation was interpreted as the work of a deity, even as evidence for the existence of God. William Paley believed that organisms were perfectly adapted to the lives they lead, an argument that shadowed Gottfried Wilhelm Leibniz, who had argued that God had brought about "the best of all possible worlds." Voltaire's Dr. Pangloss is a parody of this optimistic idea, and David Hume also argued against design. The *Bridgewater Treatises* are a product of natural theology, though some of the authors managed to present their work in a fairly neutral manner. The series was lampooned by Robert Knox, who held quasi-evolutionary views, as the *Bilgewater Treatises*. Charles Darwin broke with the tradition by emphasising the flaws and limitations which occurred in the animal and plant worlds.
- Lamarckism is a proto-evolutionary hypothesis of the inheritance of acquired characteristics, whose main purpose is to explain adaptations by natural means. Jean-Baptiste Lamarck proposed a tendency for organisms to become more complex, moving up a ladder of progress, plus "the influence of circumstances," usually expressed as *use and disuse*. His evolutionary ideas, and those of Étienne Geoffroy Saint-Hilaire, fail because they cannot be reconciled with heredity. This was known even before Gregor Mendel by medical men interested in human races (William Charles Wells, William Lawrence), and especially by August Weismann.

Many other students of natural history, such as Buffon, accepted adaptation, and some also accepted evolution, without voicing their opinions as to the mechanism. This illustrates the real merit of Darwin and Alfred Russel Wallace, and secondary figures such as Henry Walter Bates, for putting forward a mechanism whose significance had only been glimpsed previously. A century later, experimental field studies and breeding experiments by people such as E. B. Ford and Dobzhansky produced evidence that natural selection was not only the ‹engine' behind adaptation, but was a much stronger force than had previously been thought.

TYPES OF ADAPTATIONS

Changes in Habitat

Before Charles Darwin, adaptation was seen as a fixed relationship between an organism and its habitat. It was not appreciated that as the climate changed, so did the habitat; and as the habitat changed, so did the biota. Also, habitats are subject to changes in their biota: for example, invasions of species from other areas. The relative numbers of species in a given habitat are always changing. Change is the rule, though much depends on the speed and degree of the change.

When the habitat changes, three main things may happen to a resident population: habitat tracking, genetic change or extinction. In fact, all three things may occur in sequence. Of these three effects only genetic change brings about adaptation.

Habitat Tracking

When a habitat changes, the most common thing to happen is that the resident population moves to another locale which suits it; this is the typical response of flying insects or oceanic organisms, who have wide (though not unlimited) opportunity for movement. This common response is called *habitat tracking*. It is one explanation put forward for the periods of apparent stasis in the fossil record (the punctuated equilibrium theory).

Genetic Change

Genetic change is what occurs in a population when natural selection acts on the genetic variability of the population; moreover, some mutations may create genetic variation that will lead to differing characteristics of offspring and hence abet adaptation. The first pathways of enzyme-based metabolism may have been parts of purine nucleotide metabolism, with previous metabolic pathways being part of the ancient RNA world. By this means, the population adapts genetically to its circumstances. Genetic changes may result in visible structures, or may adjust physiological activity in a way that suits the habitat.

It is now clear that habitats and biota do frequently change. Therefore, it follows that the process of adaptation is never finally complete. Over time, it may happen that the environment changes little, and the species comes to fit its surroundings better and better. On the other hand, it may happen that changes in the environment occur relatively rapidly, and then the species becomes less and less well adapted. Seen like this, adaptation is a genetic *tracking process*, which goes on all the time to some extent, but especially when the population cannot or does not move to another, less hostile area. Also, to a greater or lesser extent, the process affects every species in a particular ecosystem.

Leigh Van Valen thought that even in a stable environment, competing species had to constantly adapt to maintain their relative standing. This became known as the Red Queen hypothesis. One of the manifestations of the Red Queen dynamics can be seen in host-parasite interaction.

Intimate Relationships: co-Adaptations

In coevolution, where the existence of one species is tightly over bound up with the life of another species, new or 'improved' adaptations which occur in one species are often followed by the appearance and spread of corresponding features in the other species. There are many examples of this; the idea emphasises that the life and death of living things is intimately connected, not just with the physical environment, but with the life of other species. These relationships are intrinsically dynamic, and may continue on a trajectory for millions of years, as has the relationship between flowering plants and insects (pollination).

Pollinator constancy: these honey bees selectively visit flowers from only one species, as can be seen by the colour of the pollen in their baskets:

The gut contents, wing structures, and mouthpart morphologies of fossilized beetles and flies suggest that they acted as early pollinators. The association between beetles and angiosperms during the Early Cretaceous period led to parallel radiations of angiosperms and insects into the Late Cretaceous. The evolution of nectaries in Late Cretaceous flowers signals the beginning of the mutualism between hymenopterans and angiosperms.

Mimicry

A and B show real wasps; the rest are mimics: three hoverflies and one beetle.

Bates' work on Amazonian butterflies led him to develop the first scientific account of mimicry, especially the kind of mimicry which bears his name: Batesian mimicry. This is the mimicry by a palatable species of an unpalatable or noxious species. A common example seen in temperate gardens is the hoverfly, many of which—though bearing no sting—mimic the warning colouration of hymenoptera (wasps and bees). Such mimicry does not need to be perfect to improve the survival of the palatable species.

Bates, Wallace and Fritz Müller believed that Batesian and Müllerian mimicry provided evidence for the action of natural selection, a view which is now standard amongst biologists.All aspects of this situation can be, and have been, the subject of research. Field and experimental work on these ideas continues to this day; the topic connects strongly to speciation, genetics and development.

The basic Machinery: Internal Adaptations

There are some important adaptations to do with the overall coordination of the systems in the body. Such adaptations may have significant consequences. Examples, in vertebrates, would be temperature regulation, or improvements in brain function, or an effective immune system. An example in plants would be the development of the reproductive system in flowering plants. Such adaptations may make the clade (monophyletic group) more viable in a wide range of habitats. The acquisition of such major adaptations has often served as the spark for adaptive radiation, and huge success over long periods of time for a whole group of animals or plants.

Compromise and Conflict Between Adaptations

All adaptations have a downside: horse legs are great for running on grass, but they can't scratch their backs; mammals' hair helps temperature, but offers a niche for ectoparasites; the only flying penguins do is under water. Adaptations serving different functions may be mutually destructive. Compromise and makeshift occur widely, not perfection. Selection pressures pull in different directions, and the adaptation that results is some kind of compromise.

Since the phenotype as a whole is the target of selection, it is impossible to improve simultaneously all aspects of the phenotype to the same degree.

—Ernst Mayr, The Growth of Biological Thought: Diversity, Evolution, and Inheritance

Consider the antlers of the Irish elk, (often supposed to be far too large; in deer antler size has an allometric relationship to body size). Obviously, antlers serve positively for defence against predators, and to score victories in the annual rut. But they are costly in terms of resource. Their size during the last glacial period presumably depended on the relative gain and loss of reproductive capacity in the population of elks during that time. Another example: camouflage to avoid detection is destroyed when vivid colors are displayed at mating time. Here the risk to life is counterbalanced by the necessity for reproduction.

Stream-dwelling salamanders, such as Caucasian salamander or Gold-striped salamander have very slender, long bodies, perfectly adapted to life at the banks of fast small rivers and mountain brooks. Elongated body protects their larvae from being washed out by current. However, elongated body increases risk of desiccation and decreases dispersal ability of the salamanders; it also negatively affects their fecundity. As a result, fire salamander, less perfectly adapted to the mountain brook habitats, is in general more successful, have a higher fecundity and broader geographic range.

An Indian peacock's trainin full display.

The peacock's ornamental train (grown anew in time for each mating season) is a famous adaptation. It must reduce his maneuverability and flight, and is hugely conspicuous; also, its growth costs food resources. Darwin's explanation of its advantage was in terms of sexual selection: "This depends on the advantage which certain individuals have over other individuals of the same sex and species, in exclusive relation to reproduction." The kind of sexual selection represented by the peacock is called 'mate choice,' with an implication that the process selects the more fit over the less fit, and so has survival value. The recognition of sexual selection was for a long time in abeyance, but has been rehabilitated. In practice, the Indian peafowl (*Pavo cristatus*) is a successful species, with a large natural range in India, so the overall outcome of their mating system is quite viable.

The conflict between the size of the human foetal brain at birth, (which cannot be larger than about 400 cm^3, else it will not get through the mother's pelvis) and the size needed for an adult brain (about 1400 cm^3), means the brain of a newborn child is quite immature. The most vital things in human life (locomotion, speech) just have to wait while the brain grows and matures. That is the result of the birth compromise. Much of the problem comes from our upright bipedal stance, without which our pelvis could be shaped more suitably for birth. Neanderthals had a similar problem.

As another example, the long neck of a giraffe is a burden and a blessing. The neck of a giraffe can be up to 2 m (6 ft 7 in) in length. This neck can be used for interspecies competition or for foraging on tall trees where shorter herbivores cannot reach. However, as previously stated, there is always a trade-off. This long neck is heavy and it adds to the body mass of a giraffe, so the giraffe needs an abundance of nutrition to provide for this costly adaptation.

SHIFTS IN FUNCTION

Pre-Adaptations

This occurs when a species or population has characteristics which (by chance) are suited for conditions which have not yet arisen. For example, the polyploid cordgrass *Spartina townsendii* is better adapted than either of its parent species to their own habitat of saline marsh and mud-flats. White Leghornchicken are markedly more resistant to vitamin B_1 deficiency than other breeds. On a plentiful diet there is no difference, but on a restricted diet this preadaptation could be decisive.

Pre-adaptation may occur because a natural population carries a huge quantity of genetic variability.In diploid eukaryotes, this is a consequence of the system of sexual reproduction, where mutant alleles get partially shielded, for example, by the selective advantage of heterozygotes. Microorganisms, with their huge populations, also carry a great deal of genetic variability.

The first experimental evidence of the pre-adaptive nature of genetic variants in microorganisms was provided by Salvador Luria and Max Delbrück who developed Fluctuation Test, a method to show the random fluctuation of pre-existing genetic changes that conferred resistance to bacteriophage in the bacterium *Escherichia coli.*

Co-option of Existing Traits: Exaptation

The classic example is the ear ossicles of mammals, which we know from paleontological and embryological studies originated in the upper and lower jaws and the hyoid bone of their synapsid ancestors, and further back still were part of the gill arches of early fish. We owe this esoteric knowledge to the comparative anatomists, who, a century ago, were at the cutting edge of evolutionary studies. The word *exaptation* was coined to cover these shifts in function, which are surprisingly common in evolutionary history. The origin of wings from feathers that were originally used for temperature regulation is a more recent discovery (see feathered dinosaurs).

RELATED ISSUES

Non-adaptive Traits

Some traits do not appear to be adaptive, that is, they appear to have a neutral or even deleterious effect on fitness in the current environment. Because genes have pleiotropic effects, not all traits may be functional (i.e. spandrels). Alternatively, a trait may have been adaptive at some point in an organism's evolutionary history, but a change in habitats caused what used to be an adaptation to become unnecessary or even a hindrance (maladaptations). Such adaptations are termed vestigial.

Vestigial Organs

Many organisms have vestigial organs, which are the remnants of fully functional structures in their ancestors. As a result of changes in lifestyle the organs became redundant, and are either not functional or reduced in functionality. With the loss of function goes the loss of positive selection, and the subsequent accumulation of deleterious mutations. Since any structure represents some kind of cost to the general economy of the body, an advantage may accrue from their elimination once they are not functional. Examples: wisdom teeth in humans; the loss of pigment and functional eyes in cave fauna; the loss of structure in endoparasites.

Fitness Landscapes

Sewall Wright proposed that populations occupy *adaptive peaks* on a fitness landscape. In order to evolve to another, higher peak, a population would first have to pass through a valley of maladaptive intermediate stages. A given population might be "trapped" on a peak that is not optimally adapted.

Extinction

If a population cannot move or change sufficiently to preserve its long-term viability, then obviously, it will become extinct, at least in that locale. The species may or may not survive in other locales. Species extinction occurs when the death rate over the entire species exceeds the birth rate for a long enough period for the species to disappear. It was an observation of Van Valen that groups of species tend to have a characteristic and fairly regular rate of extinction.

Coextinction

Just as we have co-adaptation, there is also coextinction. Coextinction refers to the loss of a species due to the extinction of another; for example, the extinction of parasitic insects following the loss of their hosts. Coextinction can also occur when a flowering plant loses its pollinator, or through the disruption of a food chain. Ecologist Lian Pin Koh and colleagues discuss coextinction, stating, "Species coextinction is a manifestation of the interconnectedness of organisms in complex ecosystems. . . . While coextinction may not be the most important cause of species extinctions, it is certainly an insidious one."

Flexibility, Acclimatization, Learning

Generalists, such as birds, are sometimes able to adapt to urban areas.

Flexibility deals with the relative capacity of an organism to maintain themselves in different habitats: their degree of specialization. *Acclimatization* is a term used for automatic physiological adjustments during life; *learning* is the term used for improvement in behavioral performance during life. In biology these terms are preferred, not adaptation, for changes during life which improve the performance of individuals. These adjustments are not inherited genetically by the next generation.

Adaptation, on the other hand, occurs over many generations; it is a gradual process caused by natural selection which changes the genetic make-up of a population so the collective performs better in its niche.

Flexibility

Populations differ in their phenotypic plasticity, which is the ability of an organism with a given genotype to change its phenotype in response to changes in its habitat, or to move to a different habitat.

To a greater or lesser extent, all living things can adjust to circumstances. The degree of flexibility is inherited, and varies to some extent between individuals. A highly specialized animal or plant lives only in a well-defined habitat, eats a specific type of food, and cannot survive if its needs are not met. Many herbivores are like this; extreme examples are koalas which depend on *eucalyptus*, and giant pandas which require bamboo. A generalist, on the other hand, eats a range of food, and can survive in many different conditions. Examples are humans, rats, crabs and many carnivores. The *tendency* to behave in a specialized or exploratory manner is inherited—it is an adaptation.

Rather different is *developmental flexibility:* "An animal or plant is developmentally flexible if when it is raised in or transferred to new conditions, it changes in structure so that it is better fitted to survive in the new environment," writes evolutionary biologist John Maynard Smith. Once again, there are huge differences between species, and the *capacities* to be flexible are inherited.

Acclimatization

If humans move to a higher altitude, respiration and physical exertion become a problem, but after spending time in high altitude conditions they *acclimatize* to the pressure by increasing production of Red blood cells. The *ability* to acclimatize is an adaptation, but not the acclimatization itself. Fecundity goes down, but deaths from some tropical diseases also goes down.

Over a longer period of time, some people will reproduce better at these high altitudes than others. They will contribute more heavily to later generations. Gradually the whole population becomes adapted to the new conditions. This we know takes place, because the performance of long-term communities at higher altitude is significantly better than the performance of new arrivals, even when the new arrivals have had time to make physiological adjustments.

Some kinds of acclimatization happen so rapidly that they are better called reflexes. The rapid colour changes in some flatfish, cephalopods, chameleons are examples.

Learning

Social learning is supreme for humans, and is possible for quite a few mammals and birds: of course, that does not involve genetic transmission except to the extent that the capacities are inherited. Similarly, the *capacity to learn* is an inherited adaptation, but not what is learnt; the capacity for human speech is inherited, but not the details of language.

Diversity of genome DNAs

A large diversity of genome DNAs in a species is the basis for species' adaptation and for species' differentiation. A great number of individuals are needed for carrying the great number of different genome DNAs. According to the Misrepair-accumulation aging theory, Misrepair mechanism is important in maintaining the sufficient number of individuals in a species. Misrepair is a way of repair for increasing the surviving chance of an organism when it has severe injuries. Without Misrepairs, no individual could survive to reproduction age. Thus Misrepair mechanism is an essential mechanism for the survival of a species and for maintaining the number of individuals. Although individuals die from aging, genome DNAs are being recopied and transmitted by individuals generation by generation. In addition, the DNA Misrepairs in germ cells contribute also to the diversity of genome DNAs.

FUNCTION AND TELEONOMY

Adaptation raises some issues concerning how biologists use key terms such as *function.*

Function

To say something has a function is to say something about what it does for the organism, obviously. It also says something about its history: how it has come about. A heart pumps blood: that is its function. It also emits sound, which is just an ancillary side-effect. That is not its function. The heart has a history (which may be well or poorly understood), and that history is about how natural selection formed and maintained the heart as a pump. Every aspect of an organism that has a function has a history. Now, an adaptation must have a functional history: therefore we expect it must have undergone selection caused by relative survival in its habitat. It would be quite wrong to use the word adaptation about a trait which arose as a by-product.

It is widely regarded as unprofessional for a biologist to say something like "A wing is for flying," although that is their normal function. A biologist would be conscious that sometime in the remote past feathers on a small dinosaur had the function of retaining heat, and that later many wings were not used for flying (e.g. penguins, ostriches). So, the biologist would rather say that the wings on a bird or an insect usually had the *function* of aiding flight. That would carry the connotation of being an adaptation with a history of evolution by natural selection.

Teleonomy

Teleonomy is a term invented to describe the study of goal-directed functions which are not guided by the conscious forethought of man or any supernatural entity. It is contrasted with Aristotle's teleology, which has connotations of intention, purpose and foresight. Evolution is teleonomic; *adaptation hoards hindsight rather than foresight*. The following is a definition for its use in biology:

Teleonomy: The hypothesis that adaptations arise without the existence of a prior purpose, but by chance may change the fitness of an organism.

The term may have been suggested by Colin Pittendrigh in 1958; it grew out of cybernetics and self-organising systems. Ernst Mayr, George C. Williams and Jacques Monod picked up the term and used it in evolutionary biology.

Philosophers of science have discussed the concept. Ernest Nagel analysed goal-directedness in biology; and David Hull commented on the use of teleology in biology:

...Haldane can be found remarking, 'Teleology is like a mistress to a biologist: he cannot live without her but he's unwilling to be seen with her in public.' Today the mistress has become a lawfully wedded wife. Biologists no longer feel obligated to apologize for their use of teleological language; they flaunt it. The only concession which they make to its disreputable past is to rename it 'teleonomy'.

REFERENCES

1. Mayr 1982, p. 483: "Adaptation... could no longer be considered a static condition, a product of a creative past, and became instead a continuing dynamic process."
2. Daintith, John; Martin, Elizabeth A., eds. (2010) [First published 1984 as Concise Science Dictionary]. "adaptation". A Dictionary of Science. Oxford Paperback Reference (6th ed.). Oxford; New York: Oxford University Press. p. 13. ISBN 978-0-19-956146-9. LCCN 2010287468. OCLC 444383696. Any change in the structure or functioning of successive generations of a population that makes it better suited to its environment.
3. Salzburger, Walter; Mack, Tanja; Verheyen, Erik; Meyer, Axel (February 21, 2005). "Out of Tanganyika: Genesis, explosive speciation, key-innovations and phylogeography of the haplochromine cichlid fishes"(PDF). BMC Evolutionary Biology. London: BioMed Central. 5 (17). doi:10.1186/1471-2148-5-17. ISSN 1471-2148. PMC 554777. PMID 15723698. Retrieved 2015-08-15.
4. Kornfield, Irv; Smith, Peter F. (November 2000). "African Cichlid Fishes: Model Systems for Evolutionary Biology". Annual Review of Ecology and Systematics. Palo Alto, CA: Annual Reviews. 31: 163–196. doi:10.1146/annurev.ecolsys.31.1.163. ISSN 1545-2069.
5. Hutchinson 1965. The niche is the central concept in evolutionary ecology; see especially part II: "The niche: an abstractly inhabited hypervolume." (pp. 26–78)
6. Darwin 1872, p. 397: "Rudimentary, Atrophied, and Aborted Organs"
7. See, for example, the discussion in Bowler 2003, pp. 86–95: "Whatever the true nature of Lamark's theory, it was his mechanism of adaptation that caught the attention of later naturalists." (p. 90)
8. Orr, H. Allen (February 2005). "The genetic theory of adaptation: a brief history". Nature Reviews Genetics. London: Nature Publishing Group. 6 (2): 119–127. doi:10.1038/nrg1523. ISSN 1471-0056. PMID 15716908.
9. Hogan, C. Michael (October 12, 2010). "Mutation". In Monosson, Emily. Encyclopedia of Earth. Washington, D.C.: Environmental Information Coalition, National Council for Science and the Environment. OCLC 72808636. Retrieved 2015-08-18.
10. Mayr 1982, pp. 481–483: This sequence tells how Darwin's ideas on adaptation developed as he came to appreciate it as "a continuing dynamic process."
11. Rabajante, J; et al. (2016). "Host-parasite Red Queen dynamics with phase-locked rare genotypes". Science Advances. 2: e1501548. doi:10.1126/sciadv.1501548. ISSN 2375-2548.
12. Mallet, James (November 2001). "The speciation revolution" (PDF). Journal of Evolutionary Biology. Hoboken, NJ: Wiley-Blackwell on behalf of the European Society for Evolutionary Biology. 14 (6): 887–888. doi:10.1046/j.1420-9101.2001.00342.x. ISSN 1010-061X.
13. Stebbins 1974. Contains an extensive analysis of the evolution of adaptations in the radiation of angiosperms.

14. Jacob, François (June 10, 1977). "Evolution and Tinkering". Science. Washington, D.C.: American Association for the Advancement of Science. 196 (4295): 1161–1166. doi:10.1126/science.860134. ISSN 0036-8075. PMID 860134.
15. Gould, Stephen Jay (June 1974). "The Origin and Function of ‹Bizarre' Structures: Antler Size and Skull Size in the ‹Irish Elk,' Megaloceros giganteus". Evolution. Hoboken, NJ: John Wiley & Sons for the Society for the Study of Evolution. 28 (2): 191–220. doi:10.2307/2407322. ISSN 0014-3820. JSTOR 2407322.
16. Tarkhnishvili, David N. (1994). "Interdependences between Populational, Developmental and Morphological Features of the Caucasian salamander, Mertensiella caucasica" (PDF). Mertensiella. Bonn, Germany: Deutsche Gesellschaft für Herpetologie und Terrarienkunde. 4: 315–325. ISSN 0934-6643. Retrieved 2015-08-18.
17. Rosenberg, Karen R. (1992). "The evolution of modern human childbirth". American Journal of Physical Anthropology. Hoboken, NJ: John Wiley & Sons for the American Association of Physical Anthropologists. 35 (Supplement S15): 89–124. doi:10.1002/ajpa.1330350605. ISSN 0002-9483.
18. Friedlander, Nancy J.; Jordan, David K. (October–December 1994). "Obstetric implications of Neanderthal robusticity and bone density". Human Evolution. Kluwer Academic Publishers. 9 (4): 331–342. doi:10.1007/BF02435519. ISSN 0393-9375.
19. Altwegg, Robert E.; Simmons, Res (September 2010). "Necks-for-sex or competing browsers? A critique of ideas on the evolution of giraffe". Journal of Zoology. Hoboken, NJ: Wiley-Blackwell. 282 (1): 6–12. doi:10.1111/j.1469-7998.2010.00711.x. ISSN 0952-8369.
20. Huskins, C. Leonard (1930). "The origin of Spartina Townsendii". Genetica. Martinus Nijhoff, The Hague/Kluwer Academic Publishers. 12 (6): 531–538. doi:10.1007/BF01487665. ISSN 0016-6707.
21. Lamoreux, Wilfred F.; Hutt, Frederick B. (February 15, 1939). "Breed differences in resistance to a deficiency in vitamin B_1 in the fowl" (PDF). Journal of Agricultural Research. Washington, D.C.: United States Department of Agriculture. 58 (4): 307–316. ISSN 0095-9758. Retrieved 2015-08-20.
22. Gould, Stephen Jay; Vrba, Elizabeth S. (Winter 1982). "Exaptation–A Missing Term in the Science of Form". Paleobiology. Boulder, CO: Paleontological Society. 8 (1): 4–15. ISSN 0094-8373. JSTOR 2400563.
23. Barrett et al. 1987. Charles Darwin was the first to put forward such ideas.
24. Van Valen, Leigh (July 1973). "A New Evolutionary Law" (PDF). Evolutionary Theory. Chicago, IL: University of Chicago. 1: 1–30. ISSN 0093-4755. Retrieved 2015-08-22.
25. Darwin 1872, pp. 57–58. Darwin in tells the story of "a web of complex relations" involving heartsease (Viola tricolor), red clover (Trifolium pratense), humble-bees (bumblebees), mice and cats.
26. Koh, Lian Pin; Dunn, Robert R.; Sodhi, Navjot S.; et al. (September 2004). "Species Coextinctions and the Biodiversity Crisis". Science. Washington, D.C.:

American Association for the Advancement of Science. 305 (5690): 1632–1634. doi:10.1126/science.1101101. ISSN 0036-8075. PMID 15361627.

27. Price, Trevor D.; Qvarnström, Anna; Irwin, Darren E. (July 2003). "The role of phenotypic plasticity in driving genetic evolution". Proceedings of the Royal Society B. London: Royal Society. 270 (1523): 1433–1440. doi:10.1098/rspb.2003.2372. ISSN 0962-8452. PMC 1691402. PMID 12965006.
28. Price, Trevor D. (June 2006). "Phenotypic plasticity, sexual selection and the evolution of colour patterns". The Journal of Experimental Biology. Cambridge, UK: The Company of Biologists. 209 (12): 2368–2376. doi:10.1242/jeb.02183. ISSN 0022-0949. PMID 16731813.
29. Moore, Lorna G.; Regensteiner, Judith G. (October 1983). "Adaptation to High Altitude". Annual Review of Anthropology. Palo Alto, CA: Annual Reviews. 12: 285–304. doi:10.1146/annurev.an.12.100183.001441. ISSN 0084-6570.
30. Maynard Smith 1993, p. 32. Maynard Smith uses the term physiologically versatile for such animals.
31. Wang, Jicun; Michelitsch, Thomas M.; Wunderlin, Arne; Mahadeva, Ravi (2009). "Aging as a consequence of misrepair—A novel theory of aging". arXiv:0904.0575 [q-bio.TO].
32. Wang-Michelitsch, Jicun; Michelitsch, Thomas M. (2015). "Aging as a process of accumulation of Misrepairs". arXiv:1503.07163 [q-bio.TO].
33. Wang-Michelitsch, Jicun; Michelitsch, Thomas M. (2015). "Misrepair mechanism: a mechanism essential for individual adaptation, species adaptation and species evolution". arXiv:1505.03900 [q-bio.TO].
34. Nagel, Ernest (May 1977). "Goal-Directed Processes in Biology". The Journal of Philosophy. New York: The Journal of Philosophy, Inc. 74 (5): 261–279. doi:10.2307/2025745. ISSN 0022-362X. JSTOR 2025745. Teleology Revisisted: The Dewy Lectures 1977 (first lecture).
35. Nagel, Ernest (May 1977). "Functional Explanations in Biology". The Journal of Philosophy. New York: The Journal of Philosophy, Inc. 74 (5): 280–301. doi:10.2307/2025746. ISSN 0022-362X. JSTOR 2025746. Teleology Revisisted: The Dewy Lectures 1977 (second lecture).

Index